Pure mathematics

All specifications include a 'core' of pure mathematics. This will count for two-thirds of your time during the course.

Exam themes

- → Algebra
- → Coordinate geometry
- → Sequences and series
- → Trigonometry
- → Calculus (differentiation and integration)
- → Numerical methods and vectors

You also have to understand how to construct and understand various methods of proof.

Topic checklist

	AQA	CCEA	EDEXCEL	OCR/A	OCR/MEI	WJEC
Basic algebra	MPC1	C1	C1	C1	C1	C1
Indices and surds	MPC1	C1	C1	C1	C2	C1
Polynomials	MPC1	C1	C1/C2	C2	C1	C1
Quadratic equations	MPC1	C1	C1	C1	C1	C1
Quadratic functions	MPC1	C1	C1	C1	C1	C1
Simultaneous equations	MPC1	C1	C1	C1	C1	C1
Inequalities	MPC1	C1	C1	C1	C1	C1
Functions	MPC3	C4	C3	C3	C3	C3
Graphs of functions	MPC1/MPC3	C1/C3/C4	C1/C3	C1/C3	C1/C3	C1/C3
Transformations	MPC1/MPC3	C1/C3	C1/C3	C1/C3	C1/C2/C3	C1/C3
Rational functions	MPC4	C3	C3/C4	C3	C3	C4
Exponentials and logarithms	MPC2/3/4	C2/C3	C2/C3/C4	C2	C2	C2/C3
Applications	MPC3/MPC4	C2/C3	C3/C4	C2	C2	C2/3/4
Sequences	MPC2	C2	C1/C2	C2	C2	C2
Series	MPC2	C2	C1/C2	C2	C2/P4	C2
Binomials	MPC2/MPC4	C2	C2/C4	C2/C3	C1/C3	C2/C4
Proof	MPC1/2/3/4	C1	C1/C3	C1/C4	C1	C2
The geometry of straight lines	MPC1	C1	C1	C1	C1	C1
The coordinate geometry of circles	MPC1	C2	C2	C1	C1	C2
Parameters	MPC4	C3	C4	C4	C4	C4
Trigonometry	MPC2	C2	C2	C2	C2	C2
Radians and trigonometry	MPC2	C2	C2	C2	C2	C2
Trigonometric relations	MPC3	C3/C4	C3	C3	C4	C3
Further trigonometric relations	MPC4	C4	C3	C3	C4	C4
Basic differentiation	MPC1/MPC2	C1	C1	C1	C2	C1
Applications of differentiation	MPC1	C1	C1/C2	C1	C2	C1
Rules of differentiation	MPC3	C3	C3	C3	C3	C3
Further rules of differentiation	MPC3	C3	C3	C3/C4	C3	C3
Implicit and parametric differentiation	MPC4	C4	C4	C4	C3/C4	C3
Basic integration	MPC1/2/3	C2/C3	C1/C2/C4	C2/C3	C2/C3	C2/C3
Rules of integration	MPC3	C3/C4	C4	C3/C4	C3	C3/C4
Further rules of integration	MPC3/MPC4	C4	C4	C4	C3/C4	C4
Applications of integration	MPC1/MPC3	C2/C4	C2/C4	C2/C3	C2/C4	C2/C4
Differential equations	MPC4	C4	C4	C4	C4	C4
Numerical solution of equations	MPC3	C3	C3	C3	C3	C3
Numerical integration	MPC2/MPC3	C2/C3	C2/C4	C2	C2	C2/C3
Vectors	MPC4	C4	C4	C4	C4	C4
Position vectors and lines	MPC4	C4	C4	C4	C4	C4
Applications of vectors	MPC4	C4	C4	C4	C4	C4

Basic algebra

Elementary algebra falls into the category of assumed knowledge. In other words, you won't be tested directly on the following topics at AS or A2 level, but you will be expected to use it in order to be able to answer other questions.

Multiplying out brackets

If there is one pair of brackets you must make sure the term outside is multiplied by all the terms inside.

Example: Simplify $5(3x - 7 + 4y)$.

Each term in the bracket is multiplied by the 5:

$$5 \times 3x + 5 \times (-7) + 5 \times 4y$$
$$15x - 35 + 20y$$

If there are two brackets, each term within the first bracket must be multiplied by each term in the second bracket.

Example: Expand and simplify $(x + 4)(x - 5)$.

Multiplying out: $x \times x + x \times (-5) + 4 \times x + 4 \times (-5)$
$x^2 - 5x + 4x - 20$
Simplifying: $x^2 - x - 20$

Action point
Work through this example with your book shut and make sure you have included all the terms.

Checkpoint 1
Multiply out the following:
(a) $7(x+3)$ (b) $x(x-9)$
(c) $(x-7)(x+5)$

Links
See quadratic equations on page 10.

Factorising quadratics, including $a^2 - b^2$

Factorising means to rewrite an expression in terms of its factors and this requires putting in brackets – the opposite of the process above. A quadratic expression is one containing one or more terms in which the variable's highest power is two. It can be written in the form $ax^2 + bx + c$, where a is the coefficient (the number multiplied by) of the x^2, b is the coefficient of the x and the c is the coefficient of the x^0 (i.e. a number on its own).

A quadratic expression will usually factorise into two bracketed factors.

Example: Factorise $x^2 + 7x + 6$.

To get the x^2, you need to do $x \times x$. Therefore, the brackets will start like this:

$$(x \quad)(x \quad)$$

You then need to answer two questions:

1 Which two numbers would MULTIPLY to give c?
2 Would they *also* ADD to give b?

In this case, 1 and 6 would MULTIPLY to give 6 and will also ADD to give 7.

Therefore, $x^2 + 7x + 6$ factorises to $(x + 1)(x + 6)$.

Examiner's secrets
Multiply your brackets out as a check that your factorising was correct.

A-level Study Guide

Maths

Clare Bigg
Michael Goulding
David Hodgson
Dineke Spackman
Ben Yudkin

Series Consultants: Geoff Black and Stuart Wall

Pearson Education Limited
Edinburgh Gate, Harlow
Essex CM20 2JE, England
and Associated Companies throughout the world

First published 2000
This second edition published 2005
Second impression 2006

British Library Cataloguing in Publication Data
A catalogue entry for this title is available from the British Library.

ISBN-10: 0-582-78411-5
ISBN-13: 978-0-582-78411-6

Set by 35 in Univers, Cheltenham

Printed by Ashford Colour Press, Hants

A special case is 'the difference of two squares', $a^2 - b^2$, which factorises to $(a + b)(a - b)$.

Example: Factorise $x^2 - 81$.

$$x^2 - 81 = x^2 - 9^2$$
$$= (x + 9)(x - 9)$$

Action point

Multiply out the following and simplify:
(a) $(x - 8)(x + 8)$ (b) $(f + 7)(f - 7)$
(c) $(p + 4)(p - 4)$
What has happened each time?
See notes on $a^2 - b^2$.

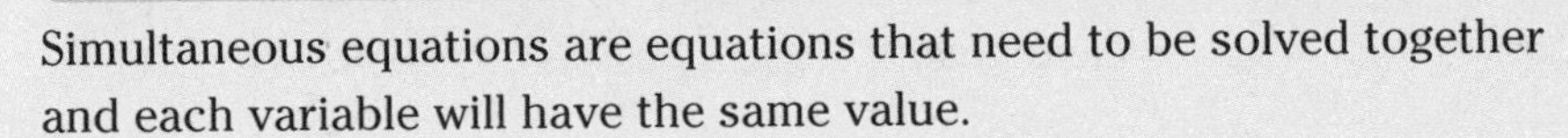

Solving simultaneous equations by eliminating a variable

Simultaneous equations are equations that need to be solved together and each variable will have the same value.

Example: Solve the simultaneous equations: $6x + 2y = 15$ and $4x + y = 11$

Label each equation: $6x + 2y = 15$ ①
$4x + y = 11$ ②

To get the y variables the same multiply Equation ② by 2: $8x + 2y = 22$

To eliminate the y variables do 'new' ② – ①:

$$8x + 2y = 22$$
$$\underline{6x + 2y = 15}$$
$$2x \quad = 7$$

Therefore, $x = 3.5$

Substituting $x = 3.5$ into equation ②: $4 \times 3.5 + y = 11$
$14 \quad + y = 11$

Rearranging: $y = -3$

Checkpoint 2

Solve this pair of simultaneous equations:
$3x - 2y = 4$
$2x + 3y = 7$

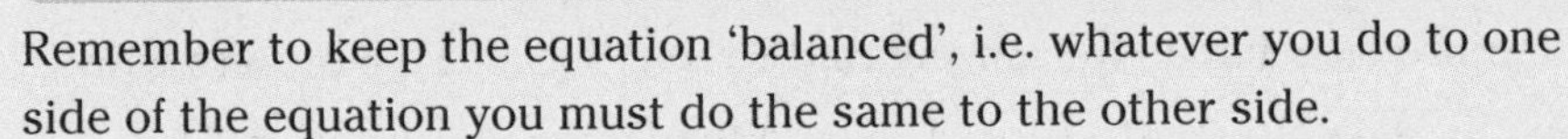

Changing the subject of a simple formula or equation

Remember to keep the equation 'balanced', i.e. whatever you do to one side of the equation you must do the same to the other side.

Example: Make u the subject of the formula $v^2 = u^2 + 2as$.

Subtract $2as$ from both sides: $v^2 - 2as = u^2$

Take the square root of both sides: $\pm\sqrt{v^2 - 2as} = u$

Exam questions

answers: page 82

1 Multiply out the following and simplify where possible:
(a) $11(f - 2g)$ (b) $5 - 2(p + 5q)$ (c) $(x + 2)(x + 13)$
(d) $(p - 3)(2p - 6)$ (e) $(3q - 4)(q + 7)$ (10 min)

2 Factorise the following:
(a) $x^2 + 11x + 18$ (b) $x^2 - 8x + 15$ (c) $x^2 + 2x - 63$
(d) $x^2 - 144$ (5 min)

3 Solve these simultaneous equations:
(a) $x + y = 4$ (b) $2x + 5y = 11$
$5x + 2y = 10$ $3x + 2y = 12$ (10 min)

4 Rearrange the following formulae to make u the subject:
(a) $v = u + at$ (b) $s = ut + \frac{1}{2}at^2$ (3 min)

Indices and surds

An index (plural *indices*) is the power to which something is raised. In the expression x^2 the index is the 2, and we call x the base. So in 10^5, 10 is the base and 5 is the index. The best way to learn these rules is through practice, so below you will find the rule, an example, and a numerical question to try – making sure that you get the final answer as a number in each case.

Rules of indices

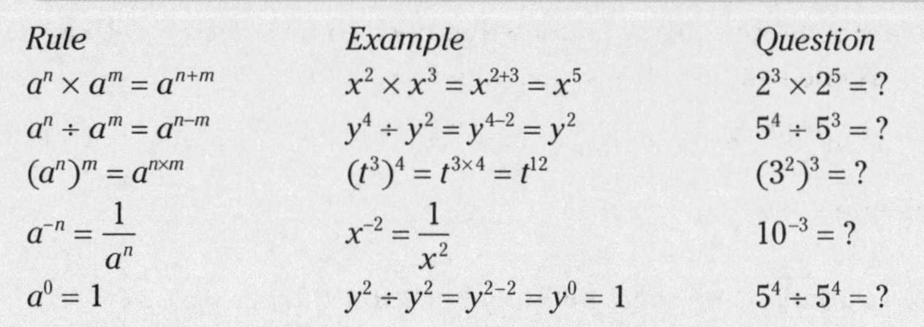

Rule	*Example*	*Question*
$a^n \times a^m = a^{n+m}$	$x^2 \times x^3 = x^{2+3} = x^5$	$2^3 \times 2^5 = ?$
$a^n \div a^m = a^{n-m}$	$y^4 \div y^2 = y^{4-2} = y^2$	$5^4 \div 5^3 = ?$
$(a^n)^m = a^{n \times m}$	$(t^3)^4 = t^{3\times 4} = t^{12}$	$(3^2)^3 = ?$
$a^{-n} = \dfrac{1}{a^n}$	$x^{-2} = \dfrac{1}{x^2}$	$10^{-3} = ?$
$a^0 = 1$	$y^2 \div y^2 = y^{2-2} = y^0 = 1$	$5^4 \div 5^4 = ?$

Watch out!

A very common mistake is to say that $x^0 = x$. In fact, $x^1 = x$ and $x^0 = 1$. Anything to the power of zero equals 1.

If that last rule doesn't make much sense then remember that anything divided by itself equals 1. It's a very useful rule as it helps to 'get rid' of things – more about that later on!

You must also be able to use fractional indices. Consider this: $x^{\frac{1}{2}} \times x^{\frac{1}{2}} = x^{\frac{1}{2}+\frac{1}{2}} = x^1$. So $x^{\frac{1}{2}}$ *multiplied by itself* gives x. That makes $x^{\frac{1}{2}}$ the *square root* of x. Also, $x^{\frac{1}{4}} \times x^{\frac{1}{4}} \times x^{\frac{1}{4}} \times x^{\frac{1}{4}} = x^1$, so $x^{\frac{1}{4}}$ is the fourth root of x. This translates into general terms as $a^{\frac{1}{n}} = \sqrt[n]{a}$ i.e. the nth root of a. So $64^{\frac{1}{2}} = \sqrt{64} = 8$, while $16^{\frac{1}{4}} = \sqrt[4]{16} = 2$.

What about something like $16^{\frac{3}{4}}$? Well, since $\frac{3}{4} = 3 \times \frac{1}{4}$, and looking at the third rule above, we can say that $16^{\frac{3}{4}} = (16^{\frac{1}{4}})^3 = (2)^3 = 8$. So to work out $16^{\frac{3}{4}}$, take the fourth root of 16, and then cube your answer. This works for any fractional index: do the rooting first to get rid of the denominator – and then treat the numerator as the index of your answer.

Checkpoint 1

Work these out:
(a) $d^2 \times d^3 = ?$ (b) $y^4 \div y^2 = ?$
(c) $w^5 \div w^7 = ?$ (d) $(t^2)^3 = ?$ (e) $9^{\frac{3}{2}} = ?$

Action point

Get to know how your calculator handles indices, especially fractional indices. For 25 to the power $\frac{3}{2}$, you should get 125, and for 64 to the power $\frac{5}{6}$ you should get 32.

Rationals, irrationals, and surds

A **rational number** is any number that can be written as an integer, or as a fraction where both numerator and denominator are integers. Examples of rational numbers are 7, 2, $\sqrt{9}$ and $\frac{3}{11}$.

An **irrational number** is any real number that cannot be expressed as the ratio of two integers. In other words, irrational numbers cannot be written as fractions. Examples are π or $\sqrt{2}$. The easy way to spot an irrational number is that no matter how many decimal places you write it out to, it never falls into a repeating pattern or sequence.

Surds come about when we want to use a number like $\sqrt{3}$ but don't want to round it off, as this means using an approximation rather than the exact value. Let's look at $\sqrt{18}$. Now, this could be written as $\sqrt{9 \times 2}$, which can then be written as $\sqrt{9}\sqrt{2}$ or plain old $3\sqrt{2}$. Try it on your calculator, $\sqrt{18}$ and $3\sqrt{2}$ – it should agree to at least 8 decimal places (accurate enough?).

Examiner's secrets

To know when you will be expected to convert a surd to a decimal, look for something in the question about the accuracy of your answer, like 'give your answer to 2 d.p.' or the more sneaky version 'give your answer to a sensible degree of accuracy'.

A few more examples are: $\sqrt{20} = \sqrt{4 \times 5} = \sqrt{4}\sqrt{5} = 2\sqrt{5}$

$$\sqrt{8} = \sqrt{4 \times 2} = \sqrt{4}\sqrt{2} = 2\sqrt{2}$$

$$\sqrt{27} = \sqrt{9 \times 3} = \sqrt{9}\sqrt{3} = 3\sqrt{3}$$

Brackets involving surds are multiplied as usual, i.e.

$$\begin{aligned}(3 + \sqrt{2})(9 - \sqrt{2}) &= 3 \times 9 + 3 \times (-\sqrt{2}) + \sqrt{2} \times 9 + \sqrt{2} \times (-\sqrt{2}) \\ &= 27 \quad - 3\sqrt{2} \quad + 9\sqrt{2} - \sqrt{2}\sqrt{2} \\ &= 27 \quad + 6\sqrt{2} \quad - 2 \\ &= 25 \quad + 6\sqrt{2}\end{aligned}$$

Here is an important case; since $(x + y)(x - y) = x^2 - y^2$, we have that $(3 + \sqrt{6})(3 - \sqrt{6}) = 3^2 - (\sqrt{6})^2 = 9 - 6 = 3$. We can use this to our advantage.

Example: To simplify $\dfrac{\sqrt{6}}{3 + \sqrt{6}}$

We get rid of the bottom term by multiplying top and bottom by $(3 - \sqrt{6})$:

$$\frac{\sqrt{6}}{(3 + \sqrt{6})} = \frac{\sqrt{6}(3 - \sqrt{6})}{(3 + \sqrt{6})(3 - \sqrt{6})} = \frac{3\sqrt{6} - \sqrt{6}\sqrt{6}}{3} = \sqrt{6} - 2$$

To make sure, check that your calculator gives the same answer for both the unsimplified and the simplified versions. This technique of simplification is called **rationalising the denominator** as you are making the denominator a rational number.

Checkpoint 2

Simplify these surds:
(a) $\sqrt{32}$, (b) $\sqrt{50}$, (c) $\sqrt{180}$.
When you have finished, check with a calculator by comparing the start number with the finished one as decimals.

Examiner's secrets

A very common mistake is to think that $\sqrt{6} \times \sqrt{6} = 6^2$ and not 6. Think about it – squaring only gets rid of the square root sign; it does not change the actual number!

Checkpoint 3

If these were on the bottom of a fraction, what would you multiply by to simplify?
(a) $4 + \sqrt{7}$, (b) $2 - \sqrt{2}$, (c) $7 + \sqrt{5}$.

Exam questions

answers: page 82

1 Simplify the following: (a) $c^5 \times c^7$ (b) $\dfrac{B^7}{B^2}$ (c) $(d^2)^3$ (d) $\dfrac{x^7 \times x^2}{x^8}$

(e) $\dfrac{\sqrt{y} \times \sqrt{y^3}}{y}$ (f) $\dfrac{z^{\frac{1}{2}} \times z^{\frac{3}{2}}}{z^{-2}}$ (10 min)

2 Evaluate each of these: (a) $\dfrac{2^7 \times 2^0}{2^3}$ (b) $64^{\frac{1}{3}}$ (c) 3^{-2} (d) $\dfrac{2^{\frac{1}{2}} \times 2^8 \times 2^{-7}}{2^{\frac{3}{2}}}$

(e) $\left(\dfrac{4}{9}\right)^{\frac{1}{2}}$ (f) $\left(\dfrac{1}{27}\right)^{\frac{1}{3}}$ (g) $\left(\left(\dfrac{16}{5}\right)^{-2}\right)^0$ (h) $\left(2\dfrac{7}{9}\right)^{\frac{3}{2}}$ (15 min)

3 Simplify: (a) $\sqrt{2}(5 - \sqrt{2})$ (b) $(3\sqrt{6} - 5)^2$ (c) $(\sqrt{p} + 1)(\sqrt{p} - 1)$ (10 min)

4 Rationalise the denominators and simplify:

(a) $\dfrac{1}{\sqrt{5}}$ (b) $\dfrac{3}{2 - \sqrt{2}}$ (c) $\dfrac{7}{3\sqrt{2} - 5}$ (d) $\dfrac{1}{(2\sqrt{3} - 1)^2}$ (15 min)

Examiner's secrets

When multiplying out surds in this fractional form *taking your time and getting it right* is better than rushing and losing easy marks. Do the top and bottom separately; give yourself plenty of space to work, and write everything down, to minimise errors!

Polynomials

This section deals firstly with multiplying out expressions which have three or more bracketed factors. We then move onto a very quick method of factorising an expression where the highest power of x is more than 2 – expressions known as polynomials.

Expanding brackets and collecting like terms

The aim of expansion is to make sure every term gets multiplied by every other term. To avoid confusion, we stick to a simple method. Suppose we had to expand something like $f(x) = (x + 2)(x + 5)(x + 10)$. We expand the last two brackets first, to get

$$(x + 2)[(x + 5)(x + 10)] = (x + 2)[x^2 + 15x + 50].$$

Now we work on these two brackets. The simplest way to avoid errors is to use the 'cover up' method. First, cover up the '+ 2' in the first bracket, and multiply everything in the second bracket by the $\boldsymbol{x}$, to get

$$(\boldsymbol{x} + 2)[x^2 + 15x + 50] = \boldsymbol{x^3} + \boldsymbol{15x^2} + \boldsymbol{50x}.$$

Next, cover up the 'x' and multiply everything in the second bracket by the **+ 2**, and add it to the end of the line, to get

$$(x + \boldsymbol{2})[x^2 + 15x + 50] = x^3 + 15x^2 + 50x + \boldsymbol{2x^2} + \boldsymbol{30x} + \boldsymbol{100}.$$

Now we have multiplied everything by everything else. It's time to collect like terms, starting with the highest powers of x. We get

$$x^3 + 15x^2 + 2x^2 + 50x + 30x + 100 = x^3 + 17x^2 + 80x + 100.$$

Expressions involving more than three brackets are done in the same way – work on the last two brackets at a time, covering up and multiplying.

Examiner's secrets

It's so easy to make a mistake with these questions – there are many things that can go wrong. *At each stage, double check* that all the signs are correct on your multiplications.

Examiner's secrets

When collecting terms, it's always a good idea to physically cross out a term once it's been included. That way, you know what's been done, and where you're up to.

Watch out

Don't forget: $(x - 4)^2$ is $(x - 4)(x - 4)$.

Example: Expand and simplify $f(x) = (x + 2)(x + 3)(x - 4)^2$

First stage: $f(x) = (x + 2)(x + 3)[x^2 - 8x + 16]$

Second stage: cover up the '+ 3', multiply by the $\boldsymbol{x}$, to get

$(x + 2)(\boldsymbol{x} + 3)[x^2 - 8x + 16] = (x + 2)[\boldsymbol{x^3} - \boldsymbol{8x^2} + \boldsymbol{16x}]$

Third stage: cover up the 'x', multiply by the + **3**, to get

$(x + 2)(x + \boldsymbol{3})[x^2 - 8x + 16] = (x + 2)[x^3 - 8x^2 + 16x + \boldsymbol{3x^2} - \boldsymbol{24x} + \boldsymbol{48}]$

Collecting terms now in the second bracket makes life easier, giving

$f(x) = (x + 2)[x^3 - 5x^2 - 8x + 48]$

Fourth stage: cover up the '+ 2', multiply through by $\boldsymbol{x}$, to get

$(\boldsymbol{x} + 2)[x^3 - 5x^2 - 8x + 48] = \boldsymbol{x^4} - \boldsymbol{5x^3} - \boldsymbol{8x^2} + \boldsymbol{48x}$

Fifth stage – nearly there: just cover the x, and multiply by + **2**, to get

$(x + \boldsymbol{2})[x^3 - 5x^2 - 8x + 48] = x^4 - 5x^3 - 8x^2 + 48x + \boldsymbol{2x^3} - \boldsymbol{10x^2} - \boldsymbol{16x} + \boldsymbol{96}$

Last stage: collecting like terms, to get

$$\begin{aligned} f(x) &= x^4 - 5x^3 + 2x^3 - 8x^2 - 10x^2 + 48x - 16x + 96 \\ &= x^4 - 3x^3 - 18x^2 + 32x + 96 \end{aligned}$$

Action point

Once you have read this through, write the question down and close the book. Work through to the end, looking back here only when you've tried everything you possibly can.

Factorisation: the factor and remainder theorems

When we go from $x^2 + 5x + 6$ to $(x + 2)(x + 3)$ it's called factorising (i.e. putting it into factors). You can think of it as expansion in reverse. With quadratics, you know that you are going to have at most two brackets. With cubics, there are at most three brackets. It follows that for a polynomial where the highest power is n then there are going to be at most n brackets.

The **remainder theorem** says that for a polynomial, $f(x)$, **f(a)** is the **remainder** when $f(x)$ is divided by $(x - a)$.

Example: Find the remainder when $f(x) = 4x^3 - 2x^2 + x - 3$ is divided by $(x - 2)$.

To find the remainder when $f(x)$ is divided through by $(x - 2)$, we find the value of $f(2)$, i.e. $4(2^3) - 2(2^2) + 2 - 3 = 32 - 8 + 2 - 3 = 23$
So 23 is the remainder when $f(x)$ is divided by $(x - 2)$.
What about dividing $f(x)$ by $(x - 1)$? Simply do

$$f(1) = 4(1^3) - 2(1^2) + 1 - 3 = 4 - 2 + 1 - 3 = 0$$

Since there is no remainder, $(x - 1)$ is a factor of $f(x)$. This important result is a case of the **factor theorem** at work. It says that if $f(a) = 0$, then $(x - a)$ is a factor of $f(x)$.

Example: Factorise fully the expression $f(x) = x^3 - 2x^2 - x + 2$.

We start by looking at the constant, 2; since it comes from all of the numbers in each bracket multiplied together, we know that they must be 1, 1, 2 in some order. Two of them might be negative. The easiest number to substitute is 1, so we'll do that first.

$$f(1) = 1^3 - 2(1^2) - 1 + 2 = 1 - 2 - 1 + 2 = 0$$

Good! $f(1) = 0$ so $(x - 1)$ is a factor of $f(x)$. Now, let's try $f(2)$, to see if one other factor is $(x - 2)$.

$$f(2) = 2^3 - 2(2^2) - 2 + 2 = 8 - 8 - 2 + 2 = 0$$

So $(x - 2)$ is a factor, making the last factor easy to work out. It must be $(x + 1)$, because then the numbers in all three brackets multiplied together give $-2 \times (-1) \times 1 = +2$, as required.
So $f(x) = (x - 2)(x - 1)(x + 1)$.

Exam questions answers: page 82

1 Expand and simplify the following:
(a) $(x - 1)(x + 2)(x + 5)$
(b) $(2x + 1)(x - 3)^2$
(c) $(x - 4)^2(2x + 3)^2$ (20 min)

2 Factorise the following expressions fully:
(a) $x^3 + 4x^2 + x - 6$
(b) $x^4 - 21x^2 - 20x$
(c) $x^3 - 3x^2 + 4$
(d) $x^3 + 3x^2 - 4x - 12$
(e) $2x^4 + 7x^3 - 2x^2 - 13x + 6$ (30 min)

Examiner's secrets

You should check all factorisations by quickly expanding them and seeing if it looks right! It doesn't take long, and *could save accuracy marks*!

Action point

Make up a polynomial, say a cubic, then find the remainder when it's divided by $(x - 1)$ by substituting 1 into it and finding the result. Try to find the remainder when it's divided by $(x - 10)$ and $(x - 5)$, too.

Examiner's secrets

When using the remainder or the factor theorems, the most common mistake made is to get the sign wrong in the bracket. Think about it – one of the brackets is $(x + a)$, and for this to be zero when $x = 2$, a must be -2.

Checkpoint 1

Try finding all the factors of $x^3 + 2x^2 - 11x - 12$.
(Hint – look at the factors of -12.)

Quadratic equations

This section shows a variety of methods used for solving quadratic equations. A quadratic f(x) is solved when we find some value, say a, such that f(a) = 0. We say that a is a root of the equation; there is usually more than one root to find.

Solution by factorisation

Links

See 'Factorising quadratics', page 4.

We begin by factorising to put the quadratic into brackets.

Example: Solve the quadratic equation: $x^2 + 3x + 2 = 0$.

First factorise the quadratic expression:

$f(x) = x^2$	$+ 3x$	$+ 2$
	Choose numbers that must add together to make 3 . . .	. . . and multiply together to make 2.

Checkpoint 1

Try getting to the other form of these: (a) $(x + 5)(x + 4) = ?$; (b) $(x + 6)(x - 5) = ?$; (c) what about $x^2 + 7x + 6 = ?$

The two numbers we require are 1 and 2, since $1 + 2 = 3$, and $1 \times 2 = 2$.
Let's check: $(x + 1)(x + 2) = x^2 + 2x + x + 2 = x^2 + 3x + 2$. It works!
To find the solutions of $(x + 1)(x + 2) = 0$ you must use the fact that either bracket or both brackets must have a value of 0. If $x + 1 = 0$, then $x = -1$ and, if $x + 2 = 0$, then $x = -2$.
Therefore, $x = -1$ and $x = -2$.

Action point

We haven't mentioned equations like $y = 2x^2 + 4x + 2$. Here, you look at factors of the constant, as usual. Factors of 2 are 1 and 2. But we know that to get the 4 in the middle, before it was added, one of the factors was multiplied by 2 (from the $2x^2$). Now $2 \times 2 + 1 = 5$, which doesn't work. $2 \times 1 + 2 = 4$ as required, so the brackets will be $(2x + 2)(x + 1)$.

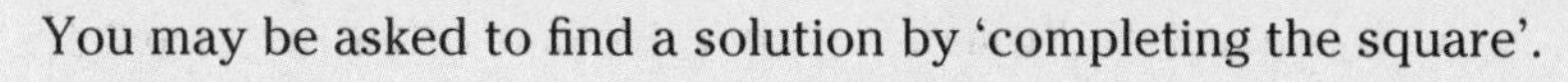

Solution by completing the square

You may be asked to find a solution by 'completing the square'.

Example: Solve $x^2 - 10x + 16 = 0$ by completing the square.

First the quadratic expression, $x^2 - 10x + 16$, has to be written in 'completed square' format: i.e. $(x + p)^2 + q$

To find the value of the 'p' divide the coefficient of x by two; here, that's $-10 \div 2 = -5$. So the bracket becomes: $(x - 5)^2$

However, if you multiplied out $(x - 5)^2$ you would get $x^2 - 10x + 25$, which has a '+ 25' on the end that we don't want. Therefore, you SUBTRACT p^2, in this case $(-5)^2$, from the bracket: $(x - 5)^2 - (-5)^2$

To work out the q, you must remember to include the number at the end of the quadratic, in this case +16: $(x - 5)^2 - 25 + 16$

The quadratic is now in 'completed square' format: $(x - 5)^2 - 9$

Now to solve the quadratic equation . . . $(x - 5)^2 - 9 = 0$

Take the 9 across to get: $(x - 5)^2 = 9$

Square root both sides to get: $(x - 5) = \pm\sqrt{9}$

Checkpoint 2

Try solving these by completing the square: (a) $x^2 + 4x - 12 = 0$; (b) $x^2 + 16x = 80$. (*Hint*: get everything on one side before you start the second one!)

So the values of x which solve the equation $x^2 - 10x + 16 = 0$ are $x = 5 + \sqrt{9} = 8$ and $x = 5 - \sqrt{9} = 2$.

This method relies on the coefficient of x^2 being 1. If the equation is $3x^2 + 8x + 5 = 0$, simply divide through by the 3 and follow the same process as before. Fractions can get tricky but time and care taken at each stage will pay off.

Solution by the formula

Try factorising $f(x) = x^2 - x - 1 = 0$. Factors of -1 are $+1$ and -1, but there's no way of making -1 by adding them together. We say that this quadratic does not factorise and turn to the formula. It looks bad, but actually using it couldn't be easier. First we need to learn our *abc*.

Our quadratic: $x^2 - x - 1$.
The general form: $ax^2 + bx + c$
In this example, $a = 1$, $b = -1$ and $c = -1$

Next we plug these values into this formula, and the resulting answers will be our solutions. Here it is:

$$x = \frac{-b \pm \sqrt{b^2 - 4ac}}{2a}$$

With our values, this becomes:

$$x = \frac{-(-1) \pm \sqrt{[(-1)^2 - 4(1)(-1)]}}{2(1)}$$

$$= \frac{1 \pm \sqrt{1+4}}{2} = \frac{1 \pm \sqrt{5}}{2}$$

So the values of x which solve $x^2 - x - 1 = 0$ are $x = \frac{1+\sqrt{5}}{2}$ and $x = \frac{1-\sqrt{5}}{2}$.

Examiner's secrets

Learning this formula is easy with practice, and saves a lot of time with formula sheets in the exam!

Action point

Close the book and try to write down the formula for solving quadratics – practice now saves lots of time later!

Checkpoint 3

When you have read this, write down the equation $4x^2 + 3x - 2 = 0$, turn over the page, and try to work through it yourself – remember, practice makes perfect!

The discriminant

The most important part of the quadratic formula is the '$\sqrt{b^2 - 4ac}$' bit. You should know that you cannot take the square root of a negative number; so what happens when '$b^2 - 4ac$' is negative? The table below shows what that means for the equation in terms of its roots.

Discriminant	*Effect*
$b^2 - 4ac < 0$	The equation has no real roots; there is no value of x that will give $f(x) = 0$. It cannot be factorised.
$b^2 - 4ac = 0$	The equation has one repeated root; there is one unique value of x giving $f(x) = 0$. It factorises to a single bracket squared, so $f(x) = (x + \alpha)^2$.
$b^2 - 4ac > 0$	The equation has two distinct roots; there are two values of x which each give $f(x) = 0$. If it does factorise, it looks like $f(x) = (x + \alpha)(x + \beta)$.

Examiner's secrets

Look at the section on quadratic functions to connect between the sign of the discriminant and where the graph of the function lies on the x- and y-axes!

Exam questions

answers: page 82

1 Factorise these quadratic expressions:
(a) $x^2 - 5x - 14$ (b) $x^2 + 9x + 20$ (c) $2x^2 - 15x + 18$ (10 min)

2 Solve these quadratic equations, by completing the square:
(a) $x^2 - 4x = 21$ (b) $x^2 + 8x - 33 = 0$ (10 min)

3 Solve these quadratic equations, where possible:
(a) $x^2 + 3x - 8 = 0$ (b) $x^2 - 5x = 24$
(c) $2x^2 + 3x - 14 = 0$ (d) $3x^2 + 7x = 18$
(e) $3x^2 + 7x + 18 = 0$ (f) $6x^2 - 4x - 3 = 0$ (40 min)

Quadratic functions

An essential skill at A-level is the ability to quickly graph quadratic functions. As long as your factorising is up to scratch, drawing the graph is actually quite easy!

Happy or sad graphs

Consider the general quadratic equation $f(x) = ax^2 + bx + c$. If a is positive, then the graph will be a ∪-shaped – or happy curve.

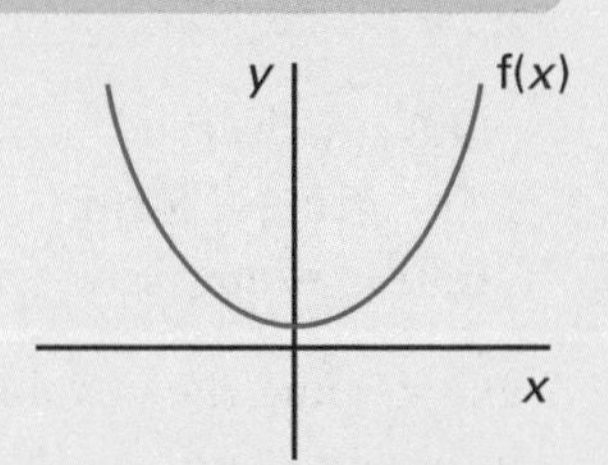

Remember: positive = happy.

Examples are $y = x^2 + 2x + 5$, $y = x^2 - 4x$, etc. If a is negative, then the graph will be a ∩-shaped – or sad curve.

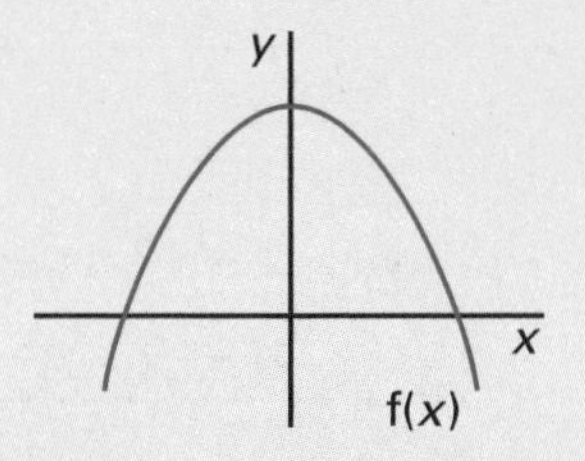

Remember: negative = sad.

Examples are $y = 3x - 4 - x^2$, $y = 16 - x^2$, etc.

Checkpoint 1

Draw rough sketches to show if these are happy or sad quadratics: (a) $x^2 - 2x - 3$; (b) $2 - 2x + x^2$; (c) $4 + 3x - x^2$.

The minimum or maximum point

Consider again the general quadratic equation $f(x) = ax^2 + bx + c$. If a is positive, it is a happy graph. Knowing what the minimum point is would be useful, since all happy quadratics are symmetrical about their minimum points. In fact, the minimum is given by $-\frac{b}{2a}$. Similarly, an unhappy graph has a maximum at the same point, $-\frac{b}{2a}$. This helps a lot when sketching curves.

If the quadratic is given in completed square format: $f(x) = (x + p)^2 + q$ then the co-ordinates of the minimum or maximum are easy to find as they will be: $(-p, q)$.

Example: To sketch the graph of the equation $y = x^2 + 4x + 3$, we first note that it's happy (∪-shaped) and then that it has a minimum point at $x = -\frac{4}{2} = -2$. We need one more point, to see where the curve goes, we so find y when $x = 0$, i.e. $y = 0^2 + 4(0) + 3 = 3$. The curve can then be sketched.

The curve of $y = x^2 + 4x + 3$ is happy; it has a minimum at $(-2,-1)$ and it goes through the point $(0,3)$.

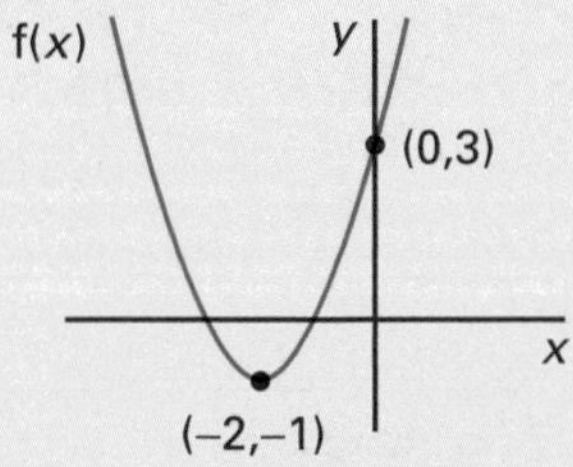

Action point

Make the sketches you did in Checkpoint 1 more accurate by finding the minimum or maximum points of each curve.

Links

See 'Solution by completing the square', page 10.

Action point

Re-write the quadratic by using the 'completing the square' method and show that the minimum is at (–2,–1).

Examiner's secrets

Although we haven't covered it here, differentiation is an acceptable technique for finding minimum or maximum points of curves – feel free to supplement these methods yourself!

Factorising, roots and graphs

Another way to find points to use is to solve the equation, hopefully by factorising it. The example above factorises to $(x + 3)(x + 1)$ – but how does that help? Well, when $(x + 3)(x + 1) = 0$, the graph of the equation crosses the x-axis, so it gives us the two points where it crosses this axis. Also, because it's symmetrical, we know that the minimum (or maximum if it's unhappy) lies exactly between these two crossing points.

Links

See 'Factorising quadratics', page 4 and 'Solution by factorisation', page 10.

Example: Sketch the graph of the equation $y = x^2 - 3x + 2$.

First factorise to get $y = (x - 2)(x - 1)$. This gives us where the graph crosses the x-axis, i.e. (1,0) and (2,0).
The a value is positive so it's happy, the graph will be ∪-shaped and will have a minimum. $-\frac{b}{2a}$ works out to 1.5, so the minimum will be at $x = 1.5$.
Finally, the graph crosses the y-axis when $x = 0$, i.e. at (0,2).
Now, you have enough information to be able to sketch the graph.

Examiner's secrets

If the quadratic is in the form: $ax^2 + bx + c$, the coordinates of the point where the graph crosses the y-axis is $(0,c)$.

The discriminant and the graph

Remember the discriminant? It's the '$b^2 - 4ac$' part of the quadratic formula. It tells us how many times the curve will cross the x-axis.

Discriminant	*Effect on the graph*	*Example*
$b^2 - 4ac < 0$	There is no value of x that will give $f(x) = 0$. The equation is said to have no real roots. The graph of $f(x)$ lies entirely above the x-axis.	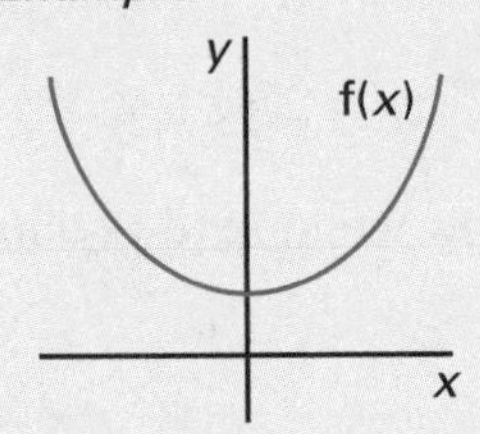
$b^2 - 4ac = 0$	There is one unique value of x which gives $f(x) = 0$. The quadratic factorises to one bracket squared (i.e. a repeated root). There is exactly one point where the curve touches the x-axis.	
$b^2 - 4ac > 0$	There are two values of x which each give $f(x) = 0$. The graph of $f(x)$ cuts the x-axis twice, once going down and once going up, at the roots of the equation $f(x) = 0$.	

Action point

Close the book and try to draw a rough sketch of a quadratic curve where the discriminant is either positive, zero or negative.

Checkpoint 2

Draw rough sketches of these equations to show where they lie on the x- and y-axes: $y = x^2 + 5x + 4$; $y = 2x^2 + x + 3$; and $y = x^2 - 6x + 9$.

Examiner's secrets

Watch out for the signs when using the discriminant, as they can get a little tricky if there are lots of minuses around! Remember that it's only one of many tools at your disposal, and if the quadratic factorises nicely then it becomes obsolete!

Exam questions answers: pages 82–3

Sketch the graphs of the following equations:

1 $y = x^2 - 7$
2 $y = x^2 + 3$
3 $y = (x + 4)^2$
4 $y = x^2 + 2x - 8$
5 $y = 2x^2 + 15x - 8$
6 $y = 8 - 2x - x^2$
7 $y = 30 - 7x - 2x^2$
8 $y = (x - 8)^2$
9 $y = (2x - 5)^2$
10 $y = 3x^2 + 10x + 8$ (45 min)

Simultaneous equations

We solve two equations simultaneously when we find values for x and y which satisfy them both individually. At GCSE, the method of eliminating one variable was the best way to achieve this. Now, however, it's not so straightforward. Our goal is to end up with an equation where only x's or y's appear. To do this, we isolate x or y in one of the equations and then substitute this into the other equation.

Solving two linear equations simultaneously

We focus on solution by substitution.

Example: To solve the equations

$$2x + 6y = 36 \quad ①$$
$$x + 2y = 13 \quad ②$$

Checkpoint 1

Solve these two equations simultaneously using the substitution method: $x + 2y = 16$ and $5x + 12y = 86$.

Let's make x the subject of Equation ②:

$$x + 2y = 13$$
$$x = 13 - 2y$$

Now we look to Equation ①, replacing x with $13 - 2y$:

$$2x + 6y = 36 \quad \Rightarrow 2(13 - 2y) + 6y = 36$$
$$\Rightarrow 26 - 4y + 6y = 36$$
$$\Rightarrow 26 + 2y = 36$$
$$\Rightarrow 2y = 36 - 26$$
$$\Rightarrow 2y = 10 \text{ so } y = 5.$$

Examiner's secrets

Laying out your solution clearly will make it very easy to see that you know what you're doing. It also makes working and spotting mistakes easier, too! *So spread it out!*

Now we know $y = 5$, we can get x. Equation ②: $x + 2(5) = 13$, so $x = 3$.

Solution of one linear and one quadratic equation

When one of the equations involves an x^2 term but the other is still linear then we can follow the same idea of isolating x in the linear equation, and then substituting that into the quadratic. The brackets can get nasty, but writing down every step will help minimise mistakes.

Example: To solve the equations

$$2x - 2y = 4 \quad ①$$
$$3x^2 - 2y^2 = -8 \quad ②$$

Action point

Once you're read through this, write down the starting equations and close the book. Try to solve them yourself!

Equation ① is linear, so make x the subject:

$$2x - 2y = 4$$
$$2x = 4 + 2y$$
$$x = 2 + y$$

Now we can replace x^2 in Equation ② with $(2 + y)^2$, to get:

$$3(2 + y)^2 - 2y^2 = -8$$

expanding: $3(4 + 4y + y^2) - 2y^2 = -8$

multiplying out: $12 + 12y + 3y^2 - 2y^2 = -8$

collecting terms: $y^2 + 12y + 12 = -8$

bring the -8 over: $y^2 + 12y + 20 = 0$

Examiner's secrets

Again, clear layout is the key to success. If you need to, write down exactly what you have done or are doing next (expanding, collecting, adding something, etc.).

We now have a quadratic in y, which factorises to $(y + 10)(y + 2)$. This has two solutions, $y = -10$ and $y = -2$. We now substitute both of these values into Equation ① to get two pairs of solutions to the simultaneous equations.

When $y = -10$, ① gives us $2x - 2(-10) = 4 \Rightarrow 2x = -16 \Rightarrow x = -8$.
When $y = -2$, ① gives us $2x - 2(-2) = 4 \Rightarrow 2x = 0 \Rightarrow x = 0$. So the solution pairs are $x = -8$, $y = -10$ and $x = 0$, $y = -2$.

Graphical solutions

It is sometimes the case that an accurately drawn graph of two equations is much more straightforward solution method than algebraic manipulation. You should remember from GCSE that the coordinates of the point where two lines meet gives the values of x and y which satisfy both equations simultaneously. We can extend this idea to cover graphs of any type – not just two linear but quadratics, cubics or reciprocals now come easily under control.

Example: Using a graph, find the values of x and y which solve $y = x^2 - 4$ and $y = (x + 4)(x + 1)(x - 3)$.

We need to do an accurate sketch of the graphs. $y = x^2 - 4$ factorises to $y = (x - 2)(x + 2)$, which cuts the x-axis at $x = -2$ and $x = 2$. The equation $y = (x + 4)(x + 1)(x - 3)$ is cubic and cuts the x-axis at $x = -4$, $x = -1$, and $x = 3$.

Once we've drawn the graphs on the same axes we can simply read off the values which solve these two equations simultaneously.

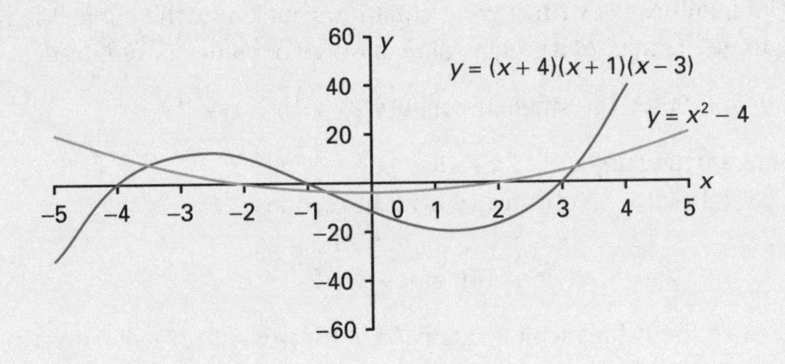

As you can see, the crossing points would not have been very friendly to calculate, so the graph is a relatively painless way to solve these equations. We state that, graphically, the solutions to these equations occur at: $x = -3.6$, $y = 9$; $x = -0.8$, $y = -5$; and $x = 3.2$, $y = 6$.

Exam questions answers: page 83

1 Solve these pairs of simultaneous equations:

(a) $6x + y = 10$, $4x + 5y = 89$ (b) $7x + 3y = 74$, $8x = 96 - 2y$ (c) $5x + 2y = 42$, $3x^2 + 4y^2 = 196$ (20 min)

2 By drawing graphs, solve these pairs of simultaneous equations, correct to 1 d.p.:

(a) $2x^2 - 12 = 2y - 2x$
$y = x(x + 5)(x - 3)$

(b) $y = x(x + 2)(x - 2)(x + 3)$
$y = x^2 - 1$ (30 min)

Examiner's secrets

If you are rusty on your graphing skills, then check out the sections on 'Quadratic functions' and 'Graphing functions'.

Action point

Make your own accurate sketch of the equations in this example, with the book closed! Can you make it more accurate than this one? Are your solutions closer than ours?

Examiner's secrets

Since you are drawing an accurate graph, make the scale sensible so that your solutions aren't too far out. As a check, substitute the values you have found back into the equations to see if they actually work – if they are miles out, take another look at your graph!

Checkpoint 2

Try solving this pair of equations simultaneously using a graph: $y = x^2 - 9$ and $y = (x + 5)(x + 2)(x - 4)$.

Inequalities

Suppose you and a friend were eating a cake. When it's your turn to take a portion, you take exactly half of what's left. Do you ever finish the cake? If you could be accurate enough with your knife, never! There is always some cake left. So the amount of cake left is greater than zero. An inequality for this situation would look like $C > 0$. If you ate the last piece, then we would have $C = 0$, an equality.

Manipulation of inequalities

The most common error with inequalities is not turning the sign around when multiplying or dividing by a negative number. Remember to ask yourself: 'Is this quantity I'm dividing by positive or negative?'

In most cases, you can treat an inequality sign just like you would an equals sign. Here are the basic rules.

Let $f < g$.	Then:	$f + c < g + c$	for any c
	and:	$fc < gc$	for positive c
	but:	$fc > gc$	for negative c.

As long as your numbers are positive, you can more or less do anything with an inequality. If you are multiplying or dividing by a negative number, however, you must turn the sign around.

Checkpoint 1

Using the inequality $10 > 1$, see if it's still true when you: (a) add 3 to both sides; (b) subtract 5 from both sides; (c) multiply both sides by 2; (d) multiply both sides by -2.

Solving single and double linear inequalities

For inequalities, as with solving equations, making x the subject is the way to go. Instead of a single value, a range of values is required.

Example: Solve the single inequality $2x + 16 < 4x + 28$.

To make x the subject of $2x + 16 < 4x + 28$ you . . .

$2x <$	$4x + 12$	subtract 16 from both sides
$-2x <$	12	subtract $4x$ from both sides
$x >$	-6	divide both sides by -2

Therefore the values which satisfy $2x + 16 < 4x + 28$ are all values x such that $x > -6$.

Action point

Try to solve these single and double inequalities on your own, with the book closed, remembering to write some words to say what you are doing at each stage. How did you do?

Example: Solve the double inequality $2x - 5 < 5x - 2 < 3x + 4$.

First you must rewrite the 'double' inequality as the *two* inequalities that it represented.

Working for $2x - 5 < 5x - 2$			*Working for* $5x - 2 < 3x + 4$		
$2x <$	$5x + 3$	add 5 to both sides	$5x <$	$3x + 6$	add 2 to both sides
$-3x <$	3	subtract $5x$ from both sides	$2x <$	6	subtract $3x$ from both sides
$x >$	-1	divide both sides by -3	$x <$	3	divide both sides by 2

We now combine our answers: x has to be bigger than -1, but less than 3, to satisfy this double inequality, i.e. $-1 < x < 3$.

Checkpoint 2

Try solving these inequalities:
(a) $10 - 3x < 15$;
(b) $4x - 3 < 3x - 6$;
(c) $3x + 9 > 5x - 1 > 4x + 1$.

Solving quadratic inequalities

Once a quadratic inequality has been factorised, finding the solution is quite straightforward so long as you are happy with graphing quadratic equations. You will either get a range of x values in which the solution must lie, or a range in which it cannot lie. The best way is to look at some examples.

Example: Find the range of x values which satisfy $x^2 - 3x - 10 \leq 0$. First factorise, to get $(x+2)(x-5) \leq 0$. Then sketch the graph of the equation $y = (x+2)(x-5)$ to show us what we need.

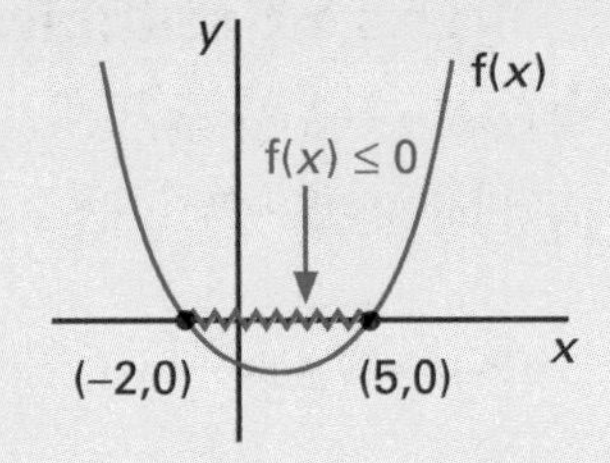

We see that between $x = -2$ and $x = 5$ the graph dips below the x-axis. Therefore, at these points $y \leq 0$. So the range of values of x which satisfy the inequality is $-2 \leq x \leq 5$.

Example: Find the ranges of values which satisfy $x^2 - x > 4 - 4x$. First, rearrange to get a zero on the right hand side: $x^2 + 3x - 4 > 0$. Then factorise, to get $(x+4)(x-1) > 0$. Again, sketching the graph of $y = (x+4)(x-1)$ shows the solution spaces quickly.

We see that the graph is above the x-axis when x is bigger than 1, or less than -4. So in these areas, $y > 0$, as required. So the ranges of x values which satisfy the inequality are $x < -4$ and $x > 1$.

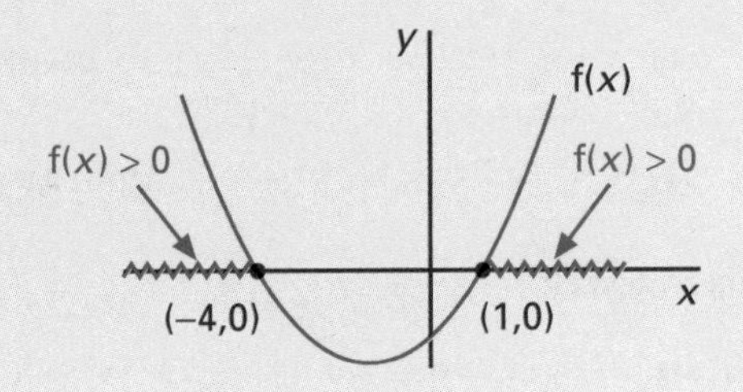

Examiner's secrets

A sketch is the best way to go with quadratic inequalities. It may take time, but sketching curves doesn't require much real thinking and the routine will give you confidence in the exam.

Checkpoint 3

With the use of a sketch, solve the inequality $x^2 + 5x + 6 < 0$.

Examiner's secrets

When giving your answer, *you will lose accuracy marks* if you use ≥ instead of >. Take a second to double check the direction of the inequality sign and whether it includes the '=' bit or not.

Checkpoint 4

With the use of a sketch, solve the inequality $x^2 - 4x - 5 \geq 0$.

Exam questions

answers: page 83

Solve these inequalities, sketching the graphs where necessary:

1 $5x - 3 > x + 5$
2 $2x + 12 < 7x - 3 < 5x + 20$
3 $x^2 + 3x - 10 \geq 0$
4 $2x^2 - 5x - 18 \leq 0$
5 $5x^2 + 4x < 15x + 12$
6 $4x^2 > 1$
7 $(x-7)(x+2) \leq x(x+9)$
8 $x(x+4)(x-3) \geq 0$
9 $3x + 10 < x + 2 < 5x - 18$
10 $3x - 14 < 2x + 5 < 4x + 20$ (45 min)

Functions

Any process that takes an input and gives an output is called a mapping. An example would be $x \to 2x$. A mapping which gives each input a unique output is called a function. So $x \to x^2$ is a function, but $x \to \sqrt{x}$ is not. The first gives only one output for any input, but the second could give two, i.e. $\pm\sqrt{x}$.

Types of mappings

There are four basic types of mapping, only two of which can be considered functions.

Function

One to one – examples are any linear function $y = 2x + 4$, or reciprocals like $y = \frac{3}{x}$.

Mapping

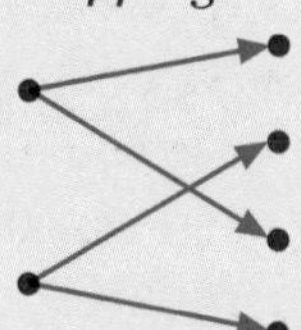

One to many – examples are any square root function, like $y = 2\sqrt{x}$.

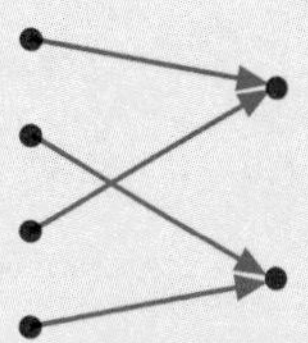

Many to one – examples are quadratic or higher polynomials, like $y = x^2 + 2x + 1$, or $y = x^3 - 8$.

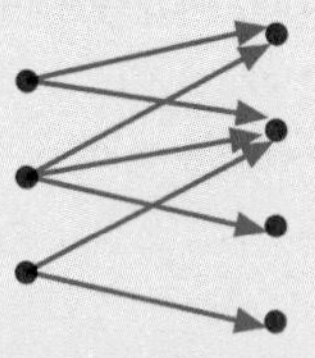

Many to many – examples of this type are not easy to find, so remember this diagram!

Action point

There are four types of mapping, two of which are also functions. Close the book and try to draw the diagrams relating to each type! Try to remember which ones are not functions, and why!

Action point

Roughly sketch the examples of equations given with each of these mappings. Can you see why only some of them are proper functions?

Types of functions

A function f is said to be even when $f(x) = f(-x)$. Even functions are symmetrical about the y-axis. Examples are $f(x) = x^2$, and the piecewise function $f(x) = -5x$ for $x \leq 0$, $5x$ for $x > 0$.

A function is said to be odd when $f(x) = -f(-x)$. Odd functions have rotational symmetry of order 2 about the origin. Examples are $f(x) = 2x$ for all x; $f(x) = x^3$ for all x; and $f(x) = \frac{1}{x}$ for $x \neq 0$.

Checkpoint 1

Make an accurate sketch of the examples given of even and odd functions, to illustrate why they get their names!

Domain and range

The place where the inputs come from, usually the x-axis, is called the **domain** of the function. When defining the domain you need to be careful, because there may be values which are excluded.

Example: To define the domain of $f(x) = \frac{1}{x}$ we note that it is undefined when $x = 0$. So its domain is simply x such that $x \neq 0$.

The range of the function is the set of outputs; usually this can be thought of as the y values you get from the function.

Example: To find the range of the function $f: x \to x^2$ think about the graph of x^2 – it never goes below zero. So the range of this function is y such that $y \geq 0$.

Examiner's secrets

A common mistake when finding domains and ranges is to forget to look for points where the function is undefined. If necessary make a quick sketch of the function and see what happens!

Inverses of functions

The inverse of a function, instead of going from the domain to the range, is a mapping which goes from the range to the domain. It is written f^{-1}. In most cases, finding the inverse is an algebraic process.

Example: Find the inverse of $f(x) = 2x$.

Set $y = 2x$; swap the x and y to get $x = 2y$; make y the subject of the formula once more, which gives $y = \frac{x}{2}$. So $f^{-1}(x) = \frac{x}{2}$.

The inverse is not always a function, as in the next example.

Example: Find the inverse of $f(x) = x^2$.

Set $y = x^2$; swap the x and y to get $x = y^2$; make y the subject of the formula again, which gives $y = \sqrt{x}$. So $f^{-1}(x) = \sqrt{x}$. Now, if $x = 4$, then $f^{-1}(x) = \sqrt{4} = \pm 2$, so the input does not lead to just one output, and the inverse is a one-to-many mapping but not a function.

Checkpoint 2

Find the inverses of these functions: (a) $f(x) = 4x$; (b) $g(x) = 5x + 3$; (c) $h(x) = 4x^2$. State, with reasons, whether the inverses are themselves functions or not.

Functions of functions

Let $f(x) = 3x + 2$ and let $g(x) = x^2$. Now, what about $fg(x)$, or $gf(x)$? These are called the composites of f and g and the order is very important. You do the last one first!

Now: $fg(x) = f(g(x)) = f(x^2) = 3x^2 + 2$. We do g first, so the input for f becomes x^2. f then takes the input, multiplies it by 3 and adds 2.

But: $gf(x) = g(f(x)) = g(3x + 2) = (3x + 2)^2$. We do f first, so the input for g becomes $3x + 2$. g then takes the input and squares it.

Examiner's secrets

Composites of functions is an area where mistakes can easily be made. Take it one step at a time, and remember that it's really just substituting one expression into another.

Piecewise functions

There are functions where the domain has been split up into pieces.

Example: Find the range and inverse of a function f where $f(x) = -2x$ for $x \le 0$ and $f(x) = x^2$ for $x > 0$. Look at the first x region: $-2x$ is a straight line sloping down towards the origin and at $x = 0$, $f(x) = 0$. Beyond this, the function becomes x^2; so y is never negative. The range of the function is therefore y such that $y \ge 0$. The inverse for the first section is $f^{-1}(x) = -\frac{x}{2}$, and of the second section is $f^{-1}(x) = \sqrt{x}$. The first is a one-to-one function. As we know the second section goes over positive x values only, we know to only take the positive root of x. It becomes a one-to-one function.

The inverses exist for both sections, and so $f^{-1}(x) = -\frac{x}{2}$ for $x \le 0$ and $f^{-1}(x) = \sqrt{x}$ for $x > 0$.

Checkpoint 3

If $f(x) = 4 - 3x$ and $g(x) = x^2 + 2x$, find $fg(x)$ and $gf(x)$.

Checkpoint 4

Find the range and inverse of the function f defined by: $f(x) = x^2 + 2$ for $x < 0$, and $f(x) = 2 - x$ for $0 \le x \le 5$.

Exam questions

answers: page 84

1 If $f(x) = 7x + 3$ and $g(x) = x^3$, find:
(a) $fg(x)$ (b) $gf(x)$ (c) $f^{-1}(x)$ (d) $g^{-1}(x)$ (e) $f^{-1}g^{-1}(x)$ (15 min)

2 Write down the domain and range of these functions:
(a) $f(x) = \frac{1}{x+3}$ (b) $g(x) = \sqrt{x+7}$ (10 min)

3 Sketch the graph for each of the following functions. Also, state whether it is an even function, an odd function, or neither:
(a) $f(x) = 3x$ (b) $f(x) = \frac{1}{x}$ where $x \ne 0$ (c) $f(x) = 4x^2$
(d) $f(x) = -x^3$ (e) $f(x) = 2x + 5$ (f) $f(x) = \frac{1}{x^2}$ where $x \ne 0$ (35 min)

Graphs of functions

One particularly necessary skill at A-level is curve sketching. Most examples are based on variations of the quadratic, cubic, reciprocal or linear theme, and it's these you should really get to grips with. For quadratic graphing, check out the section on quadratic functions.

Cubic graphs

The cubic function takes the general form $f(x) = ax^3 + bx^2 + cx + d$, where a is non-zero. Sketching these graphs is a combination of finding easy points, like y when $x = 0$, and knowledge of the general shapes.

Action point

Close the book and try to sketch the cubic curves for positive and negative x^3 coefficient values.

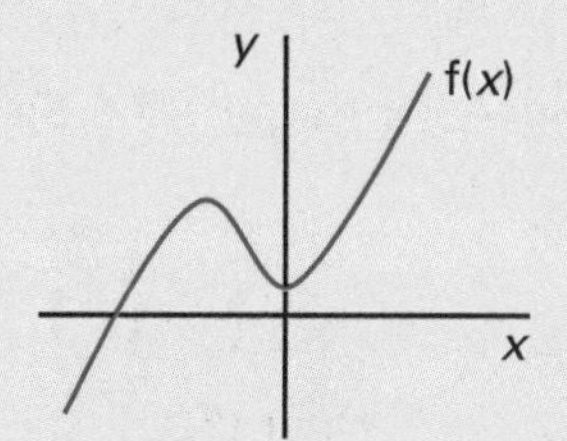

A cubic where $a > 0$.

A cubic where $a < 0$.

Checkpoint 1

Sketch the following curves:
(a) $y = (x + 2)(x + 4)(x - 3)$;
(b) $y = x(x - 2)(x - 5)$;
(c) $y = (x + 1)(x + 3)^2$.

The peaks can be found through differentiation of the function, and setting the result equal to zero. Or, if the function is given in factorised form, then the crossing points can be obtained by finding the value of x which makes each bracket zero. If there is a repeated bracket, then this will be where the curve 'bounces off' of the x-axis.

Reciprocal graphs 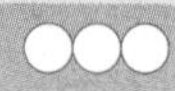

Anything that looks like $f(x) = \frac{a}{x}$ where a is any number is called a **reciprocal**.

Examples are $y = \frac{1}{x}$, $y = \frac{6}{x}$, or $y = 10 + \frac{5}{x}$.

This is a sketch of $y = \frac{1}{x}$. It can never cross the y or x axes since these involve dividing something by zero. This is what you have to look out for when graphing this kind of function.

Examiner's secrets

It is often not necessary to make your sketch very accurate if you are only drawing it to help you decide what the range of a function is, for example. The general form will do, along with any particular points which may be important, like where the curve crosses the axes, or any minimum or maximum points.

Example: Sketch the graph of $y = \frac{1}{1 - x}$. It's going to have roughly the same shape as $y = -\frac{1}{x}$, but notice that when $x = 1$, we get $1 \div 0$ which can't happen. So the line $x = 1$ cannot appear in the domain. Since y can never be zero (why?), we have two lines which the function can never cross. These lines are called **asymptotes**.

Action point

Close the book and try to draw the graph in this example. Remember to label the asymptotes.

The modulus function

The **modulus** of any number is its positive value. The modulus of 2 is 2, and the modulus of –2 is also 2. We write the modulus of x like this: $y = |x|$. The best way to sketch these graphs is to try values of x and see what you get, looking at the positive and negative parts separately.

Example: Sketch the graph of $y = |x + 2|$. Anything less than $x = -2$ will make $(x + 2)$ negative, which the modulus sign converts back into a positive number. For example, $x = -3$ gives $|-3 + 2| = |-1| = 1$. Therefore the line slopes downwards, until hitting zero at $x = -2$. After that, $x + 2$ is positive anyway, so the modulus sign has no effect.

The graph of $y = |x + 2|$ has two parts: the first is the region where $x + 2$ is negative, which the modulus sign only takes the positive value of. Part two is where $x + 2$ is positive anyway.

Examiner's secrets

This is where most graphing mistakes get made – modulus functions. The trick is to be very careful about splitting the domain up into sections where the stuff inside the modulus sign is negative and where it is positive. Treat them as two totally separate pieces, and you'll be fine!

Checkpoint 2

Sketch the graph of $y = |2 - x|$, and state the range of this function. *Hint*: start by working out the point at which $2 - x$ switches from positive to negative and work outwards!

Graphs of inverse functions

The inverse of a function is a mapping which takes the range to the domain. Nicely enough, to find the inverse of any function, all you have to do is reflect its graph in the line $y = x$. The inverse of this cubic function is shown by the dotted line.

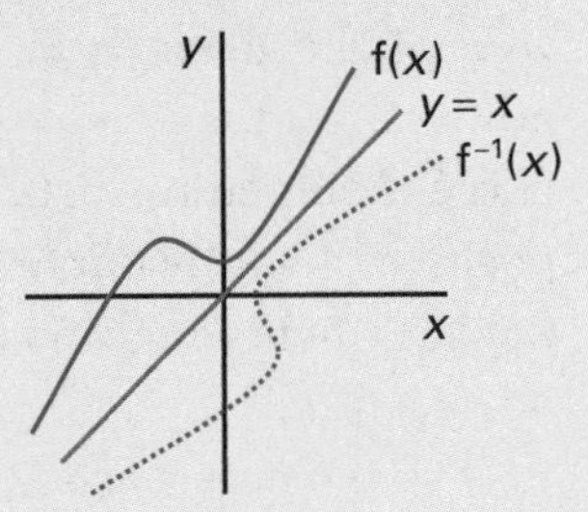

The inverse of $y = x^2$ is not a proper function: to draw the inverse function we have to restrict where the graph goes to compensate. Here, the inverse can only be for positive x.

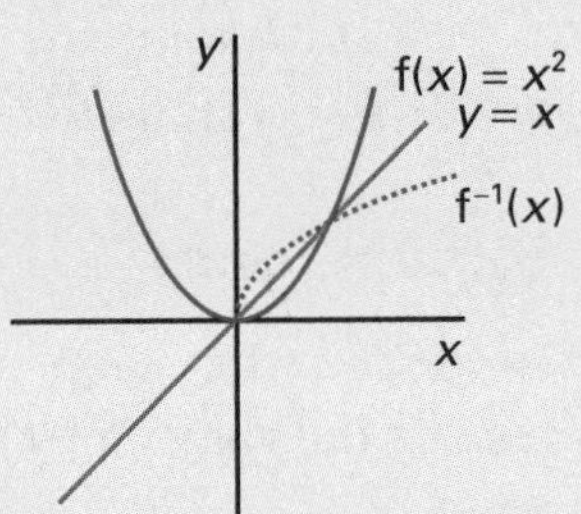

Action point

Write down the equations from these examples and try, with the book closed, to draw them and their inverses. Don't peek!

Exam questions

answers: pages 84–5

Sketch the graph of each of the following:

1 $y = 2x^2 + 3x + 4$

2 $y = (x - 3)^2 + 5$

3 $y = (x - 2)(x + 3)$

4 $y = x(x - 1)(x + 2)$

5 $y = -x^3 + 4x + 5$

6 $y = \dfrac{1}{x + 7}$

7 $y = 5 - \dfrac{2}{x}$

8 $y = |x + 3|$

9 $y = |2x - 7|$ (30 min)

Sketch the following functions and their inverses (on the same axes):

10 $y = 2x^2, x \geq 0$

11 $y = 2x - 6$

12 $y = -x^2 + 8, x \geq 0$ (30 min)

Transformations

Once you have drawn your graph, you may need to investigate what happens to it when it is transformed in some way. We split the transformations into two types: translations, where the whole curve is shifted, or expansions, where the curve is stretched or compressed.

Translations

A translation can take two directions: vertical or horizontal. We'll look at what happens to $f(x) = x^2$ under a vertical translation and then a horizontal one.

Look at $g(x) = f(x) + k$, where k is some constant. Take the point $x = 0$. Here, $f(x)$ would have given $y = 0^2 = 0$, but $f(x) + k$ would give $0^2 + k = k$. The whole curve is shifted up by a distance of k. (If k is negative, then it's shifted down.)

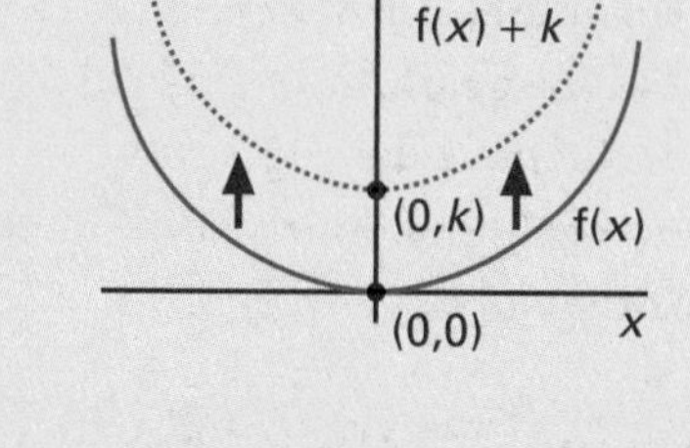

Now look at $f(x + k)$. Before, $x = 0$ would have given $y = 0$. To get $y = 0$ now, x must be $-k$. So the whole curve is shifted by distance k. If k is positive, then the curve is shifted to the left, since what was (0,0) has become $(-k,0)$.

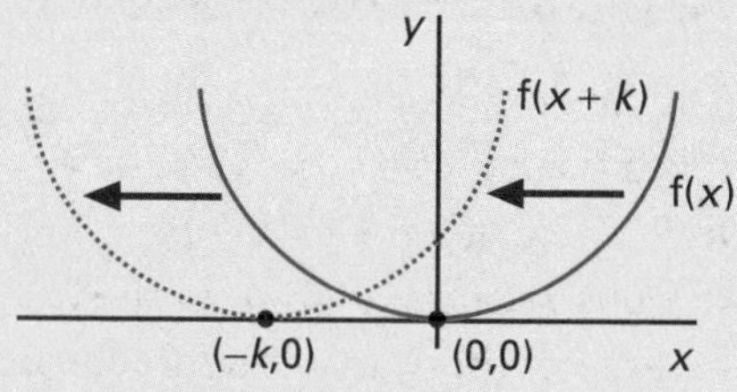

We summarise as follows:

$y = f(x) + k$ shifts the whole graph vertically in the direction of k.

$y = f(x + k)$ shifts the whole graph left for $k > 0$, right for $k < 0$.

Action point

On the same axes, sketch the graphs of $y = x^2$, $y = x^2 + 2$ and $y = (x + 2)^2$. Try plotting points from $x = -5$ to $x = 5$, and make a note about what happened after each transformation.

Examiner's secrets

The most common mistake to avoid is thinking that $f(x + 8)$ shifts the graph 8 units to the right, because you've added 8. This is not the case, as the last 'Action point' would have shown. Remember, when it goes inside the brackets, it goes the 'wrong' way, OK?

Expansions

We've seen what happens when a constant is added either to the x value or the value of $f(x)$. What about when x or $f(x)$ is multiplied by a constant?

This time, start with $f(x)$ as a cubic function. Let's look at $y = 2f(x)$. When $f(x) = 0$ there's no change. When $x = 4$, we now get $y = 2f(4)$, i.e. $f(4)$ is moved up. When $x = -3$, $y = 2f(-3)$, i.e. $f(-3)$ gets moved down. Each point is pushed out, but not by the same amount. The curve is stretched by a factor 2 in the y direction.

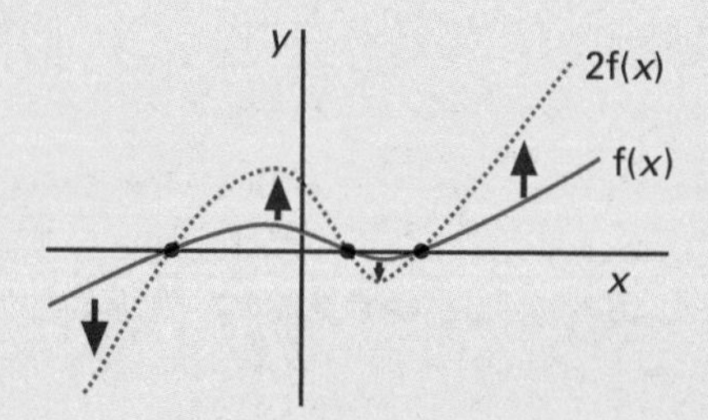

Action point

On the same axes, sketch the graphs of $y = \sin x$, $y = 2 \sin x$, $y = \sin 2x$ and $y = \sin \frac{1}{2}x$. Have x go from -4π to $+4\pi$. Make notes about what happens to the curve under each transformation.

When every x value has been multiplied by 2 we get f(2x). When $x = 0$, $y = f(2 \times 0)$ so it doesn't move. But when $x = 1$, $y = f(2 \times 1) = f(2)$, so f(2) has been moved one unit to the left. At $x = -1$, $y = f(2 \times (-1)) = f(-2)$, so f(–2) gets moved one unit to the right. The overall effect is to compress the curve by a factor of $\frac{1}{2}$ in the x direction.

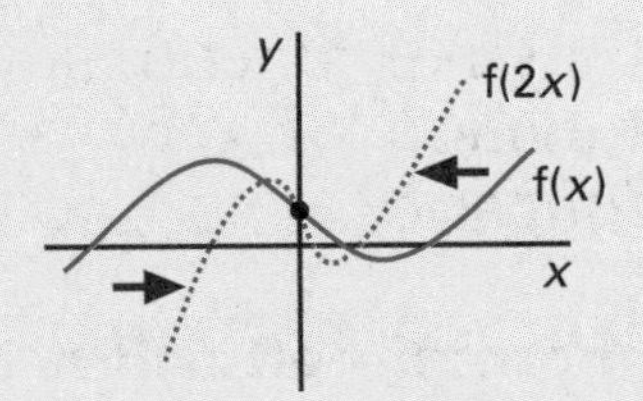

Combinations of transformations

Example: Let $f(x) = (x - 1)(x - 2)(x + 3)$. Sketch the graph of $g(x) = 3f(x - 4)$.

First, sketch f(x). It is a cubic which crosses the x-axis at $x = 1$, $x = 2$ and $x = -3$.

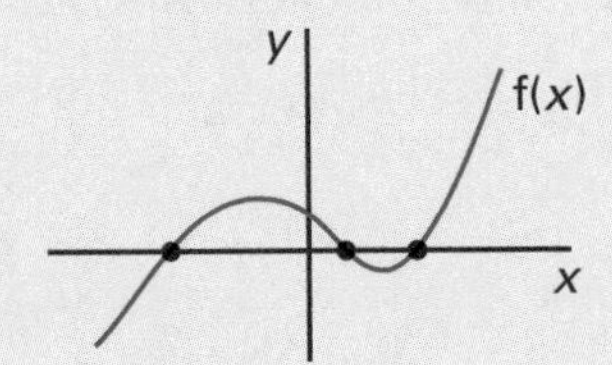

We work in stages. Multiplying by 3 will stretch the curve by a factor of 3, trebling all y coordinates (dotted line).

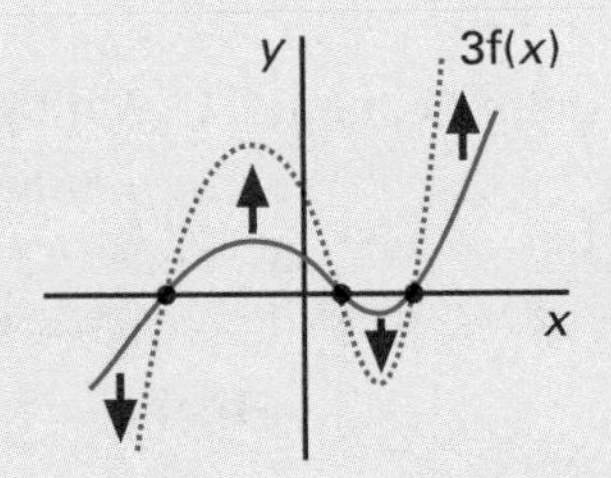

Adding –4 to every x value then shifts the stretched curve 4 units to the right (dashed line).

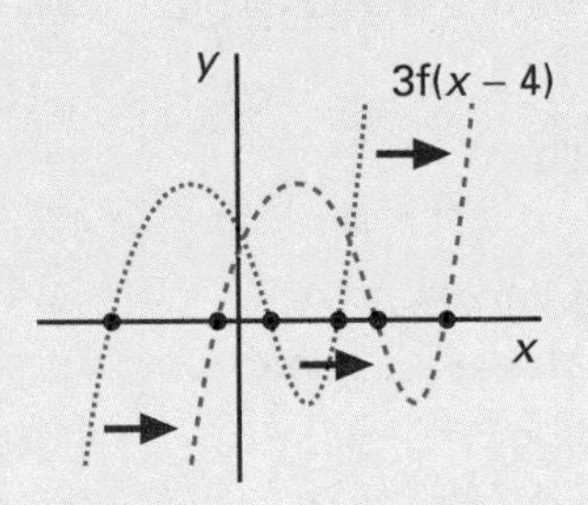

Checkpoint

This is the big one! On the same axes, with x from -4π to $+4\pi$, sketch the graphs of: $y = \cos x$, $y = 2\cos x$, $y = \cos\frac{1}{2}x$, $y = \cos(x + \frac{\pi}{2})$ and $y = 1 - \cos x$. Phew! Make it big, to fit them all on, and remember labels. Make a note about how each curve is transformed from the original, and what change in the equation brought this about.

Examiner's secrets

Always work in stages when sketching composite functions. Draw two or more graphs if you have to, to make it clearer where to go next. The best place to start is the original function, taking the transformations one at a time until you're through.

Action point

Starting with $f(x) = x^2$, draw the graphs of $g(x) = 2f(x + 3)$ and $h(x) = 10 - f(3x)$. For the second one, sketch $f(3x)$ first, then pick points and see where they go!

Exam questions

answers: page 85

1 Each of the following are transformations of the graph $y = x^2$. Describe each transformation fully.

(a) $y = x^2 + 2$ (b) $y = x^2 - 8$ (c) $y = (x - 3)^2$
(d) $y = (x + 5)^2 - 2$ (e) $y = (x + 8)^2 + 6$ (f) $y = 2x^2$ (15 min)

2 If $f(x) = (x + 1)(x - 2)(x + 4)$, sketch the graph of $g(x) = 2f(x + 5)$. (10 min)

3 Sketch the graphs of the following functions, with x from -2π to 2π.

(a) $\sin 4x$ (b) $\cos\left(x + \frac{\pi}{4}\right)$ (c) $\tan\left(x - \frac{\pi}{6}\right)$

(d) $\sin\left(2x + \frac{\pi}{2}\right)$ (e) $\cos\frac{x}{2}$ (f) $3\sin x$ (20 min)

Rational functions

Rational functions are algebraic fractions. Divide one function of x by another and you've got a rational function.

Formal definition

If $P(x)$ is one polynomial function of x and $Q(x)$ is another, $\frac{P(x)}{Q(x)}$ is a rational function.

Links

See 'Polynomials', page 8.

Simplification

You'll be expected to simplify these *algebraic* fractions, so make sure you are confident with your addition, subtraction, multiplication and division of *numerical* fractions. Here are simple reminders:

Addition/subtraction:
1. Multiply denominators together (to get a common factor).
2. Whatever you multiplied each denominator by, do the same to the numerator.
3. Add/subtract the numerators.

Multiplication:
1. Multiply the numerators.
2. Multiply the denominators.

Division:
1. Change the ÷ to a multiply.
2. Flip the second fraction over.
3. Treat as a multiplication.

Checkpoint 1

Remember that a polynomial is a function in which each term is in the form of ax^n, where a is a constant and n is a positive integer. Which of the following would be an example of a polynomial?
(a) $5x^3 + 2x^2 + 4$
(b) $\sqrt{x} + x^2$
(c) $4x^5 + 2 + \frac{1}{x}$
(d) $12x^7 + 4x^3 + x$

Checkpoint 2

Follow the steps here to calculate these:
(a) $\frac{4}{7} + \frac{5}{8}$; (b) $\frac{8}{9} + \frac{7}{10}$; (c) $\frac{5}{6} \div \frac{2}{3}$

Example: Simplify
$$\frac{2x - 10}{(x + 5)(x + 4)} + \frac{10}{x^2 + 3x - 4}$$

Factorising:
$$= \frac{2(x + 5)}{(x + 5)(x + 4)} + \frac{10}{(x - 1)(x + 4)}$$

Links

See 'Factorisation', page 9.

Cancelling the $(x + 5)$s and adding, using the first two steps stated above for addition/subtraction:
$$= \frac{2(x - 1) + 10}{(x + 4)(x - 1)}$$

Multiplying brackets:
$$= \frac{2x - 2 + 10}{(x + 4)(x - 1)}$$

Factorising:
$$= \frac{2(x + 4)}{(x + 4)(x - 1)}$$

Cancelling $(x + 4)$:
$$= \frac{2}{x - 1}$$

Checkpoint 3

Factorise:
(a) $5x + 15$ (b) $2xy + 4x^2$
Multiply out these brackets:
(c) $5(x - 4)$ (d) $x(x + 3)$

Algebraic division

Supposing you had a question like this:

Simplify $$\frac{x^3 - 4x^2 + 7x - 12}{x - 3}$$

You could try long division. Use the notes on the following page for guidance. *Note*: If you were trying to evaluate $(x^4 + 5x - 6) \div (x + 1)$, you would first need to write the polynomial so that it included all the zero terms, i.e. $(x^4 + 0x^3 + 0x^2 + 5x - 6) \div (x + 1)$.

Links

See the remainder theorem, page 9.

Action point

Calculate 45 731 ÷ 47 using long division and think about the process as you are doing it.

$$
\begin{array}{r l}
 & {}^{①}x^2 - {}^{⑤}x + 4 \\
(x-3) & \overline{)\,x^3 - 4x^2 + 7x - 12} \\
 & {}^{②}\underline{x^3 - 3x^2} \quad \downarrow ④ \\
 & {}^{③} -x^2 + 7x \qquad \text{subtract} \\
 & \underline{-x^2 + 3x} \\
 & 4x - 12 \quad \text{subtract} \\
 & \underline{\underline{4x - 12}}
\end{array}
$$

① Decide on the first term. To get x^3 we need to multiply the x in $x - 3$ by x^2. $\therefore x^2$ is the first term.
② Multiply $(x - 3)$ by x^2 and write the answer down.
③ Subtract to give the remainder.
④ Bring down the next term.
⑤ To get a $-x^2$ you must multiply $(x - 3)$ by $-x$.

Keep repeating steps ② to ④ until the division is complete.

Action point

Remember how to add fractions: 1 Find a common denominator (easily done by multiplying the two denominators); 2 The numerators then also need to be multiplied by their respective numbers.

E.g. $\frac{1}{3}+\frac{2}{5}=\frac{1\times5+2\times3}{3\times5}$

Complete the above calculation and do this one, using the same technique: $\frac{3}{10}+\frac{4}{7}$

Partial fractions

A partial fraction consists of at least two distinct algebraic fractions, which are being added. There are three main types: those with linear factors, repeated factors and quadratic factors.

1 **Linear factors** are always of the following format:

Example: $\dfrac{2x-7}{(x-3)(x+5)} = \dfrac{A}{x-3} + \dfrac{B}{x+5}$

$$\frac{2x-7}{(x-3)(x+5)} = \frac{A(x+5)+B(x-3)}{(x-3)(x+5)}$$

The denominators are the same, so, the numerators will also be equal to each other.

$$2x - 7 = A(x + 5) + B(x - 3)$$

Substituting $x = 3$ (to eliminate B):

$$2 \times 3 - 7 = A(3 + 5) + B(3 - 3)$$
$$-1 = 8A$$
$$\therefore A = -\tfrac{1}{8}$$

Substituting $x = -5$ (to eliminate A):

$$2(-5) - 7 = A(-5 + 5) + B(-5 - 3)$$
$$-17 = -8B$$
$$\therefore B = \tfrac{17}{8}$$

$$\Rightarrow \frac{2x-7}{(x-3)(x+5)} = \frac{-1}{8(x-3)} + \frac{17}{8(x+5)}$$

Watch out!

Sometimes substituting in values for x to eliminate A, B or C will not be sufficient to work them out. In which case, comparing the coefficients of x or x^2 may be necessary.

2 **Repeated factors** are of this format:

Example: $\dfrac{2x+7}{(x-3)(x+5)^2} = \dfrac{A}{x-3} + \dfrac{B}{x+5} + \dfrac{C}{(x+5)^2}$

3 **Quadratic factors** are of this format:

Example: $\dfrac{2x+7}{(x-3)(x^2+5)} = \dfrac{A}{x-3} + \dfrac{Bx+C}{(x^2+5)}$

Checkpoint 4

Work out the values of A, B and C in the 'Repeated factors' example.

Checkpoint 5

Evaluate A, B and C in the 'Quadratic factors' example.

Exam questions

answers: pages 85–6

1 Simplify: (40 min)

(a) $\dfrac{2}{3x-1} - \dfrac{1}{4x+1}$ (b) $\dfrac{3x+18}{(x+6)(x-4)} - \dfrac{9}{x^2-5x+4}$ (c) $\dfrac{x^3+x^2-22x-40}{x-5}$

2 Express these in partial fractions: (30 min)

(a) $\dfrac{4}{(x+1)(x-1)}$ (b) $\dfrac{2x-1}{(x+3)(x^2-1)}$ (c) $\dfrac{x^2+1}{(x-2)(x+1)^2}$ (d) $\dfrac{x+3}{x(x+5)^2}$

Exponentials and logarithms

Exponentials are in simple terms the powers of numbers and logarithms are their inverse functions. Logarithm is really another word for power or index. For example, $3^4 = 81$ can be read as '4 is the power to which the base 3 must be raised to get 81' and this can be written as $4 = \log_3 81$.

e^x and its graph

Checkpoint 1

Sketch the graphs of the functions shown as examples here.

Exponential functions are of the format $y = a^x$, where $a > 0$ (for example $y = 2^x$, $y = 3^x$, $y = 7^x$). Finding the gradient of some of these functions at the point (0,1) produces the results shown below. Notice that if you wanted a gradient of 1 at the point (0,1), the exponential function would be between $y = 2^x$ and $y = 3^x$. In fact, the number in the function is given a letter, e, where **e = 2.718 (to 4 s.f.)**.

Exponential function	*Gradient at (0,1)*
$y = 1^x$	0
$y = 2^x$	0.693
$y = 3^x$	1.099
$y = 4^x$	1.386

Checkpoint 2

How would you get the value of e on your calculator? Is it a rational or an irrational number?

Having the gradient 1 at the point (0,1) makes e^x a very special function as it is the only function which does not change when differentiated (or integrated).

$$\frac{d}{dx}e^x = e^x$$

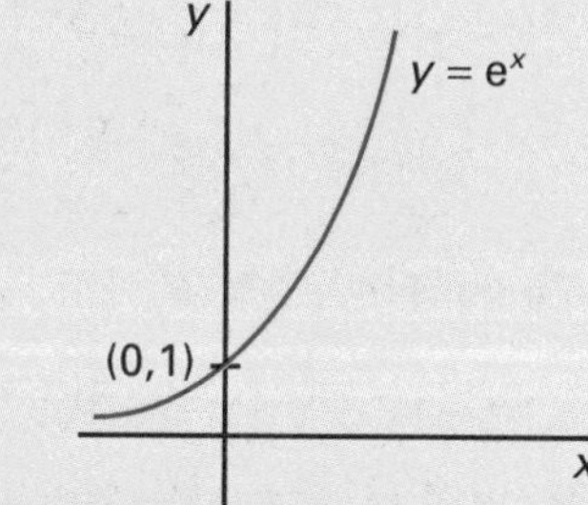

Links

See 'Basic differentiation', page 52.

This is the graph of $y = e^x$.
The gradient of $y = e^x$ at (0,1) is 1.
Note that *the* exponential function is e^x.

ln x and its graph

Links

See 'Graphs of inverse functions', page 21.

Natural logarithm is abbreviated to ln x. If you reflect the graph of a power (for example, $y = 2^x$) in the line $y = x$, you get $y = \log_2 x$. It is the inverse function. **Natural** logarithm (sometimes referred to as **Naperian** logarithm) is the inverse function of e^x. See the graph below.

Watch out!

$\ln x = \log_e x$
They can be written either way although most commonly ln x is used. ln x and log x are both on scientific and graphical calculators, however, log x is $\log_{10} x$. They are different so be careful you use the correct one.

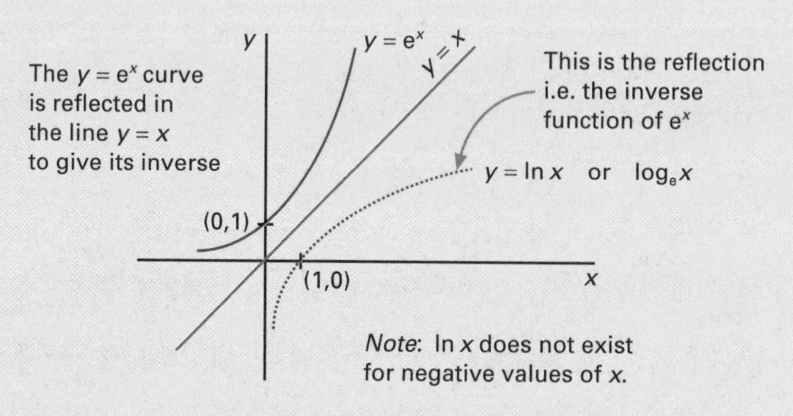

Also, note: $\frac{d}{dx}\ln x = \frac{1}{x}$

Laws of logarithms

The following law is very useful, especially for finding powers.

$$\log_a b = c \Leftrightarrow b = a^c$$

Notice with the next three laws that the logs must all have the same base.

$$\log_a x + \log_a y = \log_a xy$$
$$\log_a x - \log_a y = \log_a \frac{x}{y}$$
$$\log_a x^k = k \log_a x$$

Examples:

1 Simplify: (a) log 7 + log 3 (b) log 20 – log 10
(c) log 8 (d) log 16/log 4

2 Work out the value of:
(a) $3 \log_2 8 - \log_3 81$ (b) ln 12 (c) $\log_e 2.6$

Solutions:

1 (a) Using $\log_a x + \log_a y = \log_a xy$,
$\log 7 + \log 3 = \log 7 \times 3$
$\therefore \log 7 + \log 3 = \log 21$

(b) Using $\log_a x - \log_a y = \log_a \frac{x}{y}$
$\log 20 - \log 10 = \log \frac{20}{10}$
$\therefore \log 20 - \log 10 = \log 2$

(c) Using $\log_a x^k = k \log_a x$
$\log 8 = \log 2^3$
$\therefore \log 8 = 3 \log 2$

(d) $\frac{\log 16}{\log 4} = \frac{\log 4^2}{\log 4}$
Using $\log_a x^k = k \log_a x$
$\frac{\log 4^2}{\log 4} = \frac{2 \log 4}{\log 4} = 2$

2 (a) $\log_2 8 = 3$ (2 to the power of 3 gives 8)
$\log_3 81 = 4$ (3 to the power of 4 gives 81)
$\therefore 3 \log_2 8 - \log_3 81 = 3(3) - 4 = 5$

(b) Using a calculator
ln 12 = 2.48 (to 3 s.f.)

(c) Remember: $\log_e = \ln$
Using a calculator
$\log_e 2.6 = 0.956$ (to 3 s.f.)

Checkpoint 3

Write $9 = 3^2$ in logarithmic format.

Examiner's secrets

Learn these three laws off by heart.

Links

See 'Indices and surds' page 6.

Examiner's secrets

Get into the habit of writing down which of the laws you are using. This shows a greater understanding and will make it easier for the examiner to mark.

Examiner's secrets

Don't confuse $\frac{\log 16}{\log 4}$ with $\log \frac{16}{4}$. They are different!

Exam questions answers: page 86

1 Simplify the following: (a) log 5 + log 2 (b) 2 log 4 + log 3
(c) log 12 – log 4 (d) log 8/log 2 (e) $\frac{1}{2}\log 36 - 2 \log 5$
(f) $\frac{1}{2}\ln 25 + 2 \ln 3 - \ln 5$ (10 min)

2 Work out the value of the following, giving your answers to 3 s.f. (where appropriate):
(a) $\log_7 49$ (b) $\log_5 0.2$ (c) $\log_4 32$ (d) ln 14 (e) $\log_e 28$ (5 min)

Applications

Here are some typical applications for logarithms and exponentials.

Exponential growth and decay

This is connected with the work on natural logarithms and the exponential function. Taking the differential equation $dy/dx = y$, it can be rearranged and integrated to give $y = Ae^x$. (This is covered in more detail later on in this book.)

Also, it can be shown that:

if $dy/dx = \lambda y \Rightarrow y = Ae^{\lambda x}$

The graphs of exponential growth and decay look like this:

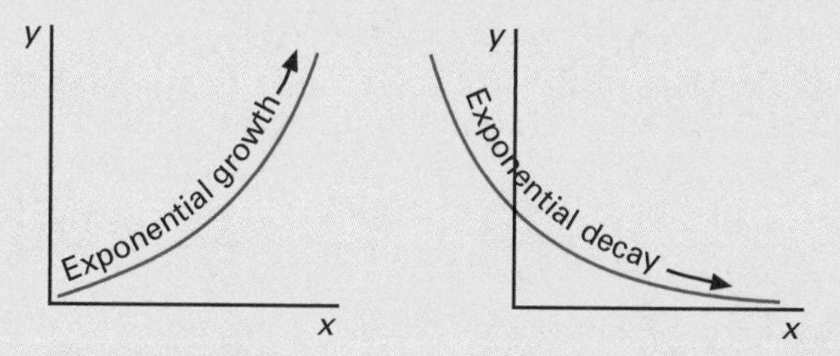

Worked examples

1 Solve $e^{3x+1} = 5$.

Taking the natural logarithm of both sides:

$$\log_e e^{3x+1} = \log_e 5$$

Using $\log_a x^k = k \log_a x$ $\quad (3x+1)\log_e e = \log_e 5$

$$3x + 1 = \log_e 5$$

$$3x + 1 = 1.609\,437\,912$$

(using a calculator)

$$\therefore x = 0.203 \text{ (to 3 d.p.)}$$

2 A cup of boiling water is left in a room and, after 7 minutes, has cooled to 85°C. If the room temperature is 28°C, how long will the water take to cool to 50°C?

The rate of change in temperature is directly proportional to the difference in the water temperature and the temperature of the room. Let t be the time in minutes and let y be the *difference* in temperature between the water and the room.

Since $dy/dt \propto y$, we can write $dy/dt = \lambda y$, where λ is a proportionality constant.

Using $dy/dx = \lambda y \Rightarrow y = Ae^{\lambda x}$ $\qquad y = Ae^{\lambda t}$

When $t = 0$ $\quad y = 100 - 28 \qquad y = 72$

$\therefore \quad 72 = Ae^0$

$\Rightarrow \quad A = 72 \qquad \therefore y = 72e^{\lambda t}$

When $t = 7$ $\quad y = 85 - 28 \qquad y = 57$

$57 = 72e^{7\lambda}$

$0.791\,6 = e^{7\lambda}$

Using $\log_a b = c \Leftrightarrow a^c = b$ $\quad \ln 0.791\,6 = 7\lambda$

$\lambda = -0.033\,4$

$\therefore y = 72e^{-0.033\,4t}$

For 50°C $\quad y = 50 - 28 \quad y = 22$

$22 = 72e^{-0.033\,4t}$

$0.305 = e^{-0.033\,4t}$

Links

See 'Differential equations', page 70.

Examiner's secrets

It would be worth your while learning this off by heart.

Checkpoint 1

Which part of the equation $y = Ae^{\lambda x}$ shows whether it represents growth or decay? And, how?

Examiner's secrets

$\log_a a = 1 \therefore \ln(e) = 1$

Checkpoint 2

Who originally discovered this rate of cooling?

Checkpoint 3

What is e^0 equal to?

Watch out!

Remember $\ln x$ is $\log_e x$.

Using $\log_a b = c \Leftrightarrow a^c = b$ $\quad \ln 0.305 = -0.033\,4t$

$\therefore t = 35.5$ minutes (to 3 s.f.)

Solution of $a^x = b$

The following worked examples involve logarithms.

1 Solve these: (a) $3^x = 12$ (b) $3^{x-1} = 5^{2x+1}$

(a) $3^x = 12$

Taking logs of both sides: $\log 3^x = \log 12$

Using $\log_a x^y = y \log_a x$: $x \log 3 = \log 12$

$$x = \frac{\log 12}{\log 3}$$

$\therefore x = 2.26$ (to 3 s.f.)

(b) $3^{x-1} = 5^{2x+1}$

Taking logs of both sides: $\log 3^{x-1} = \log 5^{2x+1}$

Using $\log_a x^y = y \log_a x$: $(x-1)\log 3 = (2x+1)\log 5$

Multiplying out the brackets: $(\log 3)x - \log 3 = (2\log 5)x + \log 5$

Solve as a linear equation: $-\log 5 - \log 3 = (2\log 5)x - (\log 3)x$

$\therefore x = -1.28$ (to 3 s.f.)

2 How long does a sum of money take to double itself when it has been invested at a compound interest rate of 7%?

Let x be the initial sum of money and n the number of years it is invested for.

$x(1.07)^n = 2x$

Cancel the x's: $1.07^n = 2$

Taking logs of both sides: $\log 1.07^n = \log 2$

Using $\log_a x^y = y \log_a x$: $n \log 1.07 = \log 2$

$$n = \frac{\log 2}{\log 1.07}$$

$\therefore n = 10.2$ years (to 3 s.f.)

Checkpoint 4

When a logarithm is written without a base, such as log, what is the value of the base?

Links

See 'Exponentials and logarithms', pages 26 and 27.

Checkpoint 5

What does the 1.07 represent in this equation?

Exam questions answers: page 86

1 A colony of ants has a population of 200 and is growing at the rate of 20 ants per day. How many ants will there be after 30 days if the rate of growth is directly proportional to the population size at that time? (10 min)

2 George is disturbed by his neighbour whilst he is making a cup of coffee. He takes his coffee without milk and has poured in the boiling water. George goes to help his neighbour and returns three minutes later. The temperature of his coffee is now 94°C. Just then the phone rings and George decides to answer it. If George only likes to drink coffee when it is 80°C or more, and the temperature in the kitchen is 20°C, how long can he spend on the phone? (10 min)

3 If £2 000 is invested in a bank at a compound interest rate of 6.5%, for how many years would it need to be left to have increased to £5 000? (10 min)

4 Solve these equations (to 3 s.f.):

(a) $7^x = 24$ (b) $5^{2x} = 12$ (10 min)

Sequences

Action point

A sequence is of the format:
$u_1, u_2, u_3, u_4, \ldots$
where each 'u' represents a different term. For example, 8, 11, 14, 17, . . .
The first term, u_1, is 8, the second term, u_2, is 11. What are the values of u_3 and u_4?

The jargon

Sometimes called 'position to term' sequences.

Checkpoint 1

Work out the nth term of the sequence shown below, using the method described in worked example 1. 6, 14, 22, 30, . . .

Examiner's secrets

Check your nth term by using it to work out the first three terms in the sequences.

Checkpoint 2

Why is $(-1)^n$ used for getting negative odd terms? What would $(-1)^{n+1}$ be used for, and why?

Test yourself

What are the values of u_1, u_2 and u_5 of the sequence in worked example 3.

Checkpoint 3

Using the technique described in worked example 4, find the nth term of this sequence: 36, 144, 576, 2 304

A sequence is a set of numbers, written out in a particular order, for which there is a rule connecting each number (usually each consecutive number).

Formula for the *n*th term

A 'term' is the name given to each member of a sequence. If the sequence does not end with a last term (usually will have '. . .' written at the end) it is known as an infinite sequence. If there is a last term (and the sequence ends with a full stop) then the sequence is said to be finite. The **general term for a sequence** is also known as the ***n*th term or the *i*th term**.

Worked examples

Here are some examples of sequences and how to find the nth term.

1 4, 7, 10, 13, . . .
$u_1 = 4$ is the first term, $u_2 = 7$ is the second term.
Taking the first difference (i.e. $7 - 4$, $10 - 7$, $13 - 10$) in this case gives us a constant '3' each time. Therefore, the nth term will involve a multiple of 3, i.e. $3n$. The first term, however, is 4. To get this you first do 3×1 (n is 1 as we are referring to the *first* term) and then, by inspection, adding 1 will make the 4. Therefore, the nth term is $3n + 1$.

2 −2, 4, −6, 8, . . .
Deal with the alternating − and + first. This is simply done by having a factor of $(-1)^n$ or $(-1)^{n+1}$ in the nth term. In this case, the odd terms are negative, therefore, we use $(-1)^n$. Ignoring the negative signs and finding the first difference we get a constant '2'. Therefore, the nth term is $(-1)^n 2n$.

3 $\frac{1}{2}, \frac{2}{3}, \frac{3}{4}, \frac{4}{5}, \frac{5}{6}, \ldots$
Sometimes it is necessary to simply look at the numbers and try to 'see' the connection. Here the nth term is clearly $\dfrac{n}{n+1}$.

4 12, 24, 48, 96, 192, 384
If the numbers are doubling then the nth term contains a power of 2. If you divide the first term by 2 it will give you the other factor in the nth term. Here, the first term is: $\frac{12}{2} = 6$. Therefore, the nth term is $6(2)^n$. The same technique applies to any 'powers of' sequences.

5 1, −8, 27, −64, 81
If numbers go up rapidly and are not obviously being multiplied by the same number, the nth term could contain a square or a cube or a. . . . Here, the nth term is $(-1)^{n+1}n^3$.

You could also be given problems where you are given the general nth term and asked to find specific terms. For example:

> The nth term for a sequence is $3n^2 + n$. Write down the first five terms.

You simply substitute in 1 for n to get the first term, 2 for the second term and so on. The terms are: 4, 14, 30, 52, 80.

Recurrence relations

Another way of giving or defining a sequence is by an **inductive definition**, which needs two items:

1 **a starting value (or values),**
2 **a recurrence relation.**

Usually the starting value is the first term(s), while the recurrence relation is a *formula* that is used to work out a term from the previous term(s).

Example: Using this sequence: 4, 7, 10, 13, we can see how an inductive definition works. You need two items.

1 The starting value. In this case, $u_1 = 4$ is the first term and the starting value.
2 The recurrence relation. To get from the first term to the second, the second to the third, etc. you add 3.

Therefore, the recurrence relation can be written as $u_{n+1} = u_n + 3$. For example, $u_2 = u_1 + 3$ and $u_1 = 4$, therefore, $u_2 = 4 + 3 = 7$.

Convergence and divergence

All sequences are either **convergent** or **divergent**.

If a *sequence is convergent*, the terms will be getting closer and closer to a fixed value. (Shown in the first diagram.) For example, 36, 18, 9, 4.5, 2.25, 1.25, . . .

If a *sequence is divergent*, the terms will keep moving away from the original value. (Shown in the second diagram.) For example, 1, 4, 9, 16, 25, . . .

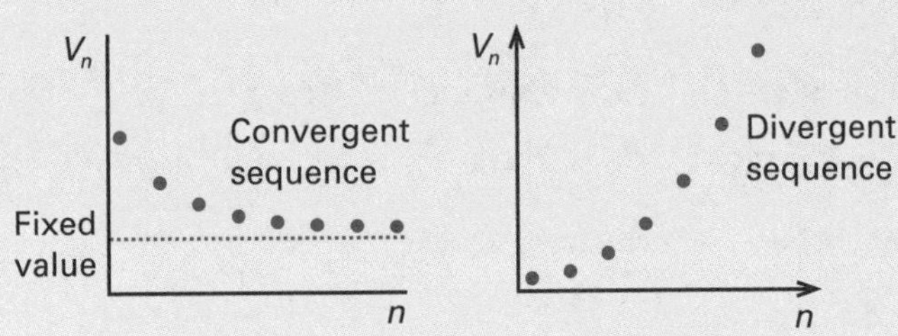

Note: both convergent and divergent sequences can also be oscillatory (i.e. they can oscillate – move or swing from side to side regularly).

Exam questions answers: page 86

1 Find the nth term of these sequences:
(a) 7, 10, 13, 16
(b) 6, 18, 54, 162
(c) 3, 6, 11, 18, 27
(d) 4, −8, 16, −32, 64
(e) 64, −32, 16, −8, 4 (15 min)

2 Write down the first five terms of the sequence X where $X_{n+1} = X_n + 6$ and $X_1 = 3$. (5 min)

3 Write down the first five terms of the sequence Y where $Y_{n+2} = Y_{n+1} + 2Y_n$ and $Y_1 = 2$ and $Y_2 = 3$. (5 min)

4 State whether the sequences for Question 1 are divergent or convergent. (5 min)

The jargon

Sometimes called 'term-to-term' sequences.

Watch out!

Sometimes there are two (or three) starting values. A very common example of a sequence with two starting values is the Fibonacci sequence. Here, $u_1 = 1$, $u_2 = 1$ are the two starting values and the recurrence relation is $u_{n+2} = u_{n+1} + u_n$.

Test yourself

Write down the first seven terms of the Fibonacci sequence as defined in the 'Watch out!' above.

Checkpoint 4

What value is this convergent sequence heading for?

Series

Adding a sequence together changes it to a series. For example, the sequence 5, 7, 9, 11, 13 becomes a series when added, i.e. 5 + 7 + 9 + 11 + 13. Σ (sigma notation) is, in simple terms, the symbol for 'the sum of' and is often used with series. For example,

$$\sum_{r=3}^{7} r^3 = 3^3 + 4^3 + 5^3 + 6^3 + 7^3$$

Links

See 'Sequences', page 30.

Watch out!

Looking at the series $\sum_{r=3}^{7} r^3$ the 3 at the bottom is the r value you start with and the 7 at the top is the number you finish with. Therefore, there are (3 → 7) 5 terms. A common misconception is that the top number is the number of terms.

Arithmetic series

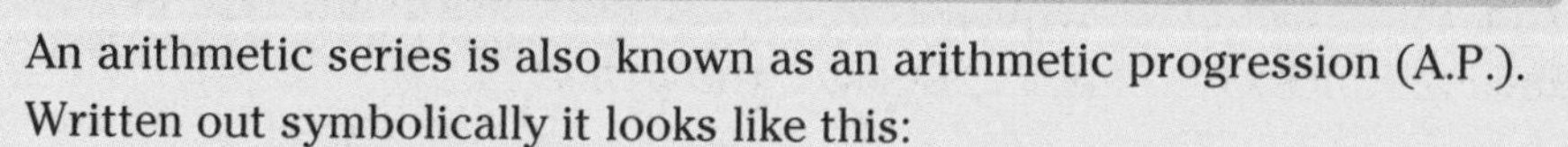

An arithmetic series is also known as an arithmetic progression (A.P.). Written out symbolically it looks like this:

$$a + (a + d) + (a + 2d) + \ldots + (a + (n - 1)d)$$

where a is the first term, d is the common difference between each term and the nth term, u_n, is $a + (n - 1)d$.

The sum of an arithmetic series is calculated using either:

$$S_n = \frac{(a + l)n}{2} \quad \text{where } l \text{ is the last term}$$

Or, by replacing l with $a + (n - 1)d$:

$$S_n = \frac{n}{2}(2a + (n - 1)d)$$

Test yourself

Substitute $a + (n - 1)d$ for l into the first formula and show that the second formula is produced.

Note: if the first difference of a series is constant, it is an **arithmetic series**.

Example: Find the sum of this arithmetic series:

$$7 + 13 + 19 + 25 + \ldots + 169$$

The last term is $a + (n - 1)d$ and is equal to 169.

$$a + (n - 1)d = 169 \quad ①$$

We also know the first term is 7. $\therefore a = 7$ ②

The difference between the terms is 6, i.e. $d = 6$ ③

Substituting a and d into ① we get $7 + (n - 1)(6) = 169$

Rearranging gives $n = 28$. Therefore, there are 28 terms.

Substituting all known information into $S_n = \frac{n}{2}(a + l)$

we get $S_{28} = (28 \div 2)(7 + 169)$

$S_{28} = 2\,464$

Checkpoint 1

Follow this method to find the sum of the A.P.: 4 + 11 + 18 + 25 + . . . + 137.

Checkpoint 2

How many terms would $\sum_{r=6}^{12} 4r$ have?

Geometric series

Geometric series are also known as geometric progressions (G.P.).

They are of the format: $a + ar + ar^2 + ar^3 + \ldots + ar^{n-1}$

where a is the first term, r is the common ratio and the nth term, u_n, is ar^{n-1}. The sum of a geometric series is worked out using:

$$S_n = \frac{a(1 - r^n)}{1 - r}$$ This can also be written as $S_n = \frac{a(r^n - 1)}{r - 1}$ when $r > 1$.

To find the common ratio: take any two consecutive terms and divide the latter by the former: e.g. 3 + 6 + 12 + 24, . . . using the first and second terms (2nd ÷ 1st) gives 6 ÷ 3 = 2, which is the common ratio.

Checkpoint 3

Follow this method to find the common ratio of this G.P.:
4 + 20 + 100 + 500
Do you get the same ratio if you choose any two consecutive terms?

Note: if the same number is multiplying each consecutive term, it is a **geometric series**.

Example: In a geometric series $u_3 = 9$ and $u_5 = 20.25$. Using this information, find the first term, the common ratio and the sum of the first eight terms of the series.

A G.P. can be written as: $a + ar + ar^2 + ar^3 + ar^4$ where $u_3 = ar^2$ and $u_5 = ar^4$

$\therefore ar^2 = 9$ ① $ar^4 = 20.25$ ②

Dividing these terms we get:

$$\frac{ar^4}{ar^2} = \frac{20.25}{9}$$

Cancelling we get $r^2 = \frac{20.25}{9}$

$r^2 = 2.25 \quad \therefore r = 1.5$

Substituting r into ①

$a \times 1.5^2 = 9 \quad \therefore a = 4$

Using $S_n = \frac{a(r^n - 1)}{r - 1}$ (This version is better for $r > 1$)

$S_8 = \frac{4(1.5^8 - 1)}{1.5 - 1} \quad S_8 = 197$ (to 3 s.f.)

The sum to infinity of a GP

Please note that the following formula **only** applies to geometric progressions where the common ratio is such that $|r| < 1$. The series will be convergent.

$$\boldsymbol{S_\infty = \frac{a}{1 - r}}$$

Example: Find the sum to infinity of the series:

$90 + 18 + 3.6 + 0.72 + \ldots$

The common ratio, $r = 18 \div 90 = 0.2$ and the first term $a = 90$.

$S_\infty = \frac{a}{1 - r} \quad S_\infty = \frac{90}{1 - 0.2}$

$S_\infty = 112.5$

Exam questions answers: pages 86–7

1 Find the sum of each of these series:
(a) $14 + 18 + 22 + \ldots + 86$ (b) $3 + 5.4 + 9.72 + \ldots + 102.036\,672$
(c) $60 + 24 + 9.6 + 3.84 + \ldots$ (15 min)

2 The sum of the first eight terms of an arithmetic series is 124 and the sum of the first 15 terms is 390. Find the sum of the first 25 terms. (20 min)

3 The sum of the first two terms of a geometric progression is 180 and the sum to infinity of the series is $182\frac{6}{7}$. Find r (given that it is a positive common ratio) and the first term. (20 min)

4 For the series below, state whether they are divergent or convergent. Work out the sum to infinity for those that are convergent.
(a) $2 + 5 + 8 + 11 + \ldots$ (b) $48 + 24 + 12 + 6 + 3 + \ldots$
(c) $80 - 20 + 5 - \frac{5}{4} + \ldots$ (15 min)

Examiner's secrets

All the formulae typed in bold on these two pages will need to be learnt off by heart.

Checkpoint 4

Find the sum of this G.P.:
$4 + 10 + 25 + \ldots + 390.625$

Watch out!

$|r| < 1$ means $-1 < r < 1$. If the common ratio is negative then put the negative ratio into the formula. For example, find the sum of the infinite series

$8 - 6 + 4\frac{1}{2} - 3\frac{3}{8} + \ldots$

The common ratio is $-\frac{3}{4}$.

$\therefore S_\infty = \frac{8}{1 - (-\frac{3}{4})} = 4\frac{4}{7}$

Binomials

Any expression that has two terms can be referred to as binomial. If you had three terms it is known as trinomial. An example of a binomial is $(e + 4f)^5$, where 'e' and '$4f$' are the two terms.

Binomial expansion

Expanding a binomial is relatively easy as long as the power is a small whole number. For example, $(x + y)^3$ could be expanded by multiplying out the three brackets (see below). However, if the power were of a higher order, the process is subject to errors and can become very messy.

$$(x + y)^3 = (x + y)(x + y)(x + y)$$
$$(x + y)^3 = x^3 + 3x^2y + 3xy^2 + y^3$$

The numbers in front of each term of the result are known as the **binomial coefficients.**

Pascal's Triangle can be used to work out these binomial coefficients.

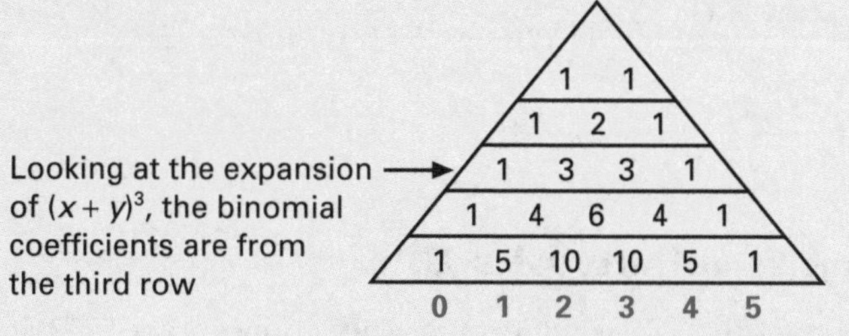

Alternatively, use ^nC_r which can also be written as $\binom{n}{r}$. These are abbreviations for $\dfrac{n!}{r!(1 - r)!}$

The binomial expansion, for positive integer n, summarised with the binomial coefficients is the **binomial theorem**.

$$(a + b)^n = \binom{n}{0}a^n + \binom{n}{1}a^{n-1}b + \binom{n}{2}a^{n-2}b^2 + \ldots + \binom{n}{r}a^{n-r}b^r + \ldots + \binom{n}{n}b^n$$

Note that the number of terms in the expansion will always be 1 more than the power. For example, $(a + b)^6$ expanded will have $6 + 1 = 7$ terms.

Worked example

Expand the following binomial: $(e + f)^5$.

$n = 5$, a is e and b is f. Substituting this information into the binomial theorem above.

$$(e + f)^5 = \binom{5}{0}e^5 + \binom{5}{1}e^{5-1}f + \binom{5}{2}e^{5-2}f^2 + \binom{5}{3}e^{5-3}f^3 + \binom{5}{4}e^{5-4}f^4 + \binom{5}{5}f^5$$

$$\therefore (e + f)^5 = e^5 + 5e^4f + 10e^3f^2 + 10e^2f^3 + 5ef^4 + f^5$$

Notice that for each term the powers add up to 5 (the power of the expansion).

Links

See 'expanding brackets', page 8.

Test yourself

Expand $(x + y)^3$ and check you get $x^3 + 3x^2y + 3xy^2 + y^3$.

Checkpoint 1

How are each of the rows of numbers in Pascal's Triangle formed?
Work out the next three rows.

Test yourself

Find ^nC_r on your calculator and check that you get $^7C_5 = 21$, $^6C_2 = 15$ and $^4C_2 = 6$. (Also, note that this is the 3rd number in the 4th row of Pascal's Triangle.)

Checkpoint 2

Find $^8C_3$ and $^8C_5$. Using Pascal's Triangle, explain your answers.

The binomial series for any rational n

We can get the binomial series from the binomial theorem. The binomial series is simply the expansion of $(1 + x)^n$, although it only applies when $-1 < x < 1$.

$$(a+b)^n = \binom{n}{0}a^n + \binom{n}{1}a^{n-1}b + \binom{n}{2}a^{n-2}b^2 + \ldots \text{ where } a = 1,\ b = x$$

$$(1+x)^n = 1^n + n\,.\,1^{n-1}\,.\,x + n(n-1)1^{n-2}\,.\,\frac{x^2}{2!} + \ldots \text{ Remember: } 1^n = 1^{n-1} = 1^{n-2} = 1$$

$$(1+x)^n = 1 + nx + \frac{n(n-1)}{2!}x^2 + \ldots$$

The binomial series is:

$$\mathbf{(1+x)^n = 1 + nx + \frac{n(n-1)}{2!}x^2 + \ldots \frac{n(n-1)(n-2)x^3}{3!} + \ldots \text{ for } |x| < 1.}$$

Example: Show that $\sqrt{1-3x} = 1 - \frac{3}{2}x - \frac{9}{8}x^2 - \ldots$

For what values of x is this expression valid?

$\sqrt{1-3x}$ can be written as $(1-3x)^{\frac{1}{2}}$

Using the binomial series: notice the 'x' is replaced with '$-3x$' and $n = \frac{1}{2}$.

$$(1-3x)^{\frac{1}{2}} = 1 + (\tfrac{1}{2})(-3x) + \frac{(\frac{1}{2})(\frac{1}{2}-1)(-3x)^2}{2!} + \ldots$$

$$\sqrt{1-3x} = 1 - \tfrac{3}{2}x - \tfrac{9}{8}x^2 - \ldots$$

This would normally be for $|x| < 1$, except the x was replaced with $-3x$. Therefore, $-1 < -3x < 1$, i.e. $-\frac{1}{3} < x < \frac{1}{3}$ are the values for which the expression is valid.

The binomial series can also be used for approximations.

Example: Find the value of $(0.98)^8$ correct to three decimal places, without using a calculator.

0.01 needs to be substituted into the binomial expansion of $(1 - 2x)^8$:

$$(1-2x)^8 = 1 + 8(-2x) + \frac{8(7)}{2!}(-2x)^2 + \ldots + (-2x)^8$$

and $(0.98)^8 = (1-0.02)^8 = (1 - 2\times 0.01)^8 = (1-2x)^8$ when $x = 0.01$

Hence $(0.98)^8 = 1 + 8(-2\times 0.01) + \frac{8(7)}{2!}(-2\times 0.01)^2 + \ldots$

$(0.98)^8 = 1 - 0.16 + 0.011\,2 = 0.851$ (to 3 d.p.)

Exam questions answers: page 87

1. Write down the first four terms in the binomial expansion of the following:
(a) $(p+q)^{10}$ (b) $(\frac{1}{5}+2x)^7$ (c) $(4-\frac{3}{2}x)^8$ (20 min)

2. Write down the term, as stated, of the following binomial expansions:
(a) $(3+5x)^4$, 3rd term (b) $(x-2y)^9$, 6th term (c) $(2c+3d)^6$, term with c^5 in
(d) $(6f-2g)^4$, term with f^2 in. (20 min)

3. Expand the following as far as the term with x^3.
(a) $\frac{1}{(1+2x)^5}$ (b) $3\sqrt{1+2x}$ (c) $\frac{1}{\sqrt{1-2x}}$ (25 min)

4. Find the value of $(1.02)^{15}$ correct to three decimal places, by making a suitable substitution into the binomial expansion of $(1+2x)^{15}$. (15 min)

Examiner's secrets

Remember to make a note of the domain for which the binomial series is valid. Marks are often allocated for it.

Links

See 'Indices', page 6.

Examiner's secrets

You always replace the 'x' with the 'value of the x' in the expansion and then divide by the number in front of the x. Here it has been divided by -3.

Checkpoint 3

For which values of x would the expansion of the binomial series $(1+4x)^7$ be valid?

Proof

Mathematics is all about what happens when rules (or axioms) are applied. These rules must be consistent, clear and without contradictions.

Construction of arguments

Examiner's secrets

If possible, explain, in words, exactly what you are doing. It will make it clearer for the examiner to follow and give a good impression.

When trying to prove something, ensure your workings are clear and that each step is logical and correct.

Example: Prove that the equation $x^2 + fx + g = 0$ has distinct real roots if, and only if, $f^2 > 4g$.

Rewrite the equation $x^2 + fx + g = 0$, using the method of completing the square:

$$\left(x + \frac{f}{2}\right)^2 - \left(\frac{f}{2}\right)^2 + g = 0$$

Rearranging: $\left(x + \frac{f}{2}\right)^2 = \frac{f^2}{4} - g$

Taking the square root of both sides of the equation:

$$x + \frac{f}{2} = \sqrt{\frac{f^2}{4} - g}$$

Links

See 'Completing the square', page 10.

For $\sqrt{\frac{f^2}{4} - g}$ to have distinct real roots $\frac{f^2}{4} - g$ must be greater than zero. (The square root of zero or a negative number will not give distinct real roots.)

The jargon

QED is the abbreviation for *quod erat demonstrandum*, which means 'which was to be proved'. It goes at the conclusion of a proof.

$$\therefore \frac{f^2}{4} - g > 0$$

Rearranging: $f^2 > 4g$ QED

Links

See 'Series', page 32.

Example: Prove $S_n = \frac{n}{2}(2a + (n - 1)d)$

The sum of an A.P. is:

$$S_n = a + (a + d) + (a + 2d) + \ldots + (a + (n - 2)d) + (a + (n - 1)d) \quad ①$$

Rewriting this in reverse we get:

$$S_n = (a + (n - 1)d) + (a + (n - 2)d) + \ldots + (a + 2d) + (a + d) + a) \quad ②$$

By adding equations ① and ② we get:

$$\begin{aligned} S_n &= a + (a + d) + (a + 2d) + \ldots + (a + (n - 1)d) \\ S_n &= (a + (n - 1)d) + \ldots + (a + 2d) + (a + d) + a \\ \hline 2S_n &= 2a + (n - 1)d + 2a + (n - 1) + \ldots \end{aligned}$$

Examiner's secrets

If possible try to align the equals signs when doing proofs.

There are n terms of $2a + (n - 1)d$

$$\therefore 2S_n = n(2a + (n - 1)d)$$

$$\therefore S_n = \frac{n}{2}(2a + (n - 1)d) \quad \text{QED}$$

Language and grammar

You will be expected to know the following (so learn them, please):

$=$	equals
$\neq$	is not equal to
$\equiv$	is identical to or is congruent to
$\approx$	is approximately equal to
~p	not p
$p \Rightarrow q$	p implies q (if p then q)
$p \Leftarrow q$	p is implied by q (if q then p)
$p \Leftrightarrow q$	p implies and is implied by q (p is equivalent to q)

If a line is struck through any of the symbols the word 'not' is inserted.

Example: If p is the statement 'Jane is English', then ~p would be 'Jane is not English'. If q is the statement 'Jane is from Kent', then ~q would be 'Jane is not from Kent'. You should be able to work out that:

$p \nRightarrow q$	(p does not imply q.) The fact that Jane is English does not imply that she is from Kent.
$p \Leftarrow q$	(p is implied by q.) The fact that Jane is from Kent does imply that she is English.
$\sim p \Rightarrow \sim q$	(Not p implies not q.) The fact that Jane is not English implies that she is not from Kent.

The jargon

In addition to those written here, you will also need to know the terms 'necessary' and 'sufficient', and the signs: $\therefore$ for therefore and $\because$ for because.

Checkpoint 1

If p is 'ABCD is a parallelogram' and q is 'ABCD is a rectangle', connect the statements using $\Rightarrow$, $\Leftarrow$ or $\Leftrightarrow$.

Checkpoint 2

Now connect p and q from checkpoint 1 using $\nRightarrow$, $\nLeftarrow$ or $\nLeftrightarrow$. Explain the counter-example to show why it is not valid.

Contradiction and counter-example

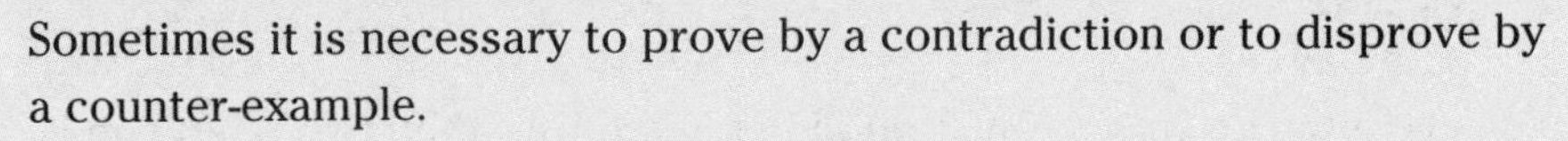

Sometimes it is necessary to prove by a contradiction or to disprove by a counter-example.

Example: $x^2 = 1 \nRightarrow x = 1$

Here, it is possible that x could be 1, but, it is equally likely to be -1. This kind of situation where there is a case which shows the implication to be invalid is called a counter-example.

Example: Prove that $\sqrt{2}$ is irrational.

For a number to be irrational it cannot be written in the form $\frac{p}{q}$.

We start by stating that $\sqrt{2} = \frac{p}{q}$ where p and q have no common factor.

Squaring both sides: $2 = \frac{p^2}{q^2} \quad \therefore 2q^2 = p^2$

As 2 is a factor of the left-hand side, it implies that p^2 must be even so p must be even. Let $p = 2k$, so $p^2 = 4k^2$. Therefore, $2q^2 = 4k^2$, so $q^2 = 2k^2$. Therefore, q^2 is even and so q is even. This shows that both p and q are even (i.e. have a common factor), which we assumed would not be the case. The supposition leads to a contradiction, therefore $\sqrt{2}$ cannot be expressed as a fraction. In other words, $\sqrt{2}$ is irrational.

Links

See irrational numbers, page 6.

Test yourself

Explain why p must be even.

Exam questions

answers: page 87

Prove $S_n = \frac{a(1 - r^n)}{1 - r}$, the sum of a finite geometric series. (10 min)

The geometry of straight lines

Note

This is an AS topic.

The line (in maths, 'line' means 'straight line') is perhaps the simplest shape there is. Yet an idea developed in 17th-century France links lines – and all the rest of geometry – with the intricate worlds of algebra and calculus. The idea? Illustrating algebraic equations as graphs using the now-familiar coordinate axes.

Coordinate geometry of a line 

Equations like $3x + 2y = 30$ have infinitely many solutions, for example $x = 4$ and $y = 9$; $x = -2$ and $y = 18$; $x = y = 6$ and so on. Each solution can be plotted as a point on a graph: (4,9), (–2,18), (6,6), etc. If you plotted all possible solutions, the points would give a line: a 'picture' of the equation $3x + 2y = 30$.

Any **linear** equation (one where the highest powers of x and y are 1) is the equation of a line graph, no matter how it is rearranged.

Checkpoint 1

Rearrange the equation $3x + 2y = 30$ into the form $y = mx + c$.

Finding the gradient and intercept of a line

A line has a slope or **gradient** (m) and an **intercept** (c) where it cuts the y-axis.

Watch out!

If the points have the same x coordinate, the line is vertical and the gradient is undefined.

→ The gradient of a line connecting two points is defined as

$$\frac{\text{difference in their } y \text{ coordinates}}{\text{difference in their } x \text{ coordinates}}$$

So given *any* two points (x_1,y_1) and (x_2,y_2) on the line,

Watch out!

It doesn't matter which point you call (x_1,y_1) and which you call (x_2,y_2) but you *must* label the x's and y's the same way round!

→ $m = \dfrac{y_2 - y_1}{x_2 - x_1}$

You are often given the equation of the line rather than two points on it. Given a linear equation in the form $ax + by + c = 0$:

Checkpoint 2

(a) What is the gradient of the line connecting (3,4) and (5,–4)? (b) What is the intercept of $3x + 2y - 5 = 0$?

→ the intercept is the y coordinate when $x = 0$. Find this by putting $x = 0$ in the equation and solving for y. Alternatively,

→ rearrange the equation into the form $y = mx + c$. This gives you the gradient and intercept directly.

It's also useful to know that:

→ the gradient of a line is equal to $\tan\theta$, where θ is the angle between the line and the positive direction of the x-axis (see graph on the next page).

Examiner's secrets

You could find the gradient first $\dfrac{4-(-1)}{3-5} = \dfrac{5}{-2}$ and then use $y - y_1 = m(x - x_1)$.

Finding the equation of a line

Given the gradient m and one point (x_1,y_1) on the line:

→ $y - y_1 = m(x - x_1)$

E.g. line with gradient 2 through (–1,3) is $y - 3 = 2(x + 1)$, i.e. $y - 2x = 5$

Checkpoint 3

(a) Which line has gradient –2 and passes through (3,4)? (b) Which line passes through (2,–5) and (–1,3)?

Given two points (x_1,y_1) and (x_2,y_2) on the line:

→ $\dfrac{y - y_1}{y_2 - y_1} = \dfrac{x - x_1}{x_2 - x_1}$

E.g. line through (3,4) and (5,–1) is $\dfrac{y-4}{-5} = \dfrac{x-3}{2}$, i.e. $5x + 2y = 23$

Given the gradient m and the intercept c:

→ $y = mx + c$

E.g. line with gradient –3 and intercept at $y = 4$ is $y = -3x + 4$

The midpoint of a line segment

The midpoint of the line segment joining (x_1,y_1) and (x_2,y_2) is simply the point whose coordinates are the average of (x_1,y_1) and (x_2,y_2).

➜ The midpoint of (x_1,y_1) and (x_2,y_2) is $\left(\frac{x_1+x_2}{2}, \frac{y_1+y_2}{2}\right)$

E.g. the midpoint of (2,3) and (–4,5) is $\left(\frac{2-4}{2}, \frac{3+5}{2}\right)$ i.e. (–1,4)

The graph below summarises some of these results

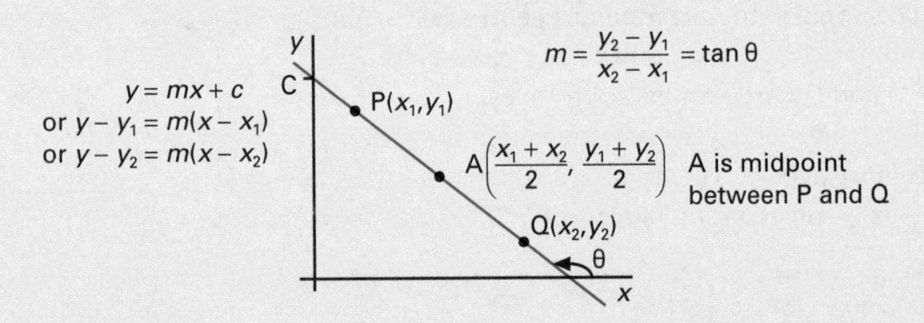

The jargon

A *line segment* is part of a line that is bounded at both ends (the whole line continues indefinitely in both directions).

Checkpoint 4

What is the midpoint of the line segment joining (–2,3) and (4,–3)?

Parallel and perpendicular lines

If you look at parallel lines on a graph, it is easy to see that

➜ parallel lines have the same gradient

Example: Find the equation of the line through (3,–1) parallel to $2y - 4x = 3$

Gradient of $2y - 4x = 3$ is 2, so our line also has gradient 2.

Line with gradient 2 through (3,–1) is $y + 1 = 2(x - 3)$, i.e. $y = 2x - 7$

The equivalent result for perpendicular lines is that

➜ perpendicular lines have gradients whose product is –1

Example: Find the equation of the line through (4,5) perpendicular to $4x + y = 9$

Gradient of $4x + y = 9$ is –4, so our line has gradient $+\frac{1}{4}$

Line with gradient $\frac{1}{4}$ through (4,5) is $y - 5 = \frac{1}{4}(x - 4)$, i.e. $4y - x = 16$

Checkpoint 5

(a) What's the equation of the line parallel to $y = 4x + 5$ passing through (1,1)?

(b) What about the line perpendicular to $2y = 4x + 5$ passing through (1,1)?

Test yourself

Make up a linear equation in the form $ax + by + c = 0$ and work out the gradient and intercept of the graph, and the equation of a line perpendicular to it passing through a point of your choice.

Exam questions

answers: pages 87–8

y
C(5,5)
A(–5,3)
x
B(7,–5)

1. A, B and C are the points (–5,3), (7,–5) and (5,5) respectively.
   (a) Find the equation of the line on which A and B lie.
   (b) D is the midpoint of AB. Show that CD is perpendicular to AB.
   (c) Show that CD crosses the x-axis when $x = \frac{5}{3}$. (15 min)

The coordinate geometry of circles

Circles and spheres are everywhere: in nature, in art and architecture and in everyday objects (plates, coins, . . .). Because of their simplicity, circles can be described without very obscure maths. The circle is the simplest of the *conic sections*, curves analysed by the ancient Greeks and still central to many branches of mathematics.

Note

This is an AS topic.

Coordinate geometry of a circle

The circle in the diagram has centre C(a,b) and radius r. P(x,y) is a point on the circumference. CQ and QP are parallel to the axes.
The length of CQ is $x - a$; the length of QP is $y - b$.
By Pythagoras' Theorem, $(x - a)^2 + (y - b)^2 = r^2$.
This is the equation of a circle.

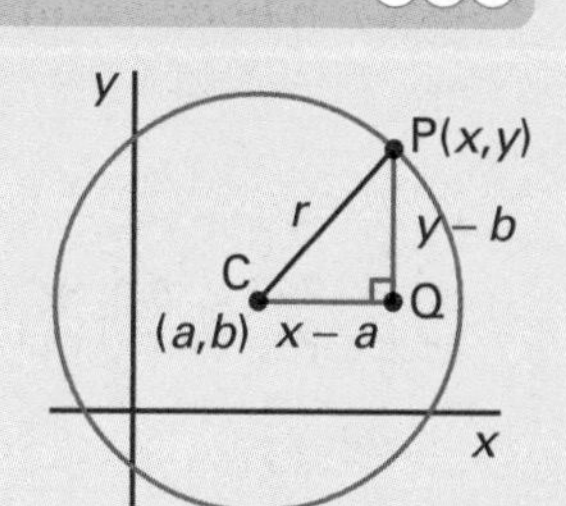

➜ $(x - a)^2 + (y - b)^2 = r^2$ is a circle with centre (a,b) and radius r

Watch out!

Sometimes a or $b = 0$ and the equation looks simpler than this. Don't get confused!
Where is the centre when $a = b = 0$?
What is the equation of the circle?

The brackets can be expanded:

$$x^2 - 2ax + a^2 + y^2 - 2by + b^2 = r^2$$
$$x^2 + y^2 - 2ax - 2by + a^2 + b^2 - r^2 = 0$$

This is another form for the equation of a circle, usually written

➜ $x^2 + y^2 + 2gx + 2fy + c = 0$

In this form, the centre is at $(-g,-f)$ and r^2 is $g^2 + f^2 - c$
If you're given the equation of a circle, to find the centre and the radius:

➜ divide or multiply the whole equation if necessary to make the coefficients of x^2 and y^2 equal to 1;
➜ group x terms and y terms for **completing the square**;
➜ rearrange into the $(x - a)^2 + (y - b)^2 = r^2$ form by completing the square. Then it's easy to find the centre (a,b) and radius r.

Checkpoint 1

Find the centre and radius of the circle
$4x^2 + 4y^2 - 16x + 8y + 19 = 0$

Normals and tangents to circles

Examiner's secrets

You need to remember the following results from GCSE:
➜ the angle in a semicircle is a right angle;
➜ the perpendicular from the centre to a chord bisects the chord;
➜ the tangent to a circle is perpendicular to the radius at its point of contact.

Once we know the equation of a circle, we can find equations for:

➜ the **normal** at a given point P on the circle, i.e. a line through P that cuts the circumference at right angles;
➜ the **tangent** at a given point P on the circle, i.e. a line just touching the circle at P but not cutting it.

To find these equations you need to know:

➜ that any radius is at right angles to the circumference, so that the normal at point P will be the line joining P to the centre;
➜ that the tangent at P is perpendicular to the radius to P;
➜ results about finding equations of lines through given points and lines perpendicular to given lines.

Action point

You'll need to know about equations of lines before revising circles.

The normal at P

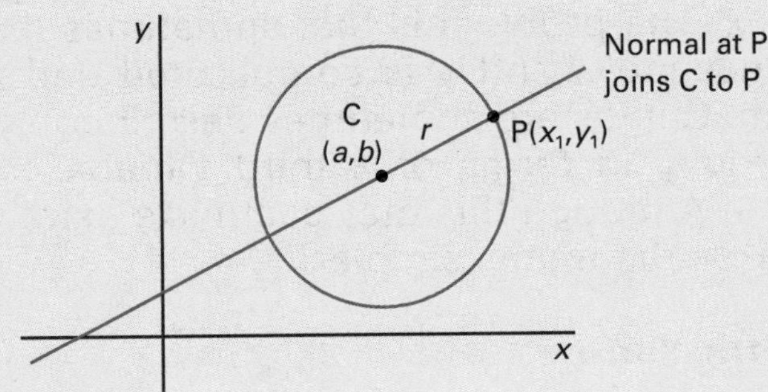

The normal at P is just the line joining $C(a,b)$ to $P(x_1,y_1)$. To find its equation, use the formula for a line through two given points (page 38).

Example: Find the normal to the circle $(x-1)^2 + (y-2)^2 = 25$ at (5,5).

The normal is the line passing through (1,2) and (5,5) so its equation is

$\frac{y-2}{5-2} = \frac{x-1}{5-1}$, i.e. $4y - 3x = 5$

The tangent at P

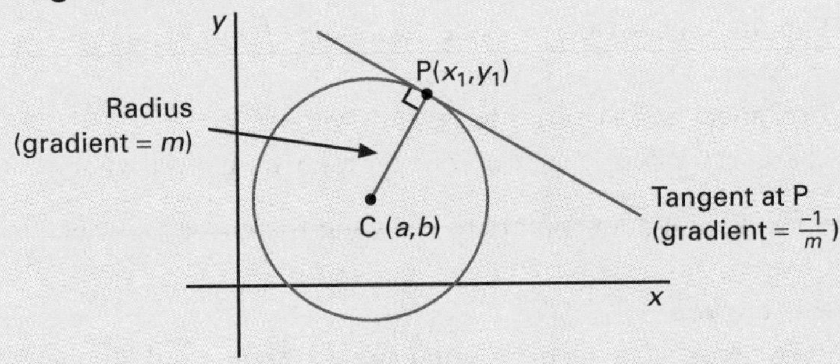

The tangent at P is a line through P at right angles to the normal at P. To find its equation, use the formula for a line through a given point and the gradient of a line perpendicular to a given line (page 39).

Example: Find the equation of the tangent to the circle $(x-10)^2 + (y-5)^2 = 100$ at (4,–3). Centre (10,5) Point (4,–3)

First find the gradient of the normal $\frac{5-(-3)}{10-4} = \frac{8}{6} = \frac{4}{3}$.

So the gradient of the tangent is $-\frac{3}{4}$ $(mm' = -1)$.

So the equation of the tangent at the point (4,–3) with gradient $-\frac{3}{4}$ is $y + 3 = -\frac{3}{4}(x-4)$, i.e. $3x + 4y = 0$.

Checkpoint 2

Find the equations of (a) the normal and (b) the tangent at the point (6,4) to the circle $(x-2)^2 + (y-1)^2 = 25$.

Exam questions answers: page 88

1 A diameter of a circle joins the points (–3,–1) and (5,5). Find the equation of the circle. (5 min)

2 A circle with centre C has equation $x^2 + y^2 + 2x - 4y - 4 = 0$
 (a) Find the coordinates of C.
 (b) Confirm that the radius of the circle is 3.
 (c) Verify that the point $A\left(\frac{4}{5}, \frac{22}{5}\right)$ lies on the circle and find the equation of the tangent to the circle at A.
 (d) Verify that the point B(4,2) lies on this tangent.
 (e) Sketch the circle, showing C, A and B and the tangent at A.
 (f) Calculate the size of angle ABC. (20 min)

Parameters

Note

This is an A2 topic.

In the Cartesian equation of a curve, y is given in terms of x. The problem is that sometimes the equation connecting x and y is complicated and hard to work with. Curves are sometimes described by giving both x and y in terms of a third variable, called a parameter. Although this may seem like extra work, it often makes the maths simpler!

Parametric form

Note

Check if sketching a curve is in your specification.

The **parameter** in terms of which x and y are given is often called t.

Sketching a curve given parametrically

Examiner's secrets

If you're asked to *sketch* a curve, don't plot individual points accurately. Draw the general shape but *always* label important features like the coordinates of points where the curve crosses the axes.

- ➜ Are there any points where x or y is undefined?
- ➜ Where does the curve cut the axes?
  - it cuts the x-axis when $y = 0$
  - it cuts the y-axis when $x = 0$
- ➜ Does the curve have any symmetry?
  - What difference does it make swapping $-t$ for t?
- ➜ What happens to x and y when t is
  - zero; small and +ve/–ve; large and +ve/–ve?
- ➜ Are there any values that x or y never take, no matter what t is?

If you're stuck, plot a few points to see what the curve looks like.

Common curves

If you come across a new curve, you can use the method above to sketch it. It's also worth learning a few common curves so you don't have to work them out each time. Here are some of the curves frequently given in parametric form:

Checkpoint 1

Where does the ellipse cut the axes? Why is it a circle when $a = b$?

➜ $x = a \sin t$, $y = b \cos t$

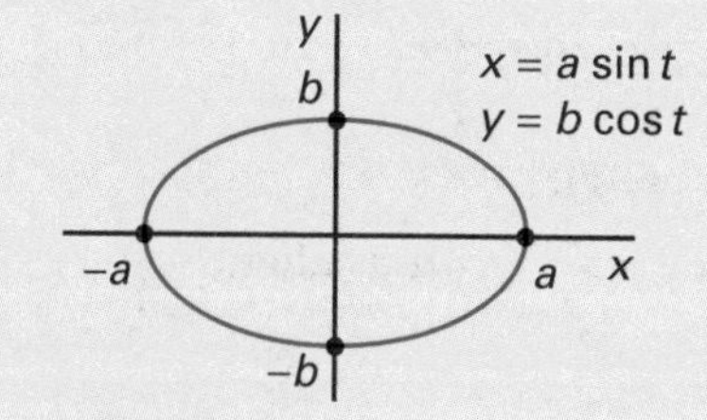

ellipse (circle if $a = b$)

Examiner's secrets

Use your graphical calculator and see how the curves build up as the parameter changes.

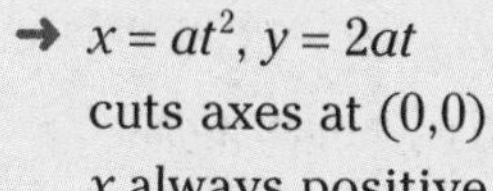

➜ $x = at^2$, $y = 2at$

cuts axes at (0,0)

x always positive whether t is +ve or –ve

y positive for +ve t and negative for –ve t

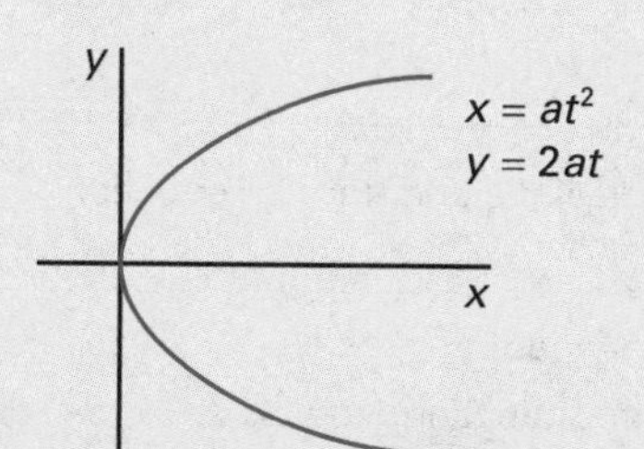

parabola

Checkpoint 2

What would be the parametric form for a parabola symmetrical about the y-axis?

→ $x = ct, y = \frac{c}{t}$

y undefined when $t = 0$

the curve doesn't cut the axes

both x and y change sign when t changes sign

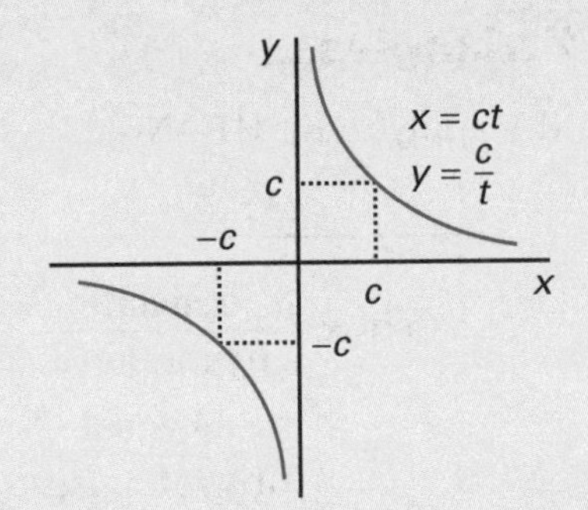

hyperbola with **asymptotes** at 90°.

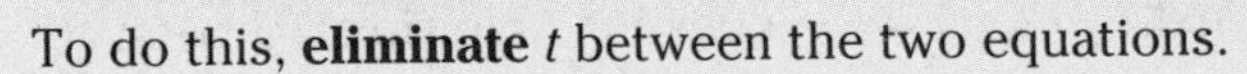

Converting from parametric to Cartesian form

To do this, **eliminate** t between the two equations.

Example: Find a Cartesian equation for $x = 6t$, $y = 3t^2$

Rearrange the x equation to give $t = \frac{6}{x}$

Eliminate t by substituting in the y equation: $y = 3\left(\frac{x}{6}\right)^2$, i.e. $y = \frac{x^2}{12}$

The process is like the elimination method for simultaneous equations. Look out for the following cases:

→ y is a power of x or similar (e.g. the example above).

→ y is tx or similar (or x is ty). Use the substitution $t = \frac{y}{x}$ or $t = \frac{x}{y}$

E.g. $x = 2t^2 + t$, $y = 2t + 1$, $t = \frac{x}{y}$ $\therefore y = 2\frac{x}{y} + 1$, i.e. $y^2 - y = 2x$

→ the parametric equations for x and y contain trig functions. Use trigonometrical identities such as $\cos^2\theta + \sin^2\theta = 1$ (page 47).

E.g. $x = 2 \sin t$, $y = 2 \cos t$ $\therefore \sin t = \frac{x}{2}$ and $\cos t = \frac{y}{2}$.

But $\cos^2 t + \sin^2 t = 1$ $\therefore \left(\frac{x}{2}\right)^2 + \left(\frac{y}{2}\right)^2 = 1$, i.e. $x^2 + y^2 = 4$ (a circle, centre at the origin, radius 2).

Checkpoint 3

Try the example $x = 4t^2$, $y = t^3$

Checkpoint 4

Try these examples:

(a) $x = t^2 - 2$, $y = t^3 - 2t$

(b) $x = a \cos t$, $y = b \sin t$

Test yourself

Write down parametric equations for an ellipse, a parabola and a hyperbola. Use specific numbers instead of a, b, etc. Then sketch the curves you have described. *Remember to label the sketches!*

Exam questions

answers: page 88

1 Sketch the curves given parametrically by the following equations and give a Cartesian equation for each curve.

(a) $x = t^2$, $y = t^3$

(b) $x = 3t^2$, $y = 6t$ (10 min)

2 Give a pair of parametric equations for the curve shown. (5 min)

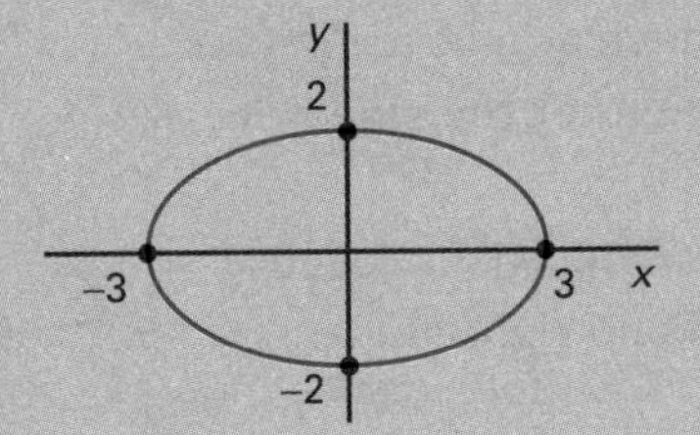

Trigonometry

Action point

Learn the three ratios by heart NOW. It'll save time in the long run.

Checkpoint

If $\sin x = \frac{3}{5}$, what are the ratios $\cos x$ and $\tan x$?

Examiner's secrets

SOHCAHTOA is an easy way of remembering the ratios. For example, the first three letters 'SOH' represent $\sin x = \frac{\text{opposite}}{\text{hypotenuse}}$

Trigonometry is based on the mathematics of triangles and the relationship between lengths and angles. The work on the right-angled triangle should be familiar from GCSE.

Trigonometry in right-angled triangles

The following ratios (or relationships) ONLY apply to RIGHT-ANGLED triangles.

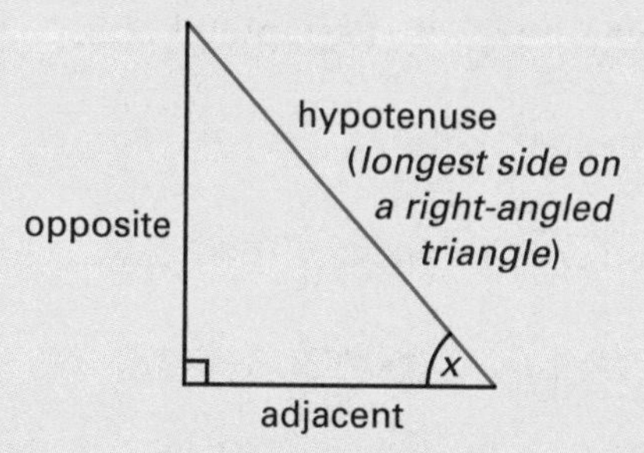

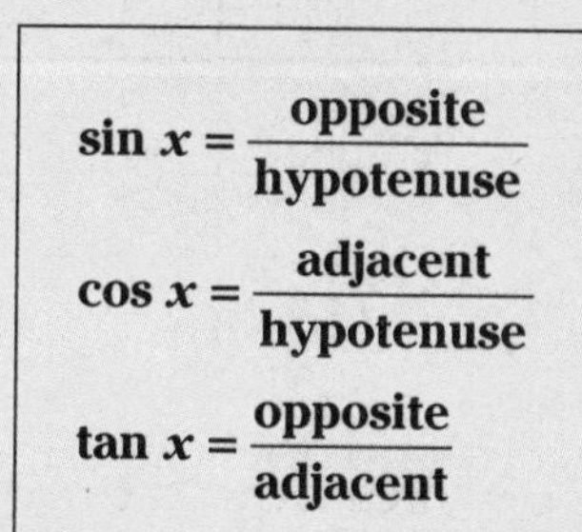

Follow these steps to help you answer a question involving right-angled triangles:

Step 1: Label the sides according to the angle. *The opposite side is always 'opposite' the angle given or required.*
Step 2: Decide which ratio should be used and put the information in.
Step 3: Re-arrange to find the answer.

Example: Work out the angle x in this triangle:

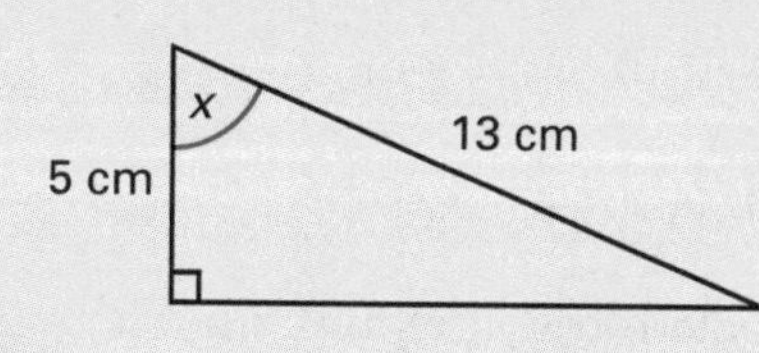

Step 1:

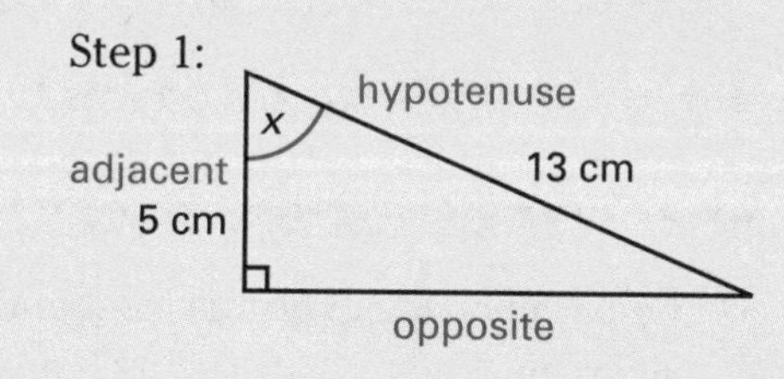

Step 2: We have the adjacent and the hypotenuse, so we need to use the cos ratio.

$$\cos x = \frac{\text{adjacent}}{\text{hypotenuse}} \qquad \cos x = \frac{5}{13}$$

Step 3: Find $\cos^{-1}$ of both sides

$$x = \cos^{-1}\frac{5}{13}$$

$$x = 67.4° \text{ (to 1 d.p.)}$$

Triangles with no right angle

If you have to solve problems involving triangles that do NOT have a right angle you will usually need to use either the sine rule or the cosine rule.

As shown here, the triangle is labelled differently when there is no right angle.

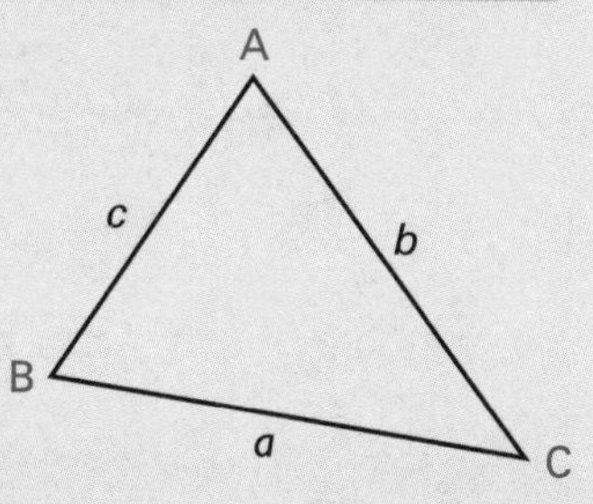

Sine rule

$\frac{a}{\sin A} = \frac{b}{\sin B} = \frac{c}{\sin C}$ which can also be written as $\frac{\sin A}{a} = \frac{\sin B}{b} = \frac{\sin C}{c}$

> **Watch out!**
> Although you ought to write out the whole formula you only need part of it.

Example: In this triangle find the side labelled with the letter a.

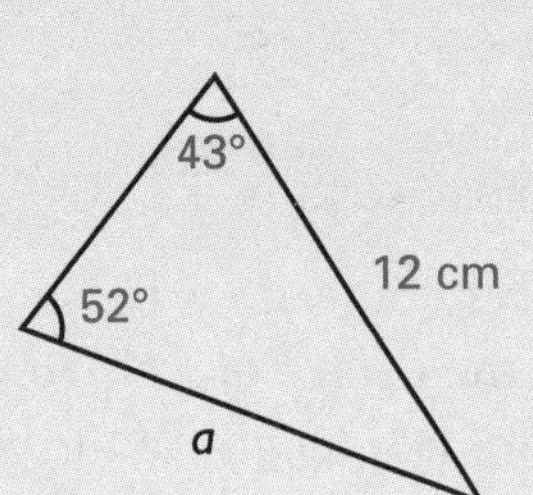

$$\frac{a}{\sin A} = \frac{b}{\sin B}$$

If a is to be found, then b is 12 cm, angle A is 43° and angle B is 52°.

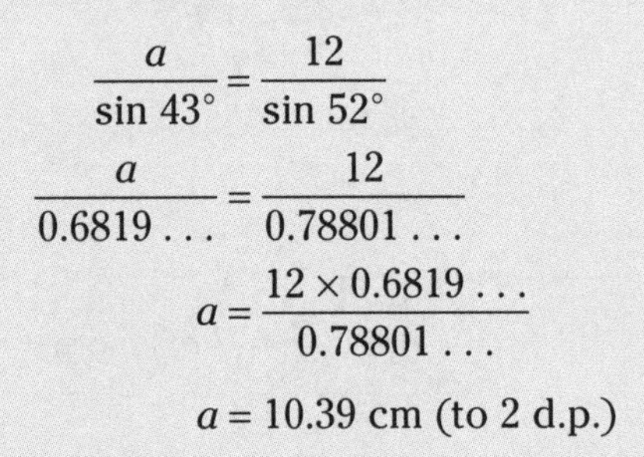

$$\frac{a}{\sin 43°} = \frac{12}{\sin 52°}$$

$$\frac{a}{0.6819\ldots} = \frac{12}{0.78801\ldots}$$

$$a = \frac{12 \times 0.6819\ldots}{0.78801\ldots}$$

$$a = 10.39 \text{ cm (to 2 d.p.)}$$

> **Watch out!**
> The sine rule doesn't work **as easily** for obtuse angles. If there is no diagram do a rough sketch. Look at the diagram – if the angle is bigger than 90°, but you've calculated a small angle, you'll need to do 180° minus your angle to get the correct angle. Why?

Cosine rule

$a^2 = b^2 + c^2 - 2bc \cos A$ which can be written as $\cos A = \frac{b^2 + c^2 - a^2}{2bc}$.

Example: Find the side labelled a in this triangle.

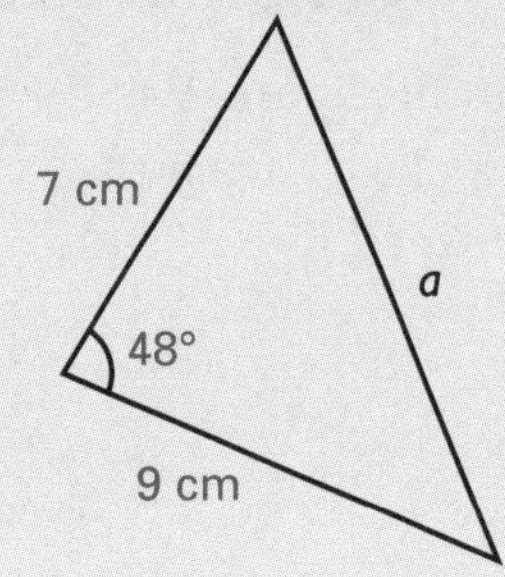

If a is to be found, then angle A is 48°, b is 7 cm and c is 9 cm.

Using cosine rule:

$$a^2 = b^2 + c^2 - 2bc \cos A$$

$$a^2 = 7^2 + 9^2 - 2 \times 7 \times 9 \times \cos 48°$$

$$a = 6.8 \text{ cm (to 1 d.p.)}$$

> **Action point**
> Work through the example using the cosine rule and make sure you get the correct answer. If you don't, make sure you had found the value of cos 48° before multiplying it by the 2, 7 and 9.

Area of a triangle

You need to know the formula Area $= \frac{1}{2}bc \sin A$ and its equivalents.

In the above triangle the area is $\frac{1}{2} \times 7 \times 9 \times \sin 48 = 23.4 \text{ cm}^2$ (to 1 d.p.)

Exam questions answers: page 88

1. In a triangle ABC, angle BAC is 90°, AB is 7 cm and AC is 8 cm. Work out angle ACB.
2. In a triangle ABC, angle BAC is 78°, angle BCA is 32° and side BC is 10 cm. Using the sine rule, work out the length of side AB.
3. In a triangle ABC, angle BAC is 47°, side BA is 8 cm and side AC is 11 cm. Using the cosine rule, work out the length of side BC. (15 min)

Radians and trigonometry

The circle was first divided into 360 degrees by Babylonian mathematicians. Other units for measuring angles exist. A particularly useful one is the radian. Using radians simplifies many calculations, particularly in calculus.

Radians

The jargon

The symbol for radians is sometimes written ^{c}. (1 rad = 1^c). Angles in radians are often shown with no symbol.

Checkpoint 1

Try to visualise one radian. How many degrees is it?

A **radian (rad)** is the angle subtended at the centre of a circle by an arc the same length as the circle's radius.

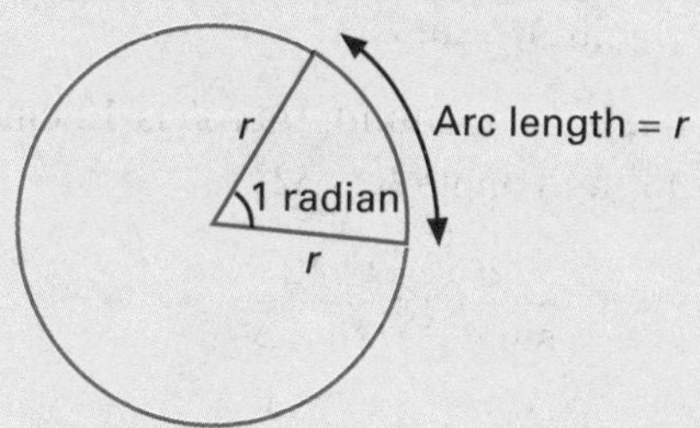

- A complete circle is 2π rad
- π rad = 180°

For a sector of radius r and angle θ radians,

- Arc length = $r\theta$
- Area = $\frac{1}{2}r^2\theta$

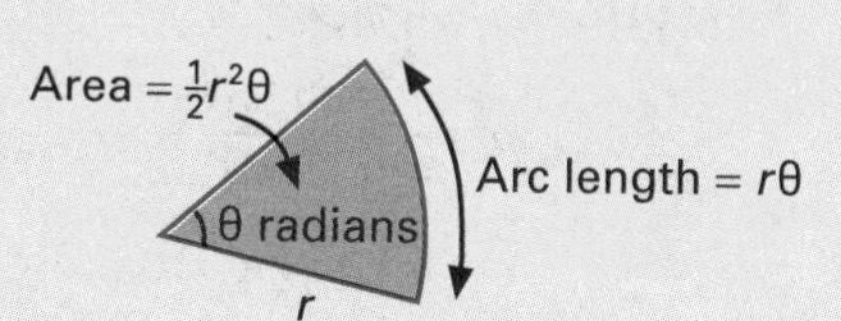

Checkpoint 2

How do the formulae for arc length and sector area work? Think about fractions of a whole circle!

E.g. for a sector of radius 3 cm and angle 1.5 radians

Arc length = $r\theta$ = 4.5 cm   Area = $\frac{1}{2}r^2\theta = 4.5 \times 1.5 = 6.75$ cm^2

Trigonometric functions and their graphs

In this figure, P is the point (x,y), OP = 1 and θ is measured anti-clockwise from the +ve direction of the x-axis. The trig ratios for θ are defined by:

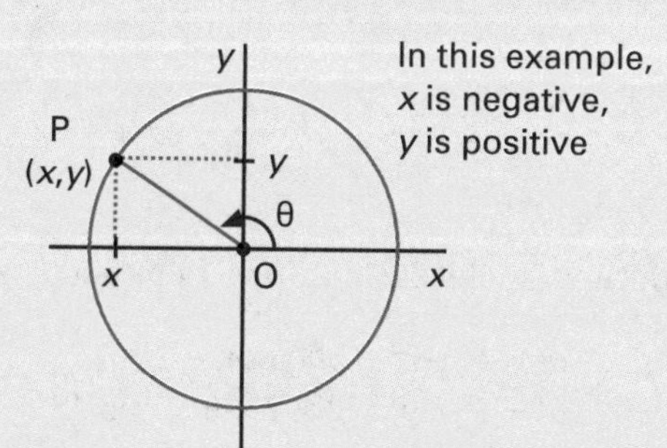

Watch out!

Note that, depending on the value of θ, x and/or y may be negative.

- $\sin\theta = y$
- $\cos\theta = x$
- $\tan\theta = \frac{y}{x}$

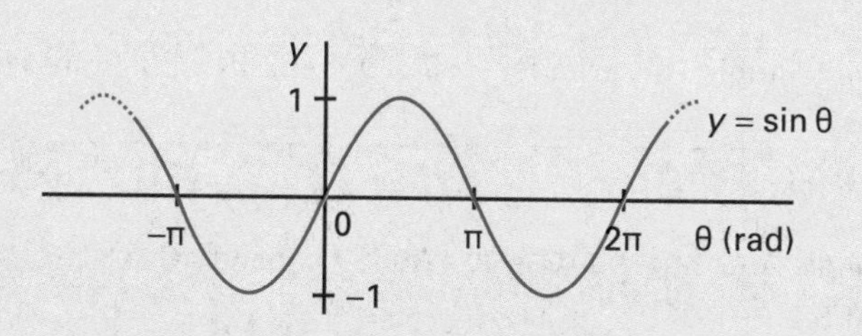

The jargon

An *odd function* is one where $f(-x) = -f(x)$. It has rotational symmetry about the origin. (Have a look at the graph!)

- $y = \sin\theta$ is an **odd** function, e.g. $\sin\frac{\pi}{2} = -\sin(-\frac{\pi}{2}) = 1$
- it has a **period** of 2π, e.g. $\sin(\frac{\pi}{4} + 2\pi) = \sin\frac{\pi}{4} = \frac{\sqrt{2}}{2}$
- $\sin(\pi - \theta) = \sin\theta$, e.g. $\sin(\pi - \frac{\pi}{6}) = \sin\frac{\pi}{6} = \frac{1}{2}$

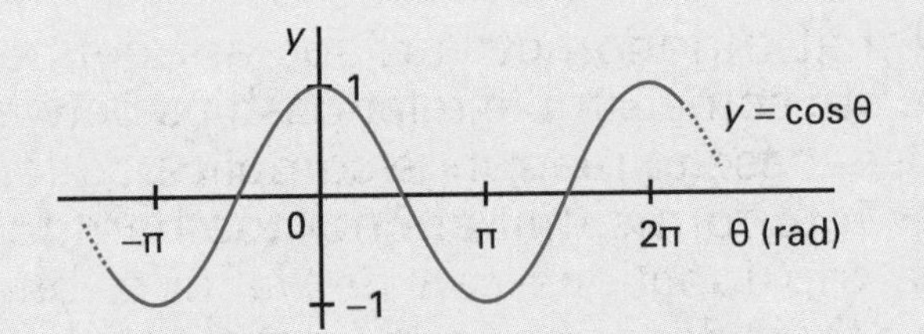

- $y = \cos\theta$ is an **even** function
- it has a period of 2π
- $\cos(2\pi - \theta) = \cos\theta$

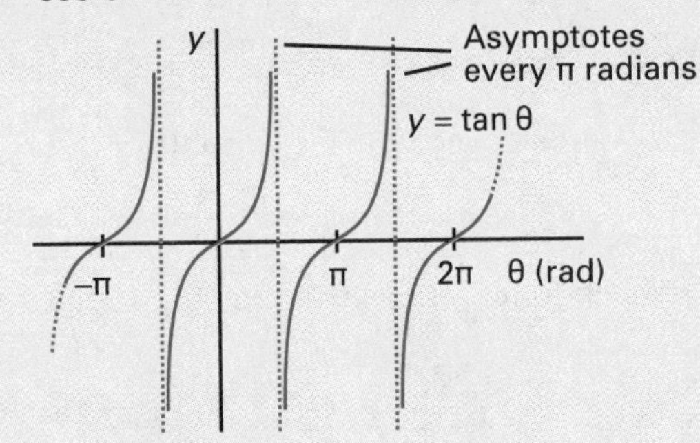

- $y = \tan\theta$ is an odd function
- it has a period of π

Example: Solve $\cos(3x + 10) = \frac{1}{2}$ for $0 \le x \le 180°$

$\cos(3x + 10) = \frac{1}{2}$. From graph +60° and −60° both have a cos of $\frac{1}{2}$

$3x + 10 = \pm 60° +$ any multiple of 360° (graph repeats every 360°)

$x + 10 = \pm 20° +$ any multiple of 120°, e.g. 20°, 100°, 140°

$x = 10°, 90°, 130°$ in the specified range

Simple trigonometric identities

Identities are true for all values of θ.

- $\cos^2\theta + \sin^2\theta \equiv 1$

E.g. $\cos^2 30° + \sin^2 30° = \left(\frac{\sqrt{3}}{2}\right)^2 + \left(\frac{1}{2}\right)^2 = \frac{3}{4} + \frac{1}{4} = 1$

- $\frac{\sin\theta}{\cos\theta} \equiv \tan\theta$

E.g. $\tan 60° = \frac{\sin 60°}{\cos 60°} = \frac{\frac{1}{2}}{\frac{\sqrt{3}}{2}} = \sqrt{3}$

Exam questions answers: pages 88–9

In the sector AXBC, AB = 10 cm and M is the midpoint of AB. MC is perpendicular to AB and MC = 10 cm. AXB is a circular arc with centre C. Find the perimeter and the area of the sector. (10 min)

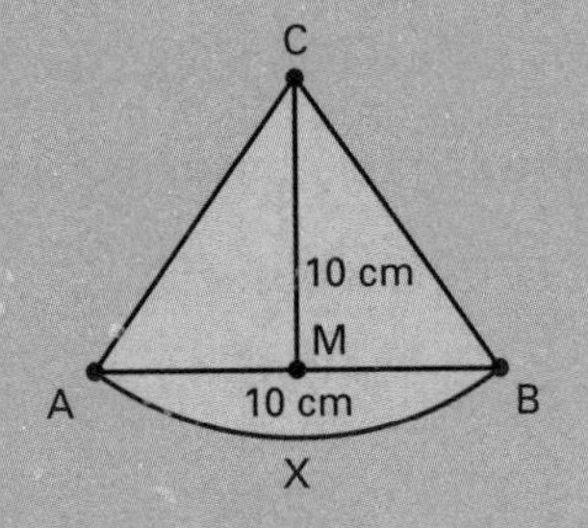

The jargon

An *even function* is one where f(−x) = f(x). It has mirror symmetry about the y-axis. (Have a look at the graph!)

The jargon

An *asymptote* is a line that the curve approaches but never touches.

Examiner's secrets

You should be really familiar with the sine, cosine and tangent of $\frac{\pi}{6}, \frac{\pi}{4}, \frac{\pi}{3}, \frac{\pi}{2}$ and related angles.

Action point

Redraw these graphs and rewrite the descriptions and the scale on the x-axis in degrees instead of radians.

Test yourself

Sketch a sector of a circle on a blank sheet of paper. Choose a radius and angle for the sector, and find its arc length and area. Draw the trig function graphs from memory, labelling key points.

Trigonometric relations

The way that trigonometric ratios are defined means that there are some simple relationships between them. We can use these to rearrange complicated-looking trig equations into forms that are not too hard to solve!

Inverse trig functions can be defined, but only by restricting their domains so they are not one-to-many relationships.

Reciprocal trig functions and trig relationships

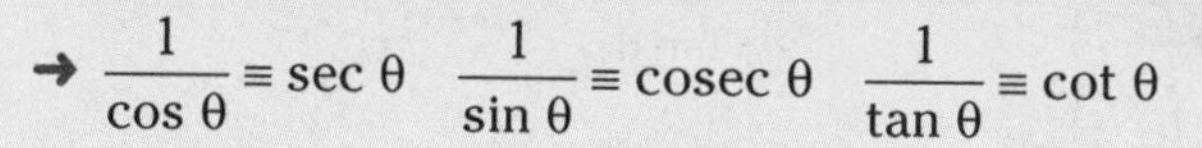

→ $\dfrac{1}{\cos\theta} \equiv \sec\theta \quad \dfrac{1}{\sin\theta} \equiv \operatorname{cosec}\theta \quad \dfrac{1}{\tan\theta} \equiv \cot\theta$

E.g. $\sec 60° = 2 \quad \operatorname{cosec} 70° = 1.064\ldots \quad \cot 30° = \sqrt{3}$

The reciprocal of each function has the same period as the function.

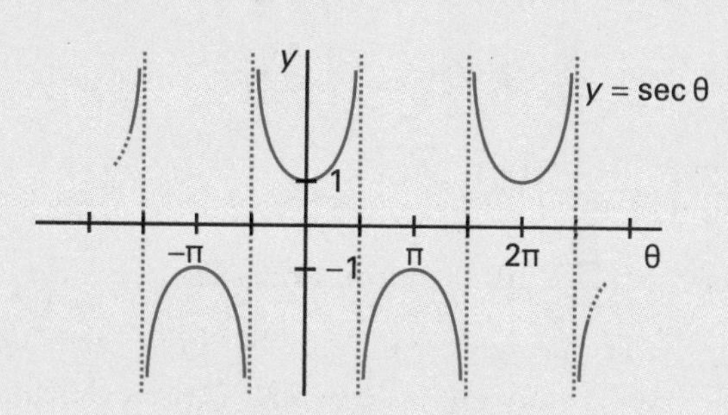

$\sec\theta \le -1$ or $\sec\theta \ge 1$ and $\sec\theta$ undefined for $\theta = (2n+1)\dfrac{\pi}{2}$

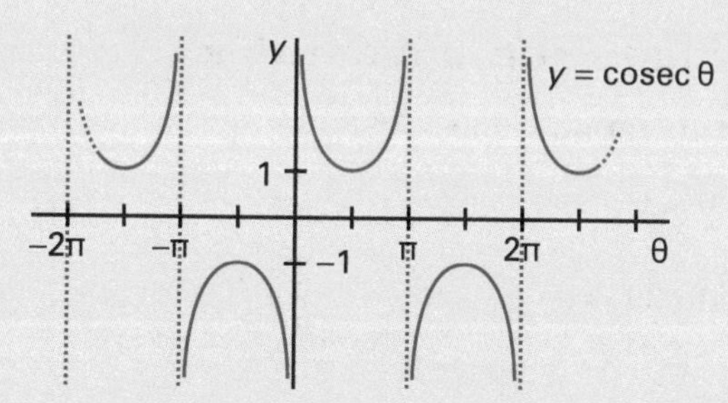

$\operatorname{cosec}\theta \le -1$ or $\operatorname{cosec}\theta \ge 1$ and $\operatorname{cosec}\theta$ undefined for $\theta = n\pi$

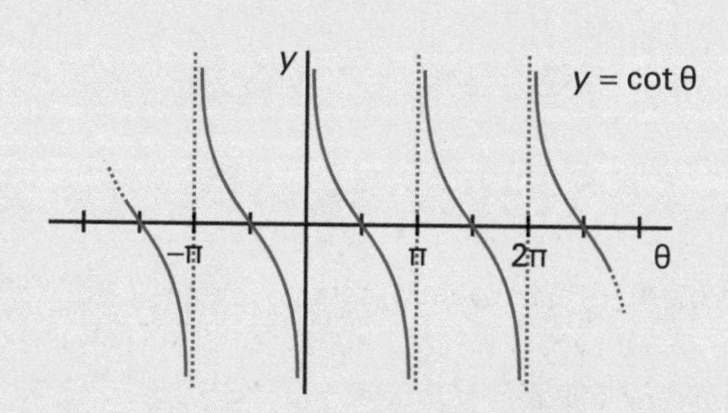

$\cot\theta$ can take any value and $\cot\theta$ undefined for $\theta = n\pi$

You should know the following relationships for reciprocal functions:

→ $1 + \tan^2\theta \equiv \sec^2\theta \qquad \cot^2\theta + 1 \equiv \operatorname{cosec}^2\theta$

→ $\dfrac{\cos\theta}{\sin\theta} \equiv \cot\theta$

Watch out!

$\dfrac{1}{\cos\theta}$ means the same as $(\cos\theta)^{-1}$, and similarly with the others. The function *must* be written in brackets. *Notation like $\cos^{-1}\theta$ means something completely different!*

Checkpoint 1

What are the periods of the trig functions and their reciprocals?

Watch out!

Be careful which function goes with which reciprocal.

cos and sec belong together.

sin and cosec belong together.

Checkpoint 2

What are the domains and ranges of sec θ, cosec θ and cot θ? Why?

Checkpoint 3

Copy the sec θ graph and add a cos θ graph on the same axes. Repeat for the other two functions and their reciprocals. What happens to cos θ, sin θ and tan θ where sec θ, cosec θ and cot θ are undefined?

Checkpoint 4

Work through the derivation of these equations from the form $\cos^2\theta + \sin^2\theta \equiv 1$.

Inverse trig functions

Inverses of sin, cos and tan are arcsin, arccos and arctan, respectively.
E.g. arcsin x is an angle whose sin is x: arcsin 'undoes' sin.
Problem: The sines of -4π, 0, 3π, etc. are all 0. Is arcsin 0, -4π, 0, 3π . . . ?
We get round this problem by defining the **principal value** of arcsin x:

→ **Arcsin x** is the angle *between* $-\frac{\pi}{2}$ and $\frac{\pi}{2}$ whose sin is x.

E.g. arcsin $0 = 0$; arcsin $\frac{1}{2} = \frac{\pi}{6}$ rad $= 30°$
By restricting the **range**, we ensure that any number can only have one arcsin, so that arcsin is a one-to-one mapping – a function. The same is true for the other inverses.

Only numbers between –1 and +1 can have an arcsin. The **domain** of arcsin x is $-1 \le x \le 1$.

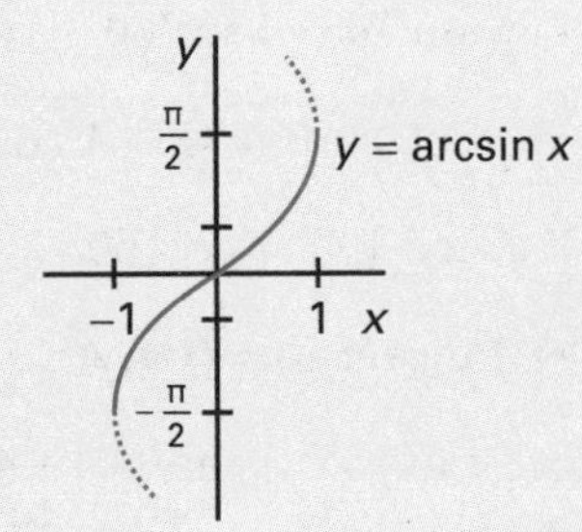

→ **Arccos x** is the angle *between* 0 *and* π whose cos is x.

E.g. arccos $0 = \frac{\pi}{2}$ or 90°; arccos$(-\frac{1}{2}) = \frac{2\pi}{3}$ rad $= 120°$
Only numbers between –1 and +1 can have an arccos. The **domain** of arccos x is $-1 \le x \le 1$.

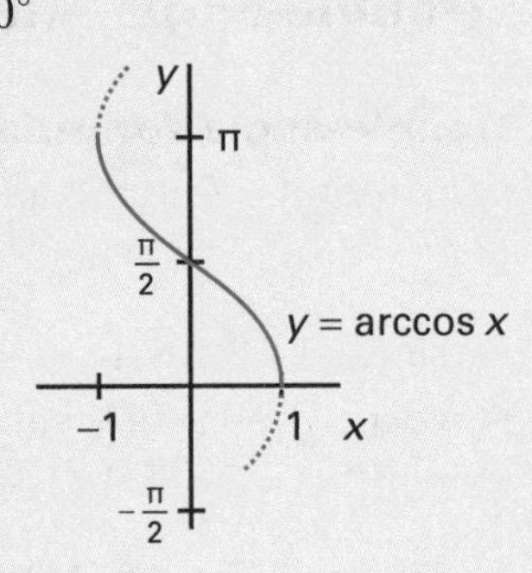

→ **Arctan x** is the angle *between* $-\frac{\pi}{2}$ *and* $\frac{\pi}{2}$ whose tan is x.

E.g. arctan $1 = \frac{\pi}{4}$ or 45°; arctan$(-1) = -\frac{\pi}{4}$ rad $= -45°$

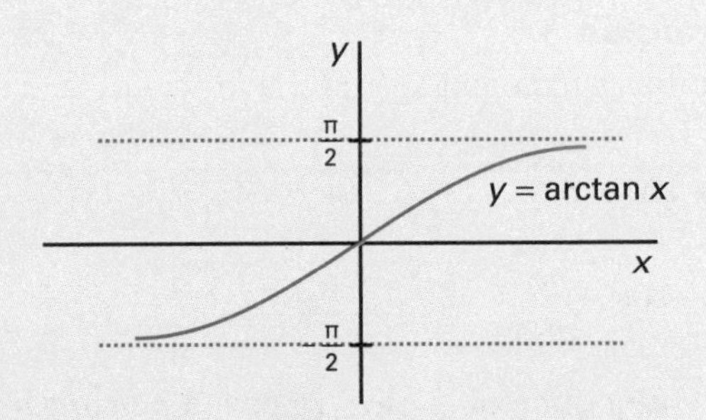

Watch out!

Many calculators and some books call the inverse functions $\sin^{-1}$, $\cos^{-1}$ and $\tan^{-1}$. *The $^{-1}$ is nothing to do with the reciprocal – it's the familiar notation for inverse functions.*

Checkpoint 5

What sort of mappings (not functions) would arcsin, arccos and arctan be if the ranges weren't restricted?

Checkpoint 6

Why are the domains of arcsin and arccos restricted? Think about why it doesn't make sense to try and find the arcsin or arccos of a number less than –1 or greater than 1.

Examiner's secrets

The examiner will be impressed if you remember that graphs of inverse functions are always reflections of the original graphs in $y = x$.

Exam questions

answers: page 89

1 Show that sec $\theta(1 - \sin\theta) = \cos\theta/(1 + \sin\theta)$ (5 min)

2 (a) Sketch on the same axes the graphs $y = x$, $y = \cos x$ and $y = \arccos x$ in the first quadrant.

(b) Show that $x - \cos x = 0$ has a root in the region of $x = 0.7$ rad and use the iterative formula $x_{n+1} = \cos x_n$ to find it to two decimal places.

(c) Hence or otherwise, solve the equation $x = \arccos(\arccos x)$. (15 min)

Test yourself

See if you can write down the trig relationships on this page – and their derivations – from memory. Sketch the graphs of the trig functions and their reciprocals and inverses, labelling any asymptotes and other features clearly.

Further trigonometric relations

The identities on this page let you express sums and products of trig functions as single functions, and *vice versa*. You can use them to rewrite complicated expressions in terms of standard ones. This is useful for simplifying trig equations and integrals.

Examiner's secrets

A candidate who is really fluent with trig identities always gives a good impression. Mastering these results also makes many exam questions quick and easy!

Compound angle formulae

These express trig functions of a compound angle (e.g. $A + B$) in terms of functions of the individual angles A and B.

The jargon

Compound angle formulae for $(A + B)$ are called *addition formulae* and those for $(A - B)$ are called *difference formulae*.

➜ $\sin(A \pm B) = \sin A \cos B \pm \cos A \sin B$

E.g. $\sin 105° = \sin(60° + 45°) = \sin 60° \cos 45° + \cos 60° \sin 45° = \dfrac{\sqrt{6} + \sqrt{2}}{4}$

➜ $\cos(A \pm B) = \cos A \cos B \mp \sin A \sin B$

E.g. $\cos 105° = \cos(60° + 45°) = \cos 60° \cos 45° - \sin 60° \sin 45° = \dfrac{\sqrt{2} - \sqrt{6}}{4}$

➜ $\tan(A \pm B) = (\tan A \pm \tan B)/(1 \mp \tan A \tan B)$

E.g. $\tan 105° = \tan(60° + 45°) = (\tan 60° + \tan 45°)/(1 - \tan 60° \tan 45°)$
$= (\sqrt{3} + 1)/(1 - \sqrt{3}) = -(2 + \sqrt{3})$

Watch out!

Be really careful about + and – signs in these formulae.

Checkpoint 1

What are the exact values of the sin, cos and tan of 75° and 15°?

Double-angle and half-angle formulae

Double-angle formulae

If you put $A = B$ in the addition formulae, you get formulae for functions of $(A + A)$, i.e. $2A$. They're called *double-angle formulae*.

➜ $\sin(2A) \equiv 2 \sin A \cos A$
➜ $\cos(2A) \equiv \cos^2 A - \sin^2 A$
➜ $\tan(2A) \equiv 2 \tan A/(1 - \tan^2 A)$

The formula for $\cos(2A)$ has two more rearrangements:

➜ $\cos(2A) \equiv 1 - 2 \sin^2 A$ $\cos(2A) \equiv 2 \cos^2 A - 1$

Speed learning

You may find it quicker just to learn the addition formulae and work out the double-angle formulae when you need them. Have a try at deriving the double-angle formulae – then decide!

Half-angle formulae

The last two can be rearranged again to give

➜ $\sin^2 \dfrac{A}{2} \equiv \dfrac{1}{2}(1 - \cos A)$

➜ $\cos^2 \dfrac{A}{2} \equiv \dfrac{1}{2}(1 + \cos A)$

Because they are functions of $\dfrac{A}{2}$, they're called *half-angle formulae*.

Checkpoint 2

How are the half-angle formulae derived from the double-angle formulae?

The form *a* cos θ + *b* sin θ

Expressions of the form $a \cos \theta + b \sin \theta$ are called *linear combinations* of $\cos \theta$ and $\sin \theta$. They can be expressed as a single sine or cosine, often written $R \cos(\theta \pm \alpha)$ or $R \sin(\theta \pm \alpha)$, with R positive and α acute. The values of R and α depend on the values of a and b in the original expression, and on whether you're using $R \cos(\theta \pm \alpha)$ or $R \sin(\theta \pm \alpha)$.

- ➜ Decide which formula you want to use. Any of them will work for any expression, but the choices suggested here avoid minus signs.
  for $a \cos\theta + b \sin\theta$ use $R\cos(\theta - \alpha)$
  for $a \cos\theta - b \sin\theta$ use $R\cos(\theta + \alpha)$
- ➜ Write down your equation, e.g. $a\cos\theta + b\sin\theta = R\cos(\theta - \alpha)$
- ➜ Expand the RHS, e.g. $a\cos\theta + b\sin\theta = R(\cos\theta\cos\alpha + \sin\theta\sin\alpha)$
- ➜ Equate coefficients of $\cos\theta$ and $\sin\theta$ on the LHS and RHS, e.g.
  $a = R\cos\alpha$ ① (coefficients of $\cos\theta$)
  $b = R\sin\alpha$ ② (coefficients of $\sin\theta$)
- ➜ Divide ② by ① to give $\tan\alpha = \dfrac{b}{a}$
- ➜ Add ①^2 and ②^2 to give $a^2 + b^2 = R^2(\cos^2\theta + \sin^2\theta) = R^2$

The method is identical if you're using the other formulae, but the expression for α will come out slightly different. This form is used for solving trigonometrical equations.

Example: Solve $4\cos\theta - 3\sin\theta = -5$ in the range $-180° \le \theta \le 180°$

Let $4\cos\theta - 3\sin\theta = R\cos(\theta + \alpha)$

$4\cos\theta - 3\sin\theta = R(\cos\theta\cos\alpha - \sin\theta\sin\alpha)$

$4 = R\cos\alpha$ (equating coefficients of $\cos\theta$)

$-3 = -R\sin\alpha$ (equating coefficients of $\sin\theta$)

Divide the second equation by the first:

$\frac{3}{4} = \tan\alpha \quad \therefore \alpha = 36.9°$ Also,

$4^2 + (-3)^2 = R^2(\cos^2\alpha + \sin^2\alpha) = R^2 \quad \therefore R = 5$

So $4\cos\theta - 3\sin\theta = 5\cos(\theta + 36.9°)$

So $5\cos(\theta + 36.9°) = -5$

$\cos(\theta + 36.9°) = -1$

$\theta + 36.9° = -180°, 180°$

$\theta = 143.1°$ (only one solution in range)

As well as being useful for equations, this form is also easier to sketch as a graph, since it is just a stretch and a translation of a standard cos or sin graph. For example, $5\cos(\theta + 36.9°)$ is a cos graph shifted 36.9° to the left and stretched by a factor of 5 parallel to the y-axis.

Action point

Work through this method using the other identities. Try making up examples with your own numbers. It's surprising how quickly it gets easier with practice!

Checkpoint 3

How is the graph of $y = R\cos(\theta - \alpha)$ related to the graph of $y = \cos\theta$? Sketch $y = \cos\theta$ and $y = 3\cos(\theta + 30°)$ on the same axes.

Exam questions

answers: page 89

1. Find all the solutions, in the interval $0 \le x \le 2\pi$, of the equation $\cos 2x + \cos x = 0$. (5 min)
2. (a) Find all the angles between 0 and 360° that satisfy the equation
   $8\sin x + 15\cos x = 10$
   (b) Two acute angles, α and β, are such that $\tan\alpha = \frac{4}{3}$ and $\tan(\alpha + \beta) = -1$. Without evaluating α or β,
   (i) show that $\tan\beta = 7$
   (ii) evaluate $\sin\alpha$ and $\sin\beta$
   (iii) evaluate $\sin^2 2\alpha$ and $\sin^2 2\beta$. (20 min)

Test yourself

On a blank sheet of paper, write down from memory the formulae from this spread. Do a sample calculation for each one.

Basic differentiation

You have seen that the gradient of a straight line is the same at every point. The gradient of a curve, however, varies from point to point. Differentiation is the process by which we can calculate the gradient at any given point. This has many applications and is a fundamental technique in mathematics.

Note

This is tested at AS level.

Derivative as a gradient; limits

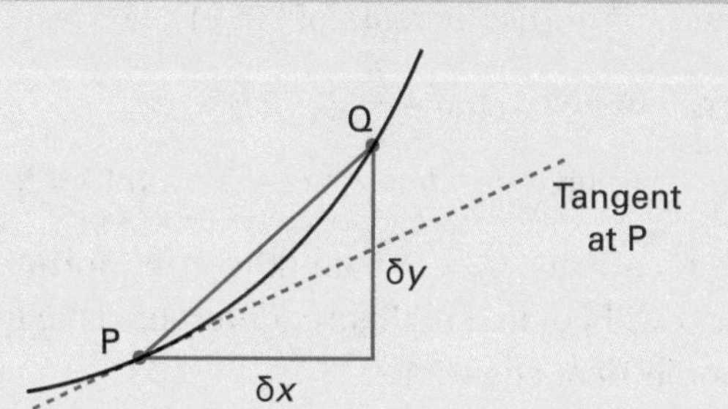

If y is written in terms of x, then $\frac{dy}{dx}$ is called the derivative or differential coefficient of y with respect to x. In this diagram, the gradient of the chord PQ $= \frac{\delta y}{\delta x}$. If you move Q closer to P, δx gets smaller and approaches zero. The ratio $\frac{\delta y}{\delta x}$ approaches a limit which is the gradient of the tangent at P. This is defined as the gradient of the curve at P.

The jargon

If $y = f(x)$, then $\frac{dy}{dx} = f'(x)$

Second derivatives

If you differentiate again, you get a second order derivative. This can be denoted by $f''(x)$ or $\frac{d^2y}{dx^2}$.

Differentiation of powers of x

Rule to know: $y = x^n, \quad \frac{dy}{dx} = nx^{n-1}$

$y = c \quad \frac{dy}{dx} = 0$

When you add or subtract functions of these types, their derivatives are similarly added or subtracted.

Example: Find $\frac{dy}{dx}$ if $y = x^7 + 4x + 5$

Answer: $\frac{dy}{dx} = 7x^6 + 4$

Example: Differentiate $\frac{1}{x^2} - \sqrt{x} + 4x^3$ with respect to x.

Answer: $= x^{-2} - x^{\frac{1}{2}} + 4x^3$

The derivative is $-2x^{-3} - \frac{1}{2}x^{-\frac{1}{2}} + 4 \times (3x^2)$

$= \frac{-2}{x^3} - \frac{1}{2\sqrt{x}} + 12x^2$

Speed learning

In words, 'decrease the power by 1 and multiply by the original power.'

Checkpoint 1

Think of a graphical reason why you get 0 when you differentiate $y = c$.

Don't forget

$y = x^7 + 4x + 5$ is a translation of $y = x^7 + 4x$.

Checkpoint 2

(a) Differentiate x^3 with respect to x.

(b) Differentiate $\frac{1}{p}$ with respect to p.

Examiner's secrets

Write as powers, differentiate, then rewrite in the original format.

Example: Differentiate $x^5 + \frac{3}{x}$ twice with respect to x

Answer: First derivative is $\frac{dy}{dx} = 5x^4 - \frac{3}{x^2}$

second derivative is $\frac{d^2y}{dx^2} = 20x^3 + \frac{6}{x^3}$

Gradients

You can find gradients using differentiation.

Example: Find (a) the gradient of the curve $y = x^2 + 3x$ where $x = 3$ and (b) the point on the curve where the gradient is 6.

Answer:

$$y = x^2 + 3x$$

$$\frac{dy}{dx} = 2x + 3$$

(a) when $x = 3$, $\frac{dy}{dx} = 2 \times 3 + 3 = 9$

(b) when the gradient is 6 $\frac{dy}{dx} = 6$

solve for x $2x + 3 = 6$ so $2x = 3$

substitute to find y $x = 1\frac{1}{2}$, $y = (1\frac{1}{2})^2 + 3 \times 1\frac{1}{2}$

$$y = 2\frac{1}{4} + 4\frac{1}{2} = 6\frac{3}{4}$$

$\therefore$ The gradient is 6 at $(1\frac{1}{2}, 6\frac{3}{4})$.

Rates of change

A derivative can also be regarded as a rate of change. For example, suppose the population, N, of a colony of rabbits at time t years is given by $N = 3t^2$.

Then $\frac{dN}{dt} = 6t$ gives you the rate of change of N at time t.

So at time $t = 5$, the population is increasing by $6 \times 5 = 30$ rabbits per year.

Checkpoint 3

If $f(x) = 5x^3 - 7x + 2$, find $f'(x)$ and $f''(x)$.

Checkpoint 4

The height, h cm, of a seedling t days after it is planted is given by $h = 4 + 6\sqrt{t}$. Find the rate at which the height is increasing 16 days after planting.

Exam questions

answers: page 90

1 Differentiate with respect to x:

(a) $3x^2 - 4x + 7$ (b) $\sqrt{x}$ (c) $x^{\frac{5}{2}} - x^{-\frac{2}{3}}$ (d) $\frac{3}{x^4}$ (e) $(x + 3)^2$ (12 min)

2 Find the second derivative with respect to x of $\left(x + \frac{1}{x}\right)^2$ (6 min)

3 (a) Find the gradient of the curve $y = x^3 - 6x^2 + 10x - 2$ at the point where $x = 2$.

(b) Find the coordinates of the points on the curve $y = x^3 - 6x^2 + 10x - 2$ where the gradient is 1. (5 min)

Examiner's secrets

When you differentiate a function like $(x + 3)^2$ multiply out the brackets first or use the chain rule (see page 56).

Applications of differentiation

You can use differentiation to find equations of tangents and normals to curves and the points where curves change direction.

Note

This is tested at AS level.

Tangents and normals 

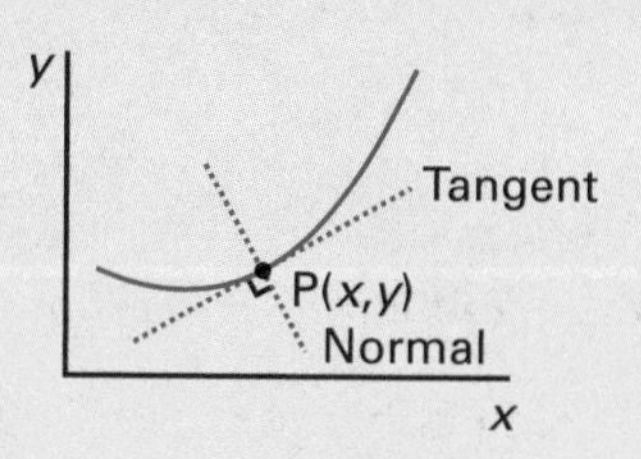

Links

See 'The geometry of straight lines', page 38.

The tangent and normal at P are perpendicular. The equations can be found by using

$$y - y_1 = m(x - x_1).$$

Example: Find the equation of the tangent to the curve $y = x^2 - 13$ at the point (4,3).

Answer: $y = x^2 - 13$

$$\frac{dy}{dx} = 2x$$

when $x = 4$, $\frac{dy}{dx} = 2x = 8$ so the gradient $m = 8$

The tangent equation: $y - y_1 = m(x - x_1)$ becomes $y - 3 = 8(x - 4)$ which simplifies to $y = 8x - 29$

Don't forget

You must substitute in the x value of your point to find the value of the gradient at that point.

Checkpoint 1

Find the equation of the tangent to the curve $y = x^3 - 5$ at the point where $x = 2$.

Example: Find the equation of the normal to the curve $y = 10 - \sqrt{x}$ at the point where $x = 9$.

Answer: $y = 10 - \sqrt{x} = 10 - x^{\frac{1}{2}}$

$$\frac{dy}{dx} = -\frac{1}{2}x^{-\frac{1}{2}} = \frac{-1}{2\sqrt{x}}$$

when $x = 9$ $\frac{dy}{dx} = -\frac{1}{2} \times 9^{-\frac{1}{2}} = \frac{-1}{2\sqrt{9}} = \frac{-1}{6}$

Gradient of normal is 6

We also need to work out the value of y when $x = 9$: $y = 10 - \sqrt{9} = 7$

Equation of normal is $y - 7 = 6(x - 9)$

This can be rewritten as $y = 6x - 47$

Don't forget

For perpendicular lines, the product of the gradients is –1 (see page 39). The gradient of the normal is the negative reciprocal of the tangent gradient.

Checkpoint 2

Find the equation of the normal to the curve $y = x^3 - 5$ at the point where $x = 2$.

Examiner's secrets

Full marks are usually gained here.

Maxima and minima 

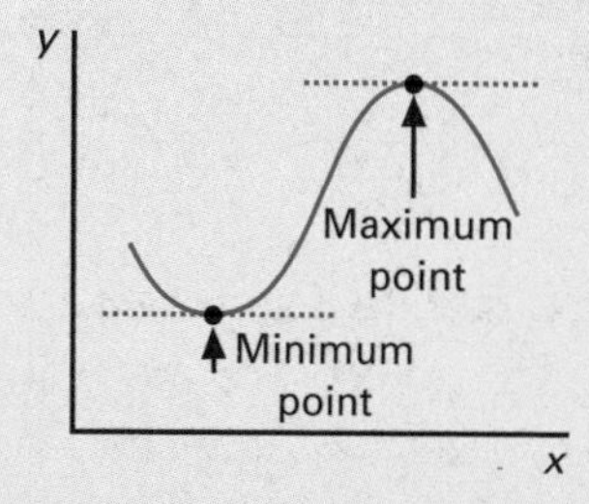

When you find the points on a curve where the gradient is zero, the turning points can be found. To distinguish between them, look at the value of $\frac{dy}{dx}$ at a point on each side of the x value that you have found.

In many cases, you may be asked to sketch the curve. This needs to be done without using a graphical calculator. The turning points and the points of intersection with the axes are needed.

Example: Find the turning points on the curve $y = 2x^3 - 3x^2 - 12x + 8$ and distinguish between them.

Answer:

$$y = 2x^3 - 3x^2 - 12x + 8$$

$$\frac{dy}{dx} = 6x^2 - 6x - 12$$

At a turning point, $\frac{dy}{dx} = 0 \quad \therefore \quad 6(x^2 - x - 2) = 0$

$$(x + 1)(x - 2) = 0$$

$$\therefore x = -1 \quad \text{or} \quad x = 2$$

x	-2	-1	0	2	3
$\frac{dy}{dx}$	24	0	-12	0	24

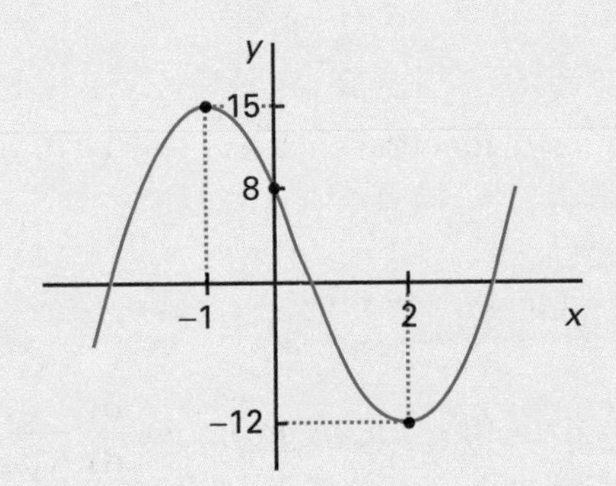

So, $(-1,15)$ is a maximum turning point and $(2,-12)$ is a minimum turning point.

Don't forget

Substitute to find y, but do it at the end.

Examiner's secrets

It is not enough just to look at the y values. It may work here, but won't with more complicated curves.

Increasing and decreasing functions

A function is increasing if $\frac{dy}{dx} > 0$. A function is decreasing if $\frac{dy}{dx} < 0$.

So in the last example, $\frac{dy}{dx}$ is negative between $x = -1$ and $x = 2$ and the function is decreasing between these values. It is therefore increasing for $x < -1$ and $x > 2$.

In the maxima and minima graph opposite, the curve is increasing between the minimum point and the maximum point. There is a point in between these two where the *gradient* of the curve reaches its maximum before decreasing to zero again at the maximum point. This point is called a point of inflection.

Don't forget

Decreasing means going from max. to min.

Exam questions answers: page 90

1. Find the gradient of the curve $y = 6x^4 - 3x^2$ at the point where $x = 3$ and hence find the equation of the tangent to the curve at that point. (6 min)
2. Find the exact values of the points on the curve $y = \left(x - \frac{2}{x}\right)^2$ where the gradient is 0. (6 min)
3. Find the turning points on the curve $y = x^4 - 2x^3 + x^2$ and distinguish between them. Sketch the curve. (8 min)

Rules of differentiation

Here are the basic rules of differentiation.

Differentiation of e^x, ln x plus sums and multiples

Rules to know: $y = e^x, \quad \frac{dy}{dx} = e^x$

$y = \ln x, \quad \frac{dy}{dx} = \frac{1}{x} \quad \text{for } x > 0$

When you add or subtract constant multiples of functions of these types, their derivatives are similarly added or subtracted.

Example: Find $\frac{dy}{dx}$ if $\quad y = 8e^x - 5 \ln x$

Answer: $\frac{dy}{dx} = 8e^x - \frac{5}{x}$

Links

$y = e^x$ and $y = \ln x$ are inverse functions. For more details see page 26.

Checkpoint 1

Find the gradient of the curve $y = 4e^x$ at the point where $x = 2$.

Checkpoint 2

Find the x coordinate of the point on the curve $y = 2 \ln x + 3$ where the gradient is 0.1.

Chain rule

If you have a composite function such as $(5x - 1)^4$ or e^{3x}, you need to use the chain rule.

Rule to know: $y = f(g(x)), \frac{dy}{dx} = f'(g(x)) \times g'(x)$

An alternative is: If $y = f(x)$ and $x = g(u)$, then $\frac{dy}{dx} = \frac{dy}{du} \times \frac{du}{dx}$

Notice how we split up the functions using another variable u.

Example: $y = (5x - 1)^4$ becomes $y = u^4$ and $u = 5x - 1$

$y = e^{3x}$ becomes $y = e^u$ and $u = 3x$

Now look how to use the rule: $\frac{dy}{dx} = \frac{dy}{du} \times \frac{du}{dx}$

Example: Differentiate $y = (5x - 1)^4$

$u = 5x - 1$ gives $\frac{du}{dx} = 5$ and $y = u^4$ gives $\frac{dy}{du} = 4u^3$

Put it together: $\frac{dy}{dx} = 4u^3 \times 5 = 20u^3 = 20(5x - 1)^3$

Example: Differentiate $y = e^{3x}$

$u = 3x$ gives $\frac{du}{dx} = 3$ and $y = e^u$ gives $\frac{dy}{du} = e^u$

Links

For more on composite functions, see page 19.

The jargon

A composite function is sometimes called *a function of a function*.

Don't forget

You must finish by getting your answer back in terms of x only, not u.

Checkpoint 3

Differentiate (a) $(7x + 2)^4$ (b) $\sqrt{x^2 + 1}$ (c) $\ln(5x + 1)$ (d) $5e^{4x}$

Put it together: $\frac{dy}{dx} = e^u \times 3 = 3e^u = 3e^{3x}$

With practice, you can reduce the amount of working you need to write down, but remember to show enough working to make your method clear.

Example: Find the equation of the normal to the curve $y = 3\sqrt{4x - 3} - 6$ at P(3,3).

Answer:

$$y = 3(4x - 3)^{\frac{1}{2}}$$

$$\frac{dy}{dx} = 3 \cdot \frac{1}{2} \cdot 4(4x - 3)^{-\frac{1}{2}}$$

$$\frac{dy}{dx} = 6(4x - 3)^{-\frac{1}{2}}$$

at $x = 3$ gradient of tangent = 2

so gradient of normal = $-\frac{1}{2}$

equation of normal is $y - 3 = -\frac{1}{2}(x - 3)$

which simplifies to $2y + x = 9$

Related rates of change

We can use the chain rule to find related rates of change.

Example: Suppose you know that the radius of a circle is increasing at a rate of 0.06 ms^{-1}, and you want to know how fast the area is increasing when the radius is 2 m.

Answer: The question says that $\frac{dr}{dt} = 0.06$, where r is the radius at time t.

We also know from GCSE that $A = \pi r^2$, so $\frac{dA}{dr} = 2\pi r$

Now use the chain rule $\frac{dA}{dt} = \frac{dA}{dr} \times \frac{dr}{dt} = 2\pi r \times 0.06 = 0.12\pi r$

So finally we deduce that the area increases at $0.12\pi \times 2 = 0.24\pi$ ms^{-1}.

Exam questions answers: page 90

1 Differentiate with respect to x:
(a) $\ln(4x - 5)$ (b) $(5x^2 + 1)^4$ (6 min)

2 (a) Use the chain rule to differentiate $\ln(x^n)$
(b) Use the laws of logarithms to write $\ln(x^n)$ in the form $a \ln b$, and hence differentiate $\ln(x^n)$ without using the chain rule. (6 min)

3 Find the equation of the tangent to the curve $y = \sqrt{4x - 3}$ at the point where $x = 7$. (5 min)

Examiner's secrets

$\dots s^{-1}$ means $\frac{d}{dt}$.

Examiner's secrets

'rate' here means $\frac{dr}{dt}$. Remember the units.

Links

For use of $\frac{dy}{dx} = \frac{1}{\frac{dx}{dy}}$ see page 59.

Checkpoint 4

A spherical balloon is being blown up so that its volume is increasing by 0.6 $m^3 s^{-1}$. Find the rate at which the radius is increasing when the radius is 0.1 m.

Further rules of differentiation

Here are some more rules to follow. You will often have to decide which rule to use. The structure of the function must be examined carefully.

Sin/cos/tan and their sums and differences

Don't forget

In all cases, x must be in radians.

Checkpoint 1

Differentiate (a) $\sin 4x$ (b) $\cos(x^2)$ (c) $-4 \tan 2x$.

Checkpoint 2

Find the equation of the tangent to the curve $y = \sin x$ at the point where $x = \frac{\pi}{3}$.

Rules to know: $y = \sin x$, $\frac{dy}{dx} = \cos x$

$y = \cos x$, $\frac{dy}{dx} = -\sin x$

$y = \tan x$, $\frac{dy}{dx} = \sec^2 x$

You can use the Chain Rule to differentiate more complex trig functions.

Examples:

$y = \cos 3x$, $\frac{dy}{dx} = -\sin 3x \times 3 = -3 \sin 3x$ (letting $u = 3x$)

$y = \sin^2 x$, $\frac{dy}{dx} = 2 \sin x \times \cos x = 2 \sin x \cos x$ (letting $u = \sin x$)

Links

This should be simplified to sin 2x – see page 50.

Product rule

When you have two different types of function multiplied together, such as $x \sin x$ or $e^x \ln x$, the product rule should be used.

Rule to know: $y = f(x)g(x)$, $\frac{dy}{dx} = f'(x)g(x) + g'(x)f(x)$

Alternatively: $y = uv$, $\frac{dy}{dx} = u\frac{dv}{dx} + v\frac{du}{dx}$

Speed learning

Differential of 1st × 2nd + differential of 2nd × 1st.

Example: Differentiate $y = x^2 \sin x$

Let $u = x^2$ so $\frac{du}{dx} = 2x$ and $v = \sin x$ so $\frac{dv}{dx} = \cos x$

Using the formula we have $\frac{dy}{dx} = x^2 \cos x + 2x \sin x$

Speed learning

(Bottom × differential of top – top × differential of bottom) divided by (bottom)^2.

Quotient rule

Similarly, if the functions are divided, such as $\frac{x}{\sin x}$ or $\frac{5 \tan x}{\sqrt{x}}$, the quotient rule should be used.

Rule to know: $y = \frac{f(x)}{g(x)}$, $\frac{dy}{dx} = \frac{f'(x)g(x) - f(x)g'(x)}{[g(x)]^2}$

Alternatively: $y = \frac{u}{v}$, $\frac{dy}{dx} = \frac{v\frac{du}{dx} - u\frac{dv}{dx}}{v^2}$

Mixed examples

Not surprisingly, at A2 level, you will be expected to figure out for yourself which rule to use.

Examples: Differentiate with respect to x:

(a) $(1 + 3x)^7$ (b) $x^2 \cos 5x$ (c) $\frac{\tan x}{x}$

Answers:

(a) This is a composite function so use the chain rule:

$$\frac{d}{dx}[(1+3x)^7] = 7(1+3x)^6 \times 3 = 21(1+3x)^6$$

(b) Use the product rule:

$$\frac{d}{dx}[x^2 \cos 5x] = 2x \times \cos 5x + (-5)\sin 5x \times x^2 = 2x\cos 5x - 5x^2 \sin 5x$$

(c) Use the quotient rule $\frac{d}{dx}\left[\frac{\tan x}{x}\right]$

$$= \frac{x \times \sec^2 x - \tan x \times 1}{x^2} = \frac{x\sec^2 x - \tan x}{x^2}$$

Example: Find the x coordinate of the turning point on the curve $y = \frac{x}{e^x}$

Differentiate using quotient rule $\frac{dy}{dx} = \frac{e^x \times 1 - x \times e^x}{(e^x)^2}$

At a turning point, $\frac{dy}{dx} = 0$ $\therefore$ $\frac{e^x - xe^x}{(e^x)^2} = 0$

Clear of fractions $e^x - xe^x = 0$

Factorise $e^x(1-x) = 0$

$e^x = 0$ or $1 - x = 0$

e^x never $= 0$ $\therefore$ $x = 1$

Examiner's secrets

First let $u = 1 + 3x$.

Don't forget

Show enough working to make your method clear.

Checkpoint 3

Differentiate with respect to x:

(a) $x \ln x$; (b) $\frac{x}{\sin x}$; (c) e^{3x-4}.

Examiner's secrets

Could rewrite as a product and use the product rule.

Checkpoint 4

Sketch the graph.

dy/dx and dx/dy

On some occasions it's much easier to find $\frac{dx}{dy}$ instead of $\frac{dy}{dx}$. So find $\frac{dx}{dy}$ then turn it upside-down to give $\frac{dy}{dx}$.

$$\frac{dy}{dx} = 1 \div \left(\frac{dx}{dy}\right)$$

Example: If $x = \tan y$, find $\frac{dy}{dx}$ in terms of x.

Answer: $\frac{dx}{dy} = \sec^2 y$ which is easily found

$$\therefore \frac{dy}{dx} = \frac{1}{\sec^2 y}$$

$$\therefore \frac{dy}{dx} = \frac{1}{1+\tan^2 y} \quad \text{using } \sec^2 y = 1 + \tan^2 y$$

$$\therefore \frac{dy}{dx} = \frac{1}{1+x^2} \quad \text{using } x = \tan y \text{ from the original equation.}$$

Don't forget

Pythagoras' Theorem (see page 40 if you have forgotten).

Exam questions answers: page 90

1 Differentiate with respect to x:
(a) $\ln(1+x^2)$ (b) $\frac{\cos x}{x}$ (c) $\sqrt{x}(8+x)$ (8 min)

2 Find the coordinates of the point on the curve $y = (3x+4)(2x-1)$ where the gradient is 13. (6 min)

3 Differentiate $y = \sin^{-1} x$ with respect to x, giving your answer in terms of x. (Hint: rewrite the equation as $x = \sin y$ first.) (5 min)

Implicit and parametric differentiation

Most of the functions that you have met so far have given y directly in terms of x, such as $y = 3x + 7$. Implicit functions are where y is given indirectly in terms of x. You will usually get a very complicated equation if you rearrange to make y the subject. Similarly, when a relationship is given in parametric form, eliminating a parameter will lead you to a more difficult task.

Implicit differentiation

Test yourself

Try to make y the subject and then differentiate.

Examples of implicit functions are $y^2 = 8x - 4$, $x^2 + y^2 = 3e^x$ and $y^2x - 3x^2 + \sin y = x$.

In each case you should differentiate each term separately with respect to x. Those with just 'x' terms are straightforward, those with just 'y' terms need a little more care, but those with 'x' and 'y' terms need great care.

$$\frac{d}{dx}(y^2) = \frac{d}{dy}(y^2) \times \frac{dy}{dx} \quad \text{Compare the chain rule}$$

$$= 2y\frac{dy}{dx}$$

Checkpoint 1

Differentiate implicitly: (a) $y^2 = 8x - 4$; (b) $x^2 + y^2 = 3e^x$; (c) $y^2x - 3x^2 + \sin y = x$.

$$\frac{d}{dx}(y^2x) = 2y\frac{dy}{dx} \times x + y^2 \times 1 \quad \text{This is a product, so use the product rule}$$

$$= 2xy\frac{dy}{dx} + y^2$$

Don't forget

This is a circle. Why? Find its centre and radius.

Example: Find $\frac{dy}{dx}$ if $x^2 - y^2 + 11x - 5y + 23 = 0$.

Answer: Differentiate each term with respect to x:

Examiner's secrets

No products here.

$$2x - 2y\frac{dy}{dx} + 11 - 5\frac{dy}{dx} = 0$$

$$(-2y - 5)\frac{dy}{dx} + 2x + 11 = 0 \quad \text{Collect like terms}$$

$$\frac{2x + 11}{2y + 5} = \frac{dy}{dx}$$

Checkpoint 2

Find the equation of the normal at this point.

Example: Find the equation of the tangent at the point (4,2) on the curve $4x^2 - 7y^2 = 36$.

Answer: Differentiate implicitly $8x - 14y\frac{dy}{dx} = 0$

Do not rearrange yet.

Substitute (4,2) $\quad 8 \times 4 - 14 \times 2\frac{dy}{dx} = 0$

$$\frac{8}{7} = \frac{dy}{dx}$$

Examiner's secrets

Full marks are usually given here, but it is safer to simplify just in case.

Equation of the tangent is $\quad y - 2 = \frac{8}{7}(x - 4)$

Which can be rewritten as $\quad 7y - 8x + 18 = 0$

Parametric differentiation

With parameters, you use a similar idea. You differentiate each parametric equation with respect to the appropriate variable and then combine using the alternative form of the chain rule.

So, if $x = 3t - 2$ and $y = 3 - 2t$, $\frac{dx}{dt} = 3$ and $\frac{dy}{dt} = -2$

use $\frac{dy}{dx} = \frac{dy}{dt} \times \frac{dt}{dx}$ and $\frac{dt}{dx} = 1 \div \left(\frac{dx}{dt}\right)$

gives $\frac{dy}{dx} = -2 \times \frac{1}{3}$

$\frac{dy}{dx} = -\frac{2}{3}$

Example: Find the equation of the normal to $x = 2 \cos \theta$, $y = \sin \theta$ at the point with parameter $\frac{\pi}{4}$. Give your answer in the form $ax + by = c$.

Answer: $\frac{dx}{d\theta} = -2 \sin \theta$

$\frac{dy}{d\theta} = \cos \theta$

$\therefore \frac{dy}{dx} = \frac{dy}{d\theta} \times \frac{d\theta}{dx}$

$= \frac{\cos \theta}{-2 \sin \theta}$ No need to simplify.

Substitute $\theta = \frac{\pi}{4}$ $= \frac{\cos \frac{\pi}{4}}{-2 \sin \frac{\pi}{4}} = -\frac{1}{2}$

Gradient of normal $= 2$

You need x and y when $\theta = \frac{\pi}{4}$

$x = 2 \cos \frac{\pi}{4} = \frac{2}{\sqrt{2}}$

$y = \sin \frac{\pi}{4} = \frac{1}{\sqrt{2}}$

Equation of the normal is $y - \frac{1}{\sqrt{2}} = 2\left(x - \frac{2}{\sqrt{2}}\right)$

Multiply by $\sqrt{2}$ $\sqrt{2}y - 1 = 2\sqrt{2}x - 4$

$2\sqrt{2}x - \sqrt{2}y = 3$

Don't forget

What shape is defined here? Eliminate t if you are not sure.

Checkpoint 3

Find the gradient of the tangent when the parameter is $\frac{\pi}{6}$.

Exam questions

answers: page 90

1 A curve is defined by $x^3 + y^3 + 3xy - 1 = 0$. Find the gradient of this curve at the point (2,–1). (7 min)

2 A curve has parametric equations $x = t^2 - 4$ and $y = 3t^4 + 8t^3$. Find the equation of the tangent to the curve at the point where $t = -1$. (10 min)

Basic integration

The jargon

'$\int \ldots dx$', means integrate the function with respect to x, i.e. x is the variable – the letter to use when integrating a constant.

Checkpoint 1

This rule works for all values of n apart from $n = -1$. Why?

Checkpoint 2

Why do we have $+ c$ at the end of each integral?

Don't forget

When you integrate x^n you increase the power by one and divide by the new power.

Checkpoint 3

Integrate with respect to x
(a) $x^4 - 5x^3$, (b) $x(3x + x^{-3})$.

Checkpoint 4

If $\frac{dy}{dx} = 2x + 3$, and $y = 8$ when $x = 2$, find y in terms of x.

Watch out!

You must be very careful when one of the limits is zero. Remember, for example, that $\cos 0° = 1$ and $e^0 = 1$.

Indefinite integration is the reverse of differentiation. If you are given the gradient of a curve, you can use integration to find its equation. You also need to be able to evaluate an integral with limits. This is called definite integration.

AS level: Integration of x^n

Rule to know: $\int x^n \, dx = \frac{x^{n+1}}{n+1} + c, \; n \neq -1$

As with derivatives, integrals can be added and subtracted just like ordinary numbers. Multiples are also worked out in a similar way.

Example: Find $\int 2x^2 + 3\sqrt{x} - \frac{7}{x^2} \, dx$

Answer: Rewrite the integral as $\int 2x^2 + 3x^{\frac{1}{2}} - 7x^{-2} \, dx$

Then integrate to get $\frac{2x^3}{3} + \frac{3x^{\frac{3}{2}}}{\frac{3}{2}} - \frac{7x^{-1}}{-1} + c$ which simplifies to $\frac{2x^3}{3} + 2\sqrt{x^3} + \frac{7}{x} + c$.

Finding the equation of a curve

Example: A curve is such that $\frac{dy}{dx} = 3x^2 + 5x$, and it passes through the point (2,8). Find the equation of the curve.

Answer: $y = \int 3x^2 + 5x \, dx$

which integrates to $y = x^3 + \frac{5x^2}{2} + c$

You are told that the curve passes through (2,8), so substitute $x = 2$ and $y = 8$ into the equation of the curve. $8 = 2^3 + \frac{5 \times 2^2}{2} + c \quad \therefore -10 = c$

Write out in full $\therefore y = x^3 + \frac{5x^2}{2} - 10$

Definite integration

Here, you are given limits. After you have integrated the function, you substitute each limit, then do (upper limit value) – (lower limit value). There is no need here to use a constant, as the subtraction will cancel it out.

Example: Find $\int_1^4 4x^3 - \frac{1}{x^2}\,dx$

Answer: First rewrite $-\frac{1}{x^2}$ as $-x^{-2}$

This integrates to $-\frac{x^{-1}}{-1}$ which is $\frac{1}{x}$

So integral is $\left[x^4 + \frac{1}{x}\right]_1^4$

$= \left[\left(4^4 + \frac{1}{4}\right) - (1 + 1)\right]$

$= 256 + \frac{1}{4} - 2 = 254\frac{1}{4}$

Further rules for A2: Integration of linear brackets, e^x and $1/x$

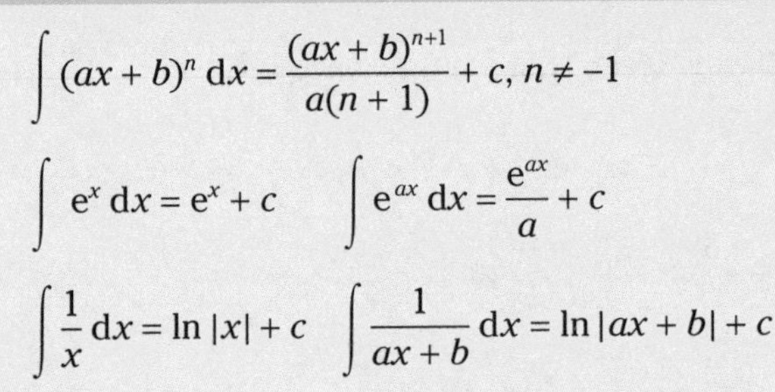

$$\int (ax+b)^n\,dx = \frac{(ax+b)^{n+1}}{a(n+1)} + c,\ n \neq -1$$

$$\int e^x\,dx = e^x + c \qquad \int e^{ax}\,dx = \frac{e^{ax}}{a} + c$$

$$\int \frac{1}{x}\,dx = \ln|x| + c \qquad \int \frac{1}{ax+b}\,dx = \ln|ax+b| + c$$

Examples: (a) $\int (3x-2)^4\,dx$ (b) $\int 8e^{2x}\,dx$ (c) $\int \frac{5}{2x-3}\,dx$

Answers: (a) $\frac{(3x-2)^5}{15} + c$ (b) $4e^{2x} + c$ (c) $2.5\ln|2x-3| + c$

Example: Evaluate exactly $\int_2^8 e^x + \frac{1}{x} + 3\,dx$

Answer: integrate $= [e^x + \ln x + 3x]_2^8$

substitute the limits $= (e^8 + \ln 8 + 3 \times 8) - (e^2 + \ln 2 + 3 \times 2)$

$= e^8 - e^2 + \ln 8 - \ln 2 + 24 - 6$

simplify using log laws $= e^8 - e^2 + \ln 2^3 - \ln 2 + 18$

$= e^8 - e^2 + 3\ln 2 - \ln 2 + 18$

$= e^8 - e^2 + 2\ln 2 + 18$

Exam questions answers: page 91

1 (AS) Integrate (a) $x^2 - \frac{1}{x^2}$ (b) $\frac{x^3 - 2}{4x^2}$ with respect to x (6 min)

2 (AS) If $\frac{dy}{dx} = 4x^3 + \frac{2}{x^3}$ and $y = 17$ when $x = 2$, find y in terms of x (5 min)

3 (AS) Evaluate exactly $\int_1^5 \frac{1}{\sqrt{x}}\,dx$ (4 min)

4 (A2) Evaluate exactly $\int_1^3 6e^{2x} + \frac{3}{4x}\,dx$ (6 min)

Links

See 'Laws of logarithms', page 27.

The jargon

c is often called the arbitrary constant.

Don't forget

The rules with the brackets only work for *linear* brackets.

Don't forget

Notice that you divide by the multiplier a (where you would have multiplied by a if you had been differentiating).

The jargon

'Exactly' means do not work out as a decimal. Leave powers of e, logs, square roots, etc. in your answer.

Checkpoint 5

Evaluate exactly (a) $\int_1^4 3x^2 + 4\sqrt{x}\,dx$.

(b) $\int_0^3 \frac{6}{2x+1}\,dx$

Rules of integration

To integrate more complicated functions, there are certain rules that you have to follow. The structure of the function must be examined carefully.

Sin/cos and their sums and differences

Rules to know:

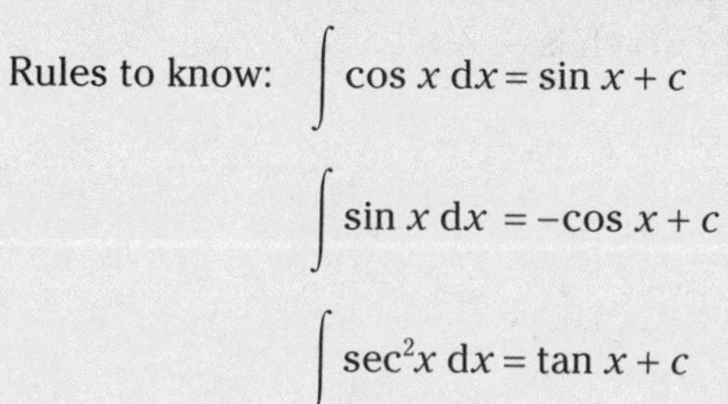

$$\int \cos x \, dx = \sin x + c$$

$$\int \sin x \, dx = -\cos x + c$$

$$\int \sec^2 x \, dx = \tan x + c$$

These are all the reverse results from differentiation and can be extended using substitution to, say, $\int \cos kx \, dx = \frac{1}{k} \sin kx + c$.

Checkpoint 1

Integrate (a) $2 \cos x - 5 \sin x$ and (b) $\sec^2 4x + \sec^2 5x$ with respect to x.

Example:

If $\frac{dy}{dx} = \sin 3x - \cos 2x$, find y, given that $y = 0$ when $x = \frac{\pi}{6}$.

Answer: $y = \int (\sin 3x - \cos 2x) \, dx$

Integrate each term $y = -\frac{1}{3} \cos 3x - \frac{1}{2} \sin 2x + c$

Substitute $\left(\frac{\pi}{6}, 0\right)$ $0 = -\frac{1}{3} \cos \frac{\pi}{2} - \frac{1}{2} \sin \frac{\pi}{3} + c$

Use exact values $0 = 0 - \frac{1}{2} \times \frac{\sqrt{3}}{2} + c$

$\therefore \quad c = \frac{\sqrt{3}}{4}$

Write out in full $y = -\frac{1}{3} \cos 3x - \frac{1}{2} \sin 2x + \frac{\sqrt{3}}{4}$

Don't forget

Add the constant.

Don't forget

Use radians during integration.

Special rule for integrating $\sin^2 x$ and $\cos^2 x$

You need to use the double-angle formulae arranged in the form:

either $\sin^2 x = \frac{1}{2}(1 - \cos 2x)$ or $\cos^2 x = \frac{1}{2}(\cos 2x - 1)$

Example: $\int \sin^2 x \, dx = \frac{1}{2} \int (1 - \cos 2x) \, dx$

$$= \frac{1}{2}\left[x - \frac{\sin 2x}{2}\right] + c$$

Links

For double-angle trig formulae see page 50.

Checkpoint 2

Integrate $\cos^2 3x$ (using a $\cos 6x$ conversion).

Substitution

Examples of functions which you can integrate by means of a substitution are $\sin(5x - 1)$, $x(x^2 + 1)^4$, $\sin^3 x \cos x$ and $\frac{x^2}{x^3 + 7}$.
You will usually be given the substitution.

The jargon

Another name is *change of variable*.

Don't forget

These are often recognised because one part of the function to be integrated is close to the differential of the other part.

Example: Find $\int x(x^2+1)^4\,dx$ by letting $u = x^2 + 1$.

Answer: You need to replace all terms in x by terms in u.

Obviously $(x^2+1)^4$ becomes u^4. You now need to differentiate $u = x^2 + 1$, to bring in 'dx' so $\frac{du}{dx} = 2x$

Rearrange this so that $\frac{du}{2} = x\,dx$ (The right-hand side is part of the required integral.)

$$\int x(x^2+1)^4\,dx = \int (x^2+1)^4\,x\,dx$$

Replace x terms with u $= \int u^4 \frac{du}{2}$

Integrate $= \frac{1}{2}\left(\frac{u^5}{5}\right) = \frac{u^5}{10}$

Put x back in $= \frac{(x^2+1)^5}{10} + c$

Example: Evaluate exactly $\int_0^2 \frac{x}{x^2+2}\,dx$ by means of the substitution $u = x^2 + 2$

Answer: $u = x^2 + 2$ differentiates to give

$$\frac{du}{dx} = 2x \text{ or } \frac{du}{2} = x\,dx$$

So the integral becomes $\int_0^2 \frac{1}{x^2+2} \times x\,dx$

$$= \int_{u=2}^{u=6} \frac{1}{u} \times \frac{du}{2}$$

$$= \frac{1}{2}\int_2^6 \frac{1}{u}\,du$$

$$= \frac{1}{2}[\ln u]_2^6 = \frac{1}{2}[\ln 6 - \ln 2]$$

$$= \frac{1}{2}\ln 3$$

Speed learning

Let u = the bracket.

Checkpoint 3

Integrate $(1 + 5x)^{11}$ letting $u = 1 + 5x$.

Checkpoint 4

Integrate $\sin^3 x \cos x$ letting $u = \sin x$.

Don't forget

Notice how the limits have become 'u' limits by working out the value of u for each value of x.

Checkpoint 5

Find $\int_0^{\frac{\pi}{6}} \frac{\cos x}{4\sin x + 1}\,dx$ letting $u = 4\sin x + 1$.

Examiner's secrets

The result $\int \frac{f'(x)}{f(x)}\,dx = \ln|f(x)| + c$ can be used to integrate this directly $(= \frac{1}{2}\ln|x^2 + 2|)$.

Exam questions answers: page 91

1 By letting $u = 1 + x^3$ or otherwise, evaluate $\int_0^1 \frac{x^2}{1+x^3}\,dx$ leaving your answer in the form of a logarithm. (6 min)

2 Find $\int x\sqrt{1+x^2}\,dx$ by letting $u = 1 + x^2$. (6 min)

Further rules of integration

Here are two more rules that you have to follow. Again, the structure of the function must be examined carefully.

Integration by parts

Examples of functions which you can integrate by parts are xe^{4x}, $x^2 \sin x$ and $x \ln x$. They are made up of different types of function multiplied together.

Action point

Compare the product rule.

The formula is $\int u\frac{dv}{dx}\,dx = uv - \int v\frac{du}{dx}\,dx$

As a general rule, if there is an x or x^2 term, let that be your u, and the other part be $\frac{dv}{dx}$.

Example: Find $\int x\,e^{4x}\,dx$

Answer:

Let $u = x$, giving $\frac{du}{dx} = 1$, and $\frac{dv}{dx} = e^{4x}$, giving $v = \frac{e^{4x}}{4}$

Examiner's secrets

Always try to differentiate the x term and integrate the other term, and show your working clearly.

Using the formula, the integral is $x\frac{e^{4x}}{4} - \int \frac{e^{4x}}{4} \times 1dx$

Simplify the second term $= \frac{x\,e^{4x}}{4} - \frac{1}{4}\int e^{4x}\,dx$

Integrate the second term $= \frac{x\,e^{4x}}{4} - \frac{e^{4x}}{16} + c$

Action point

Differentiate the answer to check you get back to the function you were integrating.

With functions such as $x^2 \cos x$ it is necessary to apply integration by parts twice. It is important that there is consistency throughout in the order that u and v are used.

The exception to the rule that you let an x term be u, is when one function is a ln term. Always let the ln term be u. This is because to integrate $\ln x$ on its own you have to use integration by parts!

Checkpoint 1

Try integrating $\ln x$ using parts. You will need to let $u = \ln x$ and $\frac{dv}{dx} = 1$.

Example: Find $\int_1^4 x \ln x\,dx$

Answer: Let $u = \ln x$, giving $\frac{du}{dx} = \frac{1}{x}$, and $\frac{dv}{dx} = x$, giving $v = \frac{x^2}{2}$.

Don't forget

For definite integrals, the uv part must also be evaluated with limits.

Using the formula, the integral becomes $\left[\ln x \times \frac{x^2}{2}\right]_1^4 - \int_1^4 \frac{x^2}{2} \times \frac{1}{x}\,dx$

$$= \left[\frac{x^2}{2}\ln x\right]_1^4 - \frac{1}{2}\int_1^4 x\,dx$$

$$= \left[\frac{x^2}{2}\ln x\right]_1^4 - \left[\frac{x^2}{4}\right]_1^4$$

$$= [8 \ln 4 - 0] - \left[4 - \frac{1}{4}\right]$$

$$= 8 \ln 4 - 3\frac{3}{4}$$

Checkpoint 2

Try $\int x^2 \sin x\,dx$. Here parts must be applied twice.

Partial fractions

You will have to split the function first and then integrate each fraction. It is likely that at least one of the integrals will lead to a logarithm.

Example: Find $\int \frac{5x^2 + 2x - 7}{(2x+3)(x+1)^2} \, dx$

Answer: When you split into partial fractions you start with

$$\frac{A}{2x+3} + \frac{B}{x+1} + \frac{C}{(x+1)^2}$$

It is easy to forget the squared term.

This splits to $\int \frac{5}{2x+3} \, dx - \int \frac{4}{(x+1)^2} \, dx$

Notice that B = 0 which just leaves two terms to integrate.

Write the second part as $\int -4(x+1)^{-2} \, dx$

which integrates to $\frac{-4(x+1)^{-1}}{-1}$

$$\therefore \int \frac{5}{2x+3} \, dx - \int \frac{4}{(x+1)^2} \, dx = \frac{5}{2} \ln |2x+3| + \frac{4}{x+1} + c$$

In some cases you may have to do long division first.

Links

Forgotten partial fractions? See page 25.

Checkpoint 3

Integrate $\frac{x+9}{(x+2)(3-x)}$ with respect to x.

Checkpoint 4

Integrate $\frac{x^2}{(x+5)(x-3)}$ with respect to x.

Note

Check in your specification if you have to integrate functions such as $\frac{1}{a^2+x^2}$ and $\frac{1}{\sqrt{a^2-x^2}}$. These integrals are in the formulae booklet.

Exam questions answers: page 91

1 Find (a) $\int x \sin x \, dx$ (b) the exact value of $\int_1^2 x^2 \ln x \, dx$.

2 Given that $f(x) = \frac{2x^2 + 10x + 7}{(1+2x)^2(1+x)}$, express f(x) in the form

$\frac{A}{(1+2x)} + \frac{B}{(1+2x)^2} + \frac{C}{(1+x)}$.

Hence evaluate $\int_0^1 f(x) \, dx$, giving your answer in the form $p + \ln q$, where p and q are to be determined. (12 min)

Applications of integration

You can use integration to find the area under a curve and the volume of revolution of a solid formed when a curve is rotated, usually about an axis. These are applications of definite integration.

Area under a curve

Note

Area under a curve is an AS topic.

Rule to know: Area $= \int_a^b y\,dx,\ y \geq 0$

You have to be careful and check that the curve does not cross the x-axis between a and b.

Action point

Sketch the curve.

Example:

Find the area under the curve $y = x^2 + 5$ between $x = 1$ and $x = 5$.

Answer: $A = \int_1^5 (x^2 + 5)\,dx$

integrate $= \left[\frac{x^3}{3} + 5x\right]_1^5$

substitute $= \left[\left(\frac{5^3}{3} + 5 \times 5\right) - \left(\frac{1^3}{3} + 5 \times 1\right)\right]$

simplify $= \left[\left(\frac{125}{3} + 25\right) - \left(\frac{1}{3} + 5\right)\right]$

$= 61\frac{1}{3}$ (square units)

Don't forget

Substitute then work out
Upper limit value – Lower limit value.

Checkpoint 1

Find the area under $y = x$, between $x = 1$ and $x = 5$. Use integration and another method to check.

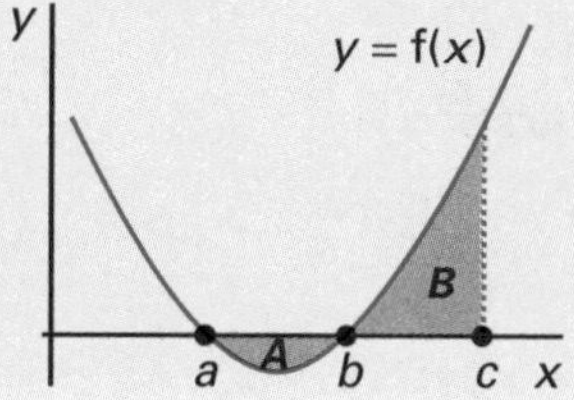

Here, area A is *below* the x-axis.

$\int_a^b y\,dx$ will be *negative*.

So, if you need the area between the curve and the x-axis from a to c, you have to do two separate integrals $\left|\int_a^b y\,dx\right| + \int_b^c y\,dx$.

The jargon

You can think of this as being in the *negative* y part of the graph.

If you were asked to evaluate the definite integral $\int_{-2}^{2} x^3\,dx$ you would get an answer of 0. This cannot be the area! As the curve is an odd function the areas above and below the x-axis cancel out. So, to calculate the area between the curve $y = x^3$, the x-axis and the lines $x = -2$ and $x = 2$, you need to find $\int_0^2 x^3 dx$ and double it. Draw a sketch to help.

Examiner's secrets

In some questions you may be asked to find the area between a curve and the x-axis from a given diagram. You have to find the limits, by finding where the curve crosses the x-axis. You may be asked to find the area between a line and a curve or between two curves. The limits here are found by working out where they intersect.

Volumes of revolution

When a curve is rotated through 360° (or 2π radians) about an axis (usually the x-axis), the volume of the solid formed is given by

$$V = \int_a^b \pi y^2\,dx$$

First you write y, then square it and then integrate.

Example: Find the exact volume formed when the curve $y = \sqrt{x}$ between $x = 1$ and $x = 7$ is rotated through 360° about the x-axis.

Answer: $V = \pi \int_1^7 y^2 \, dx \qquad y = \sqrt{x}, y^2 = x$

$$= \pi \int_1^7 x \, dx$$

integrate $\quad = \pi \left[\frac{x^2}{2} \right]_1^7$

substitute $\quad = \pi \left(\frac{7^2}{2} - \frac{1^2}{2} \right) = \pi \left(\frac{49}{2} - \frac{1}{2} \right) = 24\pi$ (cubic units)

Parameters

Similar results can be calculated for curves given parametrically. This curve is defined by

$x = \sin t, y = \cos t, 0 \le t \le \frac{\pi}{2}$.

What is the shaded area? $\frac{\pi}{4}$. Why?

Using integration, $A = \int_{t=0}^{t=\frac{\pi}{2}} y \, dx = \int_{t=0}^{t=\frac{\pi}{2}} y \frac{dx}{dt} \cdot dt$

Replace x and y terms by t terms, i.e. $y = \cos t$ and $\frac{dx}{dt} = \cos t$

$$= \int_0^{\frac{\pi}{2}} \cos t \times \cos t \, dt,$$

$$= \int_0^{\frac{\pi}{2}} \cos^2 t \, dt. \qquad \text{But } \cos^2 t = \frac{1}{2}(1 + \cos 2t)$$

Integrate $\quad = \frac{1}{2} \int_0^{\frac{\pi}{2}} (1 + \cos 2t) \, dt = \frac{1}{2} \left[t + \frac{\sin 2t}{2} \right]_0^{\frac{\pi}{2}}$

Substitute $\quad = \frac{1}{2} \left[\left(\frac{\pi}{2} + \frac{\sin \pi}{2} \right) - (0 + 0) \right] = \frac{\pi}{4}$

Exam questions answers: page 91

1. Find the area bounded by the graph of $y = \frac{1}{x^2}$, the x-axis and the lines $x = 1$ and $x = 2$. (5 min)
2. Find the area bounded by the curve $y = x^2$, the line $x + y = 2$ and the x-axis. (7 min)
3. The region R is bounded by the curve $y^2 = 4x$, the x-axis and the line $x = 4$. Find the volume generated when the curve is rotated through 360° (a) about the x-axis; (b) about the y-axis. (11 min)
4. Find the area bounded by the ellipse $x = 5 \cos t$, $y = 3 \sin t$, $0 \le t \le 2\pi$ and the positive x and y axes. (11 min)

Examiner's secrets

Take π out to get $\pi \int y^2 \, dx$.

Checkpoint 2

Find the volume when $y = 3x$ is rotated about the x-axis between 0 and 4. Check using the volume of a cone.

Don't forget

Many students lose marks because they forget to include the π *in each line.* 'Exact' means leave the π in.

Note

Check if the use of parameters is in your specification at A2.

Checkpoint 3

Find the radius of the quarter circle and hence its area.

Links

See Double-angle formulae, page 50.

Examiner's secrets

If you are asked to find the volume when a curve is rotated about the y-axis use the formula $V = \int_{y=c}^{y=d} \pi x^2 \, dy$.

Check if this is in your specification.

Differential equations

When you have an equation that contains a differential or derivative, like $\frac{dy}{dx}$, it is a differential equation. Many of the examples you will meet will be in a practical context and will refer to a rate of change, often with respect to time.

Action point

Look back at GCSE for proportion and inverse proportion (possibly called direct and indirect variation).

Formation

First look at a mathematical example. The gradient of a curve is proportional to the square root of the x coordinate at any point. What is the equation connecting y and x?

The symbol $\propto$ is used for proportional.

So, gradient $\frac{dy}{dx} \propto \sqrt{x}$.

As an equation, $\frac{dy}{dx} = k\sqrt{x}$, where k is constant.

Links

See Exponential growth and decay, page 28.

You will find differential equations where y is a function of x as above, or a function of y, or a function of x and y.

In practical situations, exponential growth and decay are common contexts that you will find. A scientific example concerns Newton's Law of Cooling. In this case, the rate of change of the temperature of a cooling body is proportional to the excess temperature over the surroundings. Although time is not stated, the rate of change is with respect to time.

Let θ be the excess temperature at time t, then $\frac{d\theta}{dt} \propto -\theta$, and $\frac{d\theta}{dt} = -k\theta$ is the differential equation. (Cooling is the clue to the negative sign.)

Solution

Checkpoint 1

Find y when $\frac{dy}{dx} = 3x + 2$, given that $y = 11$ when $x = 2$.

You have already solved many differential equations in previous sections, i.e. given $\frac{dy}{dx}$ find y. In all previous cases, y was a function of x only. The important method now is when y is a function of y or of x and y. The method is called *separating the variables*. You can think of it as collecting all the x's on one side and the y's on the other side.

So, if $\frac{dy}{dx} = 3y^2$ — divide both sides by y^2

$\frac{1}{y^2}\frac{dy}{dx} = 3$ — integrate both sides with respect to x

$\int \frac{1}{y^2}\frac{dy}{dx}\,dx = 3\int dx$ — this simplifies to

$\int \frac{1}{y^2}\,dy = 3\int dx$ — complete the integration

$-\frac{1}{y} = 3x + c$

Examiner's secrets

Remember to take the 3 outside.

Example: Given that $\frac{dy}{dx} = (y + 2)\sec^2 x$ and that $y = 1$ when $x = 0$, find an expression for y in terms of x.

Answer: $\frac{dy}{dx} = (y + 2)\sec^2 x$ separate the variables

$\frac{1}{y+2}\frac{dy}{dx} = \sec^2 x$ integrate both sides with respect to x

$\therefore \ln|y + 2| = \tan x + c$

Substitute (0,1) $\ln(1 + 2) = \tan 0 + c$

$\therefore c = \ln 3$

$\therefore \ln(y + 2) - \ln 3 = \tan x$ make y the subject using log rules

$$\ln\left(\frac{y+2}{3}\right) = \tan x$$

$$\frac{y+2}{3} = e^{\tan x}$$

$$y + 2 = 3e^{\tan x}$$

$$y = 3e^{\tan x} - 2$$

Example: A colony of micro-organisms in a liquid is growing at a rate proportional to half the number of organisms (M) present at any time. Initially there are P organisms. Given that the colony increases by 25% in 10 hours, find the time that elapses before it doubles in size.

Answer: The differential equation is:

$\frac{dM}{dt} = k\frac{M}{2}$ separate the variables

$\int \frac{1}{M}\,dM = \frac{k}{2}\int dt$ integrate both sides

$$\ln M = \frac{k}{2}t + c$$

substitute $(0,P)$ $\ln P = 0 + c \quad \therefore c = \ln P$

$$\therefore \ln M = \frac{kt}{2} + \ln P$$

$$\ln\left(\frac{M}{P}\right) = \frac{kt}{2}$$

substitute $(10,1.25P)$ $\ln\left(\frac{1.25P}{P}\right) = k\left(\frac{10}{2}\right) \quad \frac{1}{5}\ln 1.25 = k$

$$\therefore \ln\left(\frac{M}{P}\right) = \frac{1}{10}\ln 1.25 \times t$$

Need t when $M = 2P$ $\therefore \ln\left(\frac{2P}{P}\right) = \frac{1}{10}\ln 1.25 \times t$

$\therefore t = 31$ hours (2 s.f.)

Exam questions answers: pages 91–2

1 The variables x and y are related by the differential equation $x^2\frac{dy}{dx} = \sec y$. Given that $x = 1$ when $y = \frac{\pi}{4}$, find the value of x when $y = 0$. (8 min)

2 If the half-life of a radioactive element that is decaying naturally is 300 years, find how many years it will take for the original mass of the element to be reduced by 80%. (12 min)

Checkpoint 2

Show that the solution of the differential equation $\frac{dy}{dx} = \frac{y}{x}$ can be written in the form $y = cx$.

Examiner's secrets

Growth means that k is positive.

Note

Check in your specification if you need to know that $\frac{d}{dx}(a^x) = a^x \ln a$.

Numerical solution of equations

Most equations, including higher-order polynomials and all but the simplest trigonometric equations, can't be solved analytically (i.e. by algebraic manipulation.) Iterative methods are used to find approximate solutions. The process consists of finding two points between which a root exists and using iteration to find a more accurate solution.

Note

This is an A2 topic.

Checking that a solution exists

You can find out whether there is a solution using a graphical method.

→ Rearrange the equation into the form $f(x) = 0$.
→ Sketch the graph $y = f(x)$.
→ The **roots** (solutions) of the equation are points where the graph intersects the x-axis, because at these points $f(x)$ is zero.

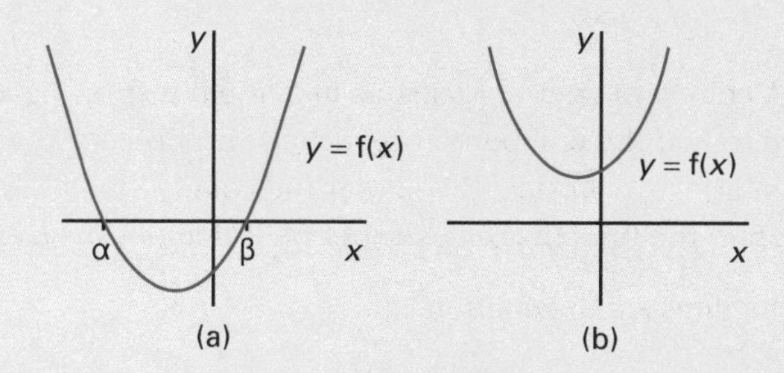

(a) $x = \alpha$ and $x = \beta$ are roots of $x^4 + x - 4 = 0$.
(b) $x^2 + x + 1 = 0$ has no real roots (graph doesn't intersect the x-axis).

Watch out!

These two methods do exactly the same thing – but don't get confused about which one you're using! The first (graphs (a) and (b)) goes with the rearrangement $f(x) = 0$. The second (graphs (c) and (d)) goes with $g(x) = x$.

An equivalent graphical method:

→ Rearrange the equation into the form $g(x) = x$.
→ Sketch the graphs $y = g(x)$ and $y = x$ on the same axes.
→ The roots of the equation are the x-coordinates of points where the graphs intersect, because at these points $g(x) = x$.

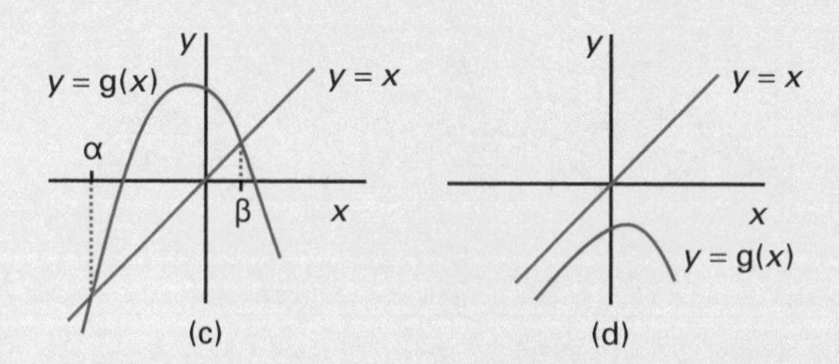

(c) $x = \alpha$ and $x = \beta$ are roots of $4 - x^4 = x$ (these are the same α and β as in graph (a)).
(d) $-x^2 - 1 = x$ has no real roots (graph doesn't intersect $y = x$).

Locating roots

Look at the first pair of graphs.
In the region of each root, $f(x)$ changes sign.

If $f(x)$ goes from positive to negative (as it does around α) or from negative to positive (as it does around β), then it must be zero in between.

So to locate a root,

- rearrange the equation into the form $f(x) = 0$
- either by sketching the graph $y = f(x)$ or by thinking about it, find a place where $f(x)$ changes sign between two values of x, say a and b. It doesn't matter whether the change is from + to – or – to +
- there is a root between a and b
- by seeing whether $f(a)$ or $f(b)$ is closer to zero, you can guess whether the root is nearer to a or b – but this is only a guess!

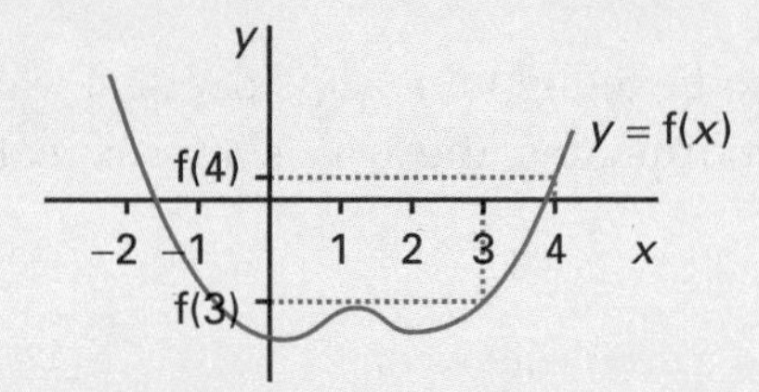

$f(x) = 0$ has a root between $x = -2$ and $x = -1$ and another root between $x = 3$ and $x = 4$. The second root is closer to 4 than to 3.

> **Checkpoint 1**
>
> For $f(x) = e^x - 3$, find two x values to 1 d.p. between which $f(x) = 0$.

> **Examiner's secrets**
>
> As you are allowed a graphics calculator in the exam, it can be a useful way of *checking* roots that you've already found. But you won't get any marks unless you show you understand how the method works!

Iterative methods

An *iterative formula* is a rule to get from an approximate solution x_n to the next, hopefully better one x_{n+1}. Applying the rule is *iterating*.

- Locate an **approximate solution** a; then . . .
  If you're given an iterative formula:
- **iterate**, using a as your first approximation in the formula
  If you're not given an iterative formula:
- rearrange the equation into the form $x = f(x)$
- iterate as above, using $x_{n+1} = f(x_n)$ as your iterative formula.

Continue till you have the required degree of accuracy. You must be sure that the next iteration won't change the last significant figure!

Example: Use the formula $x_{n+1} = -\frac{1}{5}e^{x_n}$ to solve $e^x + 5x = 0$ to 3 d.p.

Answer: There is a solution between 0 and –1 so use $x_1 = 0$. Then $x_2 = -0.2$, $x_3 = -0.163\,7\ldots$, $x_4 = -0.169\,8\ldots$, $x_5 = -0.168\,8\ldots$, $x_6 = -0.168\,9\ldots$, $x_7 = -0.168\,9\ldots$, so the solution is –0.169 to 3 d.p.

The formula $x_{n+1} = f(x_n)$ converges if $|f'(x)| < 1$ where $y = f(x)$ intersects with $y = x$. You can show this with a 'staircase' or 'cobweb' diagram.

> **Watch out!**
>
> Write down *all* the decimal places of each approximation from your calculator and use them in your working. That way, you avoid rounding errors. Of course, when writing the final answer, just give a sensible number of decimal places!

> **Checkpoint 2**
>
> Use $x_{n+1} = \frac{1}{x_n^2+1}$ to solve $x^3 + x - 1 = 0$ to 1 d.p.

> **Watch out!**
>
> Not all iterative formulae converge. If yours doesn't, *don't panic* – try another rearrangement of the equation in the form $x = f(x)$.

> **Checkpoint 3**
>
> How do staircase and cobweb diagrams illustrate the condition for convergence?

> **Note**
>
> Check if you need to know about the Newton-Raphson method.

Exam questions

answers: page 92

1 (a) Sketch the graph $y = 1 - \sin x$ for $0 \le x \le \pi$ radians. By adding a suitable line to your graph and explaining its significance, show that $1 - \sin x - x = 0$ has a root between $x = 0$ and $x = 1.5$ radians.

(b) Show by a non-graphical method that the root you have found lies between 0.5 and 0.6 radians. State with reasons whether you expect the root to lie closer to 0.5 or 0.6. (15 min)

2 Show graphically or otherwise that the equation $x^3 - x - 1 = 0$ has only one real root X, and that it lies between 1 and 2. By rearranging the equation in the form $x = f(x)^{\frac{1}{3}}$ for a suitable choice of $f(x)$, find an iterative formula for the solution X. Starting with the approximation $x_1 = 1$, estimate X to 2 decimal places. By drawing a labelled sketch on the same axes of the graphs $x = f(x)^{\frac{1}{3}}$ and $y = x$, illustrate how the iteration converges from $x_1 = 1$ towards the solution X. (15 min)

Numerical integration

There are many functions which cannot be integrated – either because they are too complicated for this specification, or they just cannot be integrated!

Trapezium rule

This is the most straightforward method to find an approximate value for the area under a curve. The method joins adjacent y-coordinates with a straight line to form trapezia. The formula is as follows:

- divide the area to be found into strips of equal width h
- if there are n y-coordinates, the area is approximately
  $\frac{h}{2}[y_1 + y_n + 2(y_2 + y_3 + \ldots + y_{n-1})]$
- in words, the trapezium rule says:
  - add the first and last y-coordinates
  - add them to twice the sum of all the other y-coordinates
  - multiply by half the width of the strips.

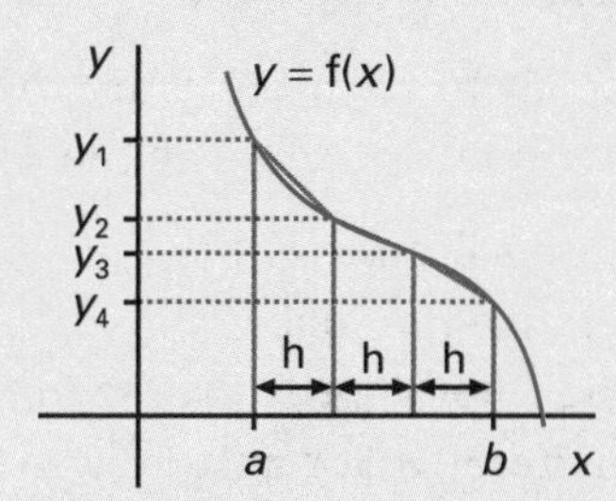

Trapezium rule approximation to $\int_b^a f(x)\,dx$

Look at each trapezium to see if it overestimates the area (e.g. the first one in the figure) or underestimates it (e.g. the last one in the figure).

Example: Find the approximate area under the curve $y = 3^x$ between $x = 0$ and $x = 6$. Use the trapezium rule with three strips.

x	0	2	4	6
3^x	1	9	81	729

Answer: In this case the width of each strip is 2 (so $h = 2$).

$$\int_0^6 3^x\,dx \approx \frac{2}{2}[1 + 729 + 2(9 + 81)]$$
$$\approx [730 + 2 \times 90]$$
$$\approx 730 + 180$$
$$\approx 910$$

Note

The trapezium rule is an AS topic, but you may be asked to use it at A2 with more complicated functions.

Watch out!

Don't forget that n y-coordinates means $n - 1$ strips!

Checkpoint 1

x	0	1	2	3	4
f(x)	3	8	17	12	4

The graph of $y = f(x)$ is continuous between $x = 0$ and $x = 4$. Use the trapezium rule to estimate the area between the curve, the x-axis and the lines $x = 0$ and $x = 4$.

Checkpoint 2

Why do you double all the y-coordinates apart from the first and the last ones?

Checkpoint 3

Work out the percentage error using strips of width 0.5.

It can be seen from the graph that this is an overestimate. You will often be asked to work out the percentage error. In this case the actual area is 662.654 156 98 . . . , so the percentage error is

$$\frac{910 - 662.654\ 156\ 98}{662.654\ 156\ 98} \times 100 = 37.33\% \text{ (to 4 s.f.)}.$$

In order to reduce this error it is sensible to take more strips. By using strips of width 1 the area is

$$\frac{1}{2}[1 + 729 + 2(3 + 9 + 27 + 81 + 243)]$$
$$= \frac{1}{2}[730 + 2 \times 363]$$
$$= \frac{1}{2} \times 1\ 456$$
$$= 728$$

This reduces the percentage error to 9.86% (to 4 s.f.).

Simpson's Rule

This method is a little more complicated in that adjacent y-coordinates are joined by a parabola. The formula is as follows:

$$\int_a^b y\,dx \approx \frac{1}{3}h\{(y_0 + y_n + 4(y_1 + y_3 + \ldots + y_{n-1}) + 2(y_2 + y_4 + \ldots + y_{n-2})\}$$

In the above example with strips of width 1 the approximation would be:

$$\frac{1}{3} \times 1 \times \{1 + 729 + 4(3 + 27 + 243) + 2(9 + 81)\}$$
$$= \frac{1}{3} \times \{730 + 4 \times 273 + 2 \times 90\}$$
$$= \frac{1}{3} \times 2002$$
$$= 667\frac{1}{3}$$

This has reduced the error to 0.7061% (to 4 s.f.).

Note

Check in your specification if you need to know Simpson's Rule.

Note

Check in your specification if you need to know the mid-ordinate rule.

Test yourself

When you have finished revising this page, write down from memory what you know about numerical methods. Use sketches to illustrate the techniques.

Exam questions

answers: page 92

1 Use the trapezium rule with four strips to find an approximation to $\int_1^3 \sqrt{1 + x^2}\,dx$. Show graphically that your answer is an overestimate. (10 min)

2 A curve has equation $y = \frac{x}{\sqrt[3]{x + 3}}$. The region R is bounded by the curve, the x-axis and the line $x = 1$. Use Simpson's rule with 5 ordinates (4 strips) to approximate the area of R to 6 significant figures. (10 min)

Vectors

> **Note**
> Vectors is an A2 topic.

Many of the concepts in which mathematics is interested are vector quantities. We use vector maths to study them. The maths of vectors is similar in many ways to the maths of ordinary numbers. Vectors can be added and subtracted, for example. Because we can draw diagrams of vectors, these operations can be represented geometrically.

Vector quantities and properties of vectors

Scalar quantities such as area, mass and age can be described by just giving their size. To describe **vector** quantities we have to give a size or **magnitude** and a direction. For example, 'the car's *velocity* is 10 m s^{-1} due east,' or 'the *displacement* of A from B is 5 m to the right.'

> **The jargon**
> *Velocity* is speed in a given direction; *displacement* is distance in a given direction.

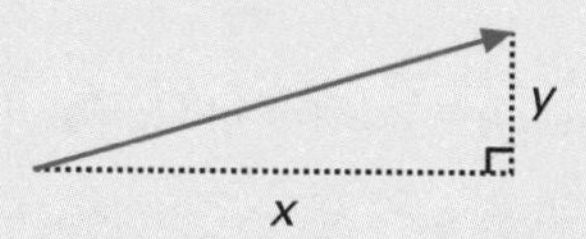

This vector has a magnitude (its length, which is 3.4 cm) and a direction (to the right and up a bit). (Always show the direction by drawing vectors with an arrow.) It can be described in different ways:

- by a letter, say **a** (in bold when printed; underlined when written)
- using the notation $\begin{pmatrix} x \\ y \end{pmatrix}$. This means a movement x in the direction of the x-axis and y in the direction of the y-axis
- using notation such as $\vec{A}$. This means the vector from point A to point B (see page 78).

> **The jargon**
> The numbers x and y in this notation are called the *components of the vector.*

In three dimensions, the notation is similar, but there are three axes:

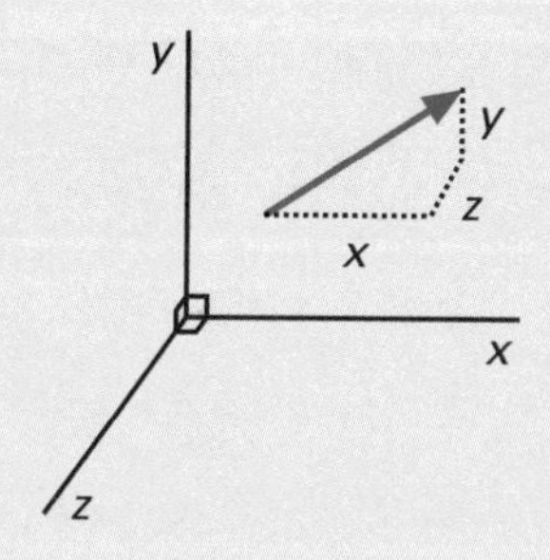

> **Examiner's secrets**
> It looks good (and makes marking easier!) if you are familiar with the convention for drawing three-dimensional axes. Practise drawing them as shown: the x- and y-axes are the same as in two dimensions and the z-axis comes *out* of the page.

In this example, the vector goes *into* the page, so z is negative.

The magnitude of a vector

- The magnitude (length) of a vector **a** is called $|\mathbf{a}|$ or just a. Look at the figure at the top of the page. The dotted lines have lengths x and y and the vector has length a. Since the three lines form a right-angled triangle,
- $|\mathbf{a}|^2 = x^2 + y^2$ (Pythagoras) In general,
- the magnitude of $\begin{pmatrix} x \\ y \end{pmatrix}$ is $\sqrt{x^2+y^2}$
- the magnitude of $\begin{pmatrix} x \\ y \\ z \end{pmatrix}$ is $\sqrt{x^2+y^2+z^2}$

> **Checkpoint 1**
> What are the magnitudes of $\begin{pmatrix} 6 \\ 8 \end{pmatrix}$ and $\begin{pmatrix} -3 \\ 6 \\ 2 \end{pmatrix}$?

Manipulating vectors

Multiplying a vector by a scalar

Simply multiply each component of the vector by the scalar:

→ If $\mathbf{a} = \begin{pmatrix} x \\ y \\ z \end{pmatrix}$ then $t\mathbf{a} = \begin{pmatrix} tx \\ ty \\ tz \end{pmatrix}$ (Remember: a scalar is just a number.)

Geometrically, $t\mathbf{a}$ is a vector in the direction of **a** but t times the length.

Unit vectors and the base vectors

A unit vector is any vector whose magnitude is 1. The vector **a** has magnitude $|\mathbf{a}|$ so it must be divided by $|\mathbf{a}|$ to give it a magnitude of 1.

→ the unit vector in the direction of **a** is $\dfrac{\mathbf{a}}{|\mathbf{a}|}$

The base vectors **i**, **j** and **k** are defined as unit vectors parallel to the x-, y- and z-axes respectively.

→ $\mathbf{i} = \begin{pmatrix} 1 \\ 0 \\ 0 \end{pmatrix}; \mathbf{j} = \begin{pmatrix} 0 \\ 1 \\ 0 \end{pmatrix}; \mathbf{k} = \begin{pmatrix} 0 \\ 0 \\ 1 \end{pmatrix}$, so the vector $\begin{pmatrix} x \\ y \\ z \end{pmatrix} \equiv x\mathbf{i} + y\mathbf{j} + z\mathbf{k}$

Vector addition and subtraction

Simply add or subtract the components of the vectors:

→ $\begin{pmatrix} u \\ v \end{pmatrix} \pm \begin{pmatrix} x \\ y \end{pmatrix} = \begin{pmatrix} u \pm x \\ v \pm y \end{pmatrix}$

To visualise a vector addition, imagine starting at point X and travelling along $\overrightarrow{XY}$ to point Y. Next travel along $\overrightarrow{YZ}$ to point Z. You've gone from X to Y to Z – in other words $\overrightarrow{XY} + \overrightarrow{YZ}$. But you could just as well have gone from X to Z all in one go. Therefore

→ $\overrightarrow{XY} + \overrightarrow{YZ} = \overrightarrow{XZ}$

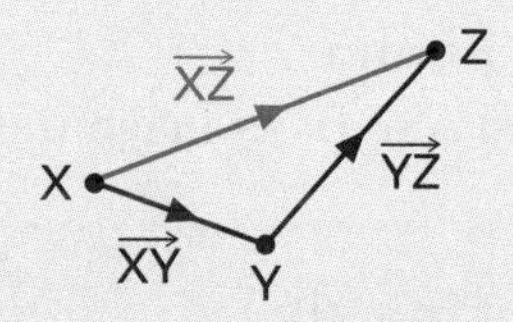

→ Because of the shape of the diagram, vectors are said to follow a 'triangle' rule of addition. (The geometrical interpretation of vector subtraction is shown on page 78.)

Checkpoint 2

What is $3\begin{pmatrix} 2 \\ 0 \\ -1 \end{pmatrix}$?

Checkpoint 3

What is the unit vector in the direction of $\begin{pmatrix} 4 \\ 3 \\ -4 \end{pmatrix}$?

Checkpoint 4

What are the base vectors in two dimensions? How would you express $\begin{pmatrix} -3 \\ -1 \end{pmatrix}$ in terms of the two-dimensional base vectors?

Checkpoint 5

Addition and subtraction are just the same in three-dimensions. What is $\begin{pmatrix} 2 \\ 1 \\ 0 \end{pmatrix} - \begin{pmatrix} 3 \\ -2 \\ 1 \end{pmatrix}$?

Test yourself

Write down two three-dimensional vectors and express each one in two different ways. Find the unit vectors in the same directions (leave surds in your answers). Add the vectors and illustrate the addition with a sketch.

Exam question answer: page 93

$\mathbf{a} = 3\mathbf{i} - 2\mathbf{j}$; $\mathbf{b} = \mathbf{i} + 7\mathbf{j}$

(a) Sketch the following vectors: **a**, **b**, $-3\mathbf{a}$, $2\mathbf{b}$.

(b) $\mathbf{c} = \mathbf{a} + 2\mathbf{b}$. Evaluate **c** and find its magnitude; illustrate the triangle rule by a sketch representing the equation $\mathbf{c} = \mathbf{a} + 2\mathbf{b}$. (5 min)

Position vectors and lines

All vectors have size and direction, but most do not have a fixed starting point. Position vectors are an important exception. A position vector must start at the origin, so it finishes at a specific point. By using position vectors, we can describe any point as a vector. We can then use vector methods to describe the lines between points.

Position vectors and displacement vectors

The **position vector a** is the vector from the origin to point A.

→ Position vectors always start at the origin

→ $\mathbf{a} = \overrightarrow{OA}$

Similarly the vector from B to the origin is $\overrightarrow{BO} = -\mathbf{b}$.

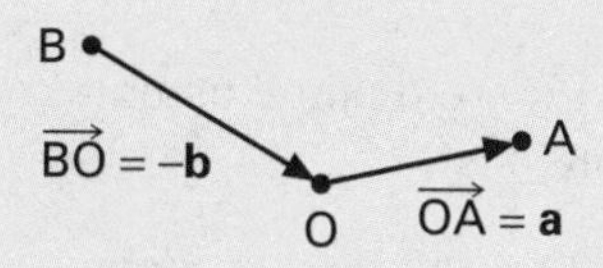

→ $\overrightarrow{AB}$ means the vector going from A (x_1,y_1,z_1) to B (x_2,y_2,z_2)

→ $\overrightarrow{AB}$ is called the **displacement vector** of B from A.

To get from A to B, imagine going from A back to the origin and then out to B. Then $\overrightarrow{AB} = \overrightarrow{AO} + \overrightarrow{OB} = -\mathbf{a} + \mathbf{b} = \mathbf{b} - \mathbf{a}$

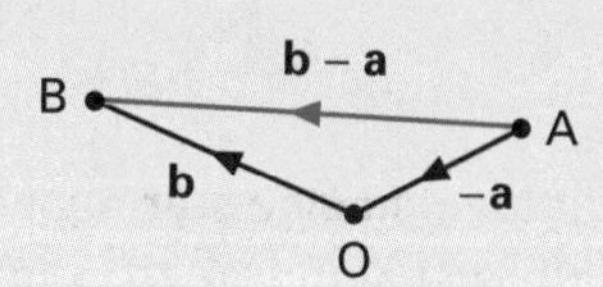

→ $\overrightarrow{AB} = \mathbf{b} - \mathbf{a}$

This is the geometrical interpretation of vector subtraction. (For the geometrical interpretation of vector addition, see page 77.)

The distance between points

Since $\overrightarrow{AB}$ is the vector from A to B, its magnitude $|\overrightarrow{AB}|$ is the distance between A and B.

→ Distance between A and B = $|\overrightarrow{AB}|$.

$$|\overrightarrow{AB}| = |\mathbf{b} - \mathbf{a}| = \left| \begin{pmatrix} x_2 - x_1 \\ y_2 - y_1 \\ z_2 - z_1 \end{pmatrix} \right| = \sqrt{(x_2 - x_1)^2 + (y_2 - y_1)^2 + (z_2 - z_1)^2}$$

→ Distance AB $= \sqrt{(x_2 - x_1)^2 + (y_2 - y_1)^2 + (z_2 - z_1)^2}$

The result in two dimensions is similar (but without z-coordinates).

Checkpoint 1

What is the relationship between any vector **a** and its negative –**a**?

Watch out!

Remember **b** – **a** takes you *from* A *to* B! The way to remember is that **b** means going forwards *to* B and –**a** means going backwards *from* A.

Checkpoint 2

What is the distance between the following pairs of points:

(a) $\begin{pmatrix} 4 \\ -1 \\ 3 \end{pmatrix}$ and $\begin{pmatrix} 3 \\ 0 \\ 5 \end{pmatrix}$?

(b) $\begin{pmatrix} 1 \\ 1 \end{pmatrix}$ and $\begin{pmatrix} 4 \\ 5 \end{pmatrix}$?

The vector equation of a line

A line through a point A and parallel to a given vector b

To get to any point R on the line

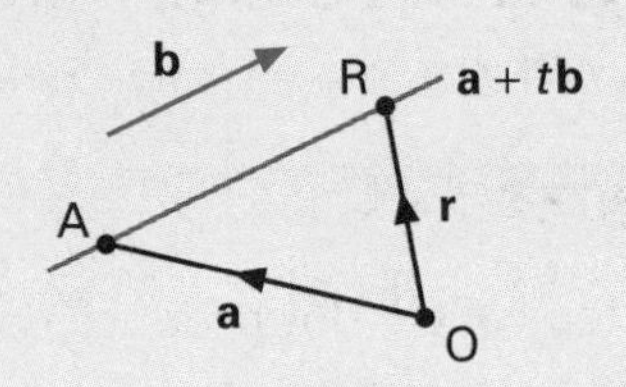

- ➜ go along **a** to get from the origin to A
- ➜ go along the line, i.e. any distance you like in the direction of **b**.

Hence the position vector **r** of a general point R on the line is:

➜ $\mathbf{r} = \mathbf{a} + t\mathbf{b}$, where t can take any value

A line through two points A and B

To get to any point R on the line

- ➜ go along **a** to get from the origin to A
- ➜ go along the line, i.e. any distance you like in the direction of $\overrightarrow{AB}$.

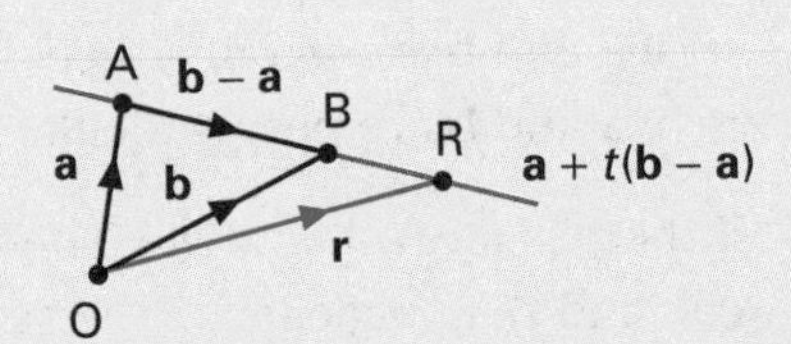

Hence the position vector **r** of a general point R on the line is

➜ $\mathbf{r} = \mathbf{a} + t\overrightarrow{AB}$, where t can take any value.

Alternative notation:

➜ $\mathbf{r} = \mathbf{a} + t(\mathbf{b} - \mathbf{a})$, where t can take any value.

Note: all the vector equations for lines are non-unique: more than one equation can describe the same line.

Checkpoint 3

What is the equation of the line through the point with position vector $\begin{pmatrix} 0 \\ 0 \\ 2 \end{pmatrix}$ and parallel to $\begin{pmatrix} -1 \\ 3 \\ -2 \end{pmatrix}$?

Checkpoint 4

(a) What is the equation of the line through the points with position vectors $\begin{pmatrix} 0 \\ 0 \\ 2 \end{pmatrix}$ and $\begin{pmatrix} -1 \\ 3 \\ -2 \end{pmatrix}$?

(b) Where is the point R relative to A and B when t takes the values $-2, 0, \frac{1}{2}, 1, 2$?

Exam questions answers: page 93

1 A and B have position vectors $\begin{pmatrix} 1 \\ 0 \\ 9 \end{pmatrix}$ and $\begin{pmatrix} -3 \\ 4 \\ 2 \end{pmatrix}$ respectively.

(a) Find a vector equation of the line on which A and B lie.

(b) Find the unit vector parallel to this line. (5 min)

2 A, B and C have position vectors $3\mathbf{i} + 2\mathbf{j} - \mathbf{k}$, $4\mathbf{i} + 4\mathbf{j}$ and $3\mathbf{i} + 4\mathbf{k}$ respectively.

(a) Find an equation for the line parallel to $\overrightarrow{AB}$; that passes through C.

(b) Show that the point D, position vector $\mathbf{i} - 4\mathbf{j} + 2\mathbf{k}$, lies on this line.

(c) Find the distance between A and D.

(d) A fifth point E is such that $\overrightarrow{CE} = 3\mathbf{i} + 8\mathbf{j} - 2\mathbf{k}$. Find $\overrightarrow{AE}$ without evaluating the position vector of E; hence show that E lies on the line AB. (15 min)

Test yourself

Write down the formulae for working out the distance between two points in two and three dimensions. Go back to the spread on straight lines (page 38) and find an example for the equation of a line through two points. Redo the example expressing points in terms of their position vectors rather than their coordinates.

Applications of vectors

The scalar product of two vectors tells us about their relative directions, among other things. The relationship between two lines in space also depends on their directions. They may intersect or be parallel if they lie in the same plane. Non-parallel non-intersecting lines are called skew.

The scalar product

The **scalar product** (often called 'dot product') of vectors **a** and **b** is

→ $\mathbf{a} \cdot \mathbf{b} = |\mathbf{a}||\mathbf{b}| \cos \theta$, where θ is the angle between the directions of the vectors.

Calculate the scalar product like this:

→ $\begin{pmatrix} x_1 \\ y_1 \\ z_1 \end{pmatrix} \cdot \begin{pmatrix} x_2 \\ y_2 \\ z_2 \end{pmatrix} = x_1x_2 + y_1y_2 + z_1z_2$ Learn this method!

Since $\cos 0 = 1$ and $\cos 90° = 0$, a useful property of the scalar product is

→ if **a** and **b** are parallel their scalar product is $|\mathbf{a}||\mathbf{b}|$
→ if **a** and **b** are perpendicular their scalar product is 0.

If there is any other angle θ between the directions of the vectors, you can find that angle using the scalar product:

→ $\theta = \arccos \dfrac{\mathbf{a} \cdot \mathbf{b}}{|\mathbf{a}||\mathbf{b}|}$

E.g. $\begin{pmatrix} -1 \\ 2 \\ 2 \end{pmatrix} \cdot \begin{pmatrix} 4 \\ -4 \\ 2 \end{pmatrix} = -1 \times 4 + 2 \times (-4) + 2 \times 2 = -8$; angle between them

$$= \arccos \frac{-1 \times 4 + 2 \times (-4) + 2 \times 2}{\sqrt{(-1)^2 + 2^2 + 2^2}\sqrt{4^2 + (-4)^2 + 2^2}} = \arccos \frac{-4}{9} = 116.4° \text{ (to 1 d.p.)}$$

Checkpoint 1

What is $\begin{pmatrix} 1 \\ 2 \\ -3 \end{pmatrix} \cdot \begin{pmatrix} -2 \\ 0 \\ 2 \end{pmatrix}$?

Watch out!

If $\mathbf{a} \cdot \mathbf{b} = 0$, they are not *necessarily* perpendicular since one of the vectors itself may be zero.

Checkpoint 2

What is the angle between the directions of $\begin{pmatrix} 2 \\ -1 \\ -2 \end{pmatrix}$ and $\begin{pmatrix} 4 \\ 4 \\ -7 \end{pmatrix}$?

Pairs of lines

Two lines $\mathbf{r}_1 = \mathbf{a}_1 + t\mathbf{b}_1$ and $\mathbf{r}_2 = \mathbf{a}_2 + s\mathbf{b}_2$ may be **parallel**, **intersect** or (in three dimensions) be **skew**.
Parallel lines

→ have the same direction, i.e. $\mathbf{b}_1$ and $\mathbf{b}_2$ are multiples of each other
→ $\mathbf{b}_1 \cdot \mathbf{b}_2 = |\mathbf{b}_1||\mathbf{b}_2|$

The lines will intersect if $\mathbf{a}_1 + t\mathbf{b}_1 = \mathbf{a}_2 + s\mathbf{b}_2$.

E.g. $\mathbf{r}_1 = \begin{pmatrix} 6 \\ 3 \\ 1 \end{pmatrix} + s\begin{pmatrix} 1 \\ 2 \\ 1 \end{pmatrix}$ and $\mathbf{r}_2 = \begin{pmatrix} 9 \\ 10 \\ 5 \end{pmatrix} + t\begin{pmatrix} 1 \\ 3 \\ 2 \end{pmatrix}$ intersect when

$\begin{pmatrix} 6 \\ 3 \\ 1 \end{pmatrix} + s\begin{pmatrix} 1 \\ 2 \\ 1 \end{pmatrix} = \begin{pmatrix} 9 \\ 10 \\ 5 \end{pmatrix} + t\begin{pmatrix} 1 \\ 3 \\ 2 \end{pmatrix}$. Treat as three simultaneous equations:

$$\left.\begin{array}{lr} 6 + s = 9 + t & ① \\ 3 + 2s = 10 + 3t & ② \\ 1 + s = 5 + 2t & ③ \end{array}\right\} ① - ③ \text{ gives } 5 = 4 - t \qquad \therefore t = -1$$

Examiner's secrets

Always check whether lines are parallel first; if they are, don't bother to look for an intersection!

Watch out!

In three dimensions, you will always be able to find values for s and t satisfying two equations. *Always check whether they fit the third one too* – if not, the lines are skew!

Checkpoint 3

What is the point of intersection of $\mathbf{r}_1 = \begin{pmatrix} 2 \\ -1 \\ -1 \end{pmatrix} + s\begin{pmatrix} -1 \\ 3 \\ 4 \end{pmatrix}$ and $\mathbf{r}_2 = \begin{pmatrix} 4 \\ 1 \\ 2 \end{pmatrix} + t\begin{pmatrix} 3 \\ -1 \\ -1 \end{pmatrix}$?

Substitute in ①: $6 + s = 9 - 1 = 8 \quad \therefore s = 2$

Check in ②, which has not been used yet: $3 + 4 = 10 - 3 = 7$

So $s = 2$ and $t = -1$, i.e. the point with position vector $\begin{pmatrix}6\\3\\1\end{pmatrix} + 2\begin{pmatrix}1\\2\\1\end{pmatrix}$ or coordinates (8,7,3). (Check that this works using the other line!). If the simultaneous equations have a unique solution, the lines intersect at a point given by your values of s and t. If the simultaneous equations have no solution, the lines don't intersect: they are parallel or skew. To visualise skew lines, imagine a straight road bridge crossing a straight railway track: they aren't parallel but they do not intersect.

The ratio theorem

Let P be a point on the line joining A and B. Its position vector is

➜ $\mathbf{p} = \dfrac{m}{m+n}\mathbf{a} + \dfrac{n}{m+n}\mathbf{b}$ (The coefficients of **a** and **b** sum to 1.)

Conversely, if any vector **p** can be written as

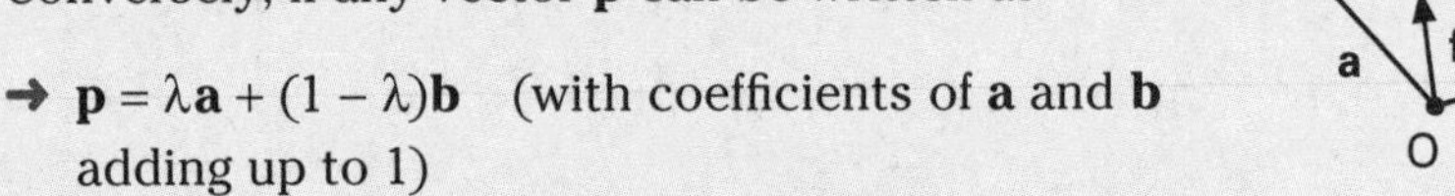

➜ $\mathbf{p} = \lambda\mathbf{a} + (1 - \lambda)\mathbf{b}$ (with coefficients of **a** and **b** adding up to 1)

then P lies on the line AB and AP : PB = $(1 - \lambda) : \lambda$. This is another form of the vector equation of a line. A, P and B are said to be **collinear**.

Planes

A plane through a point and perpendicular to a given vector

The equation of a plane through a point with position vector **a** and perpendicular to a vector **n** is

➜ $\mathbf{r} \cdot \mathbf{n} = \mathbf{a} \cdot \mathbf{n}$ The number $\mathbf{a} \cdot \mathbf{n}$ is often written d, so $\mathbf{r} \cdot \mathbf{n} = d$.

A plane through three points

Any three points must lie in the same plane. Let P be a point on the plane containing A, B and C. Its position vector is

➜ $\mathbf{p} = \lambda\mathbf{a} + \mu\mathbf{b} + (1 - \lambda - \mu)\mathbf{c}$

Conversely, if any vector **p** can be written as

➜ $\mathbf{p} = \lambda\mathbf{a} + \mu\mathbf{b} + (1 - \lambda - \mu)\mathbf{c}$ (coefficients of **a**, **b** and **c** sum to 1)

then P lies on the plane containing A, B and C.

Exam question answer: page 93

Points A, B, C and D have respective position vectors **a**, **b**, **c** and **d**, **2i** + **3j** + **6k**, **2i** + **6j** + **9k**, **2i** + μ**j** + **8k** and **6i** – **4j** + **2k**.

(a) Find μ: (i) if A, B and C are collinear; (ii) if **c** is perpendicular to $\overrightarrow{AB}$.

(b) Find equations for lines (i) L_1, which passes through B and is parallel to a; (ii) L_2, which passes through A and D, and prove that they are skew.

(c) Find the angle between the directions of L_1 and L_2. (15 min)

Action point

Show that the lines $\mathbf{r}_1 = \begin{pmatrix}3\\5\\7\end{pmatrix} + s\begin{pmatrix}1\\2\\1\end{pmatrix}$ and $\mathbf{r}_2 = \begin{pmatrix}1\\2\\3\end{pmatrix} + t\begin{pmatrix}2\\3\\5\end{pmatrix}$ do not meet.

Watch out!

Don't get confused about which coefficient goes with **a** and which with **b**! The way to remember is that if P is closer to A then the coefficient of **a** is larger (there is 'more of **a**') and vice versa. For example, in the diagram m could be 5 and n could be 2. For given **a** and **b**, all equations of the form $\mathbf{p} = \lambda\mathbf{a} + (1 - \lambda)\mathbf{b}$ (i.e. coefficients of **a** and **b** add up to 1) represent the same line. The equation is not unique.

Watch out!

Just like the equation of a line given above, this equation for a plane is not unique. For given **a**, **b** and **c**, all equations of the form $\mathbf{p} = \lambda\mathbf{a} + \mu\mathbf{b} + (1 - \lambda - \mu)\mathbf{c}$ (i.e. coefficients of **a**, **b** and **c** add up to 1) represent the same plane.

Note

Check in your specification whether you have to know about the perpendicular from a point to a line.

Note

Check in your specification whether you have to learn about planes.

Test yourself

When you've finished revising this topic, write down what you know about the following: scalar product; pairs of lines; the ratio theorem.

Answers
Pure mathematics

Basic algebra

Checkpoints

1 (a) $7x + 21$ (b) $x^2 - 9x$ (c) $x^2 + 5x - 7x - 35 = x^2 - 2x - 35$
2 $x = 2$ and $y = 1$

Exam questions

1 (a) $11f - 22g$ (b) $5 - 2p - 10q$ (c) $x^2 + 15x + 26$
(d) $2p^2 - 12p + 18$ (e) $3q^2 + 17q - 28$
2 (a) $(x + 2)(x + 9)$ (b) $(x - 3)(x - 5)$ (c) $(x + 9)(x - 7)$
(d) $(x + 12)(x - 12)$
3 (a) $x = \frac{2}{3}$ and $y = 3\frac{1}{3}$ (b) $x = 3\frac{5}{11}$ and $y = \frac{9}{11}$
4 (a) $u = v - at$ (b) $u = (s - \frac{1}{2}at^2)/t$

Indices and surds

Checkpoints

1 (a) d^5 (b) y^2 (c) w^{-2} (d) t^6 (e) 27
2 (a) $\sqrt{16 \times 2} = 4\sqrt{2}$ (b) $\sqrt{25 \times 2} = 5\sqrt{2}$
(c) $\sqrt{36 \times 5} = 6\sqrt{5}$
3 (a) $4 - \sqrt{7}$ (b) $2 + \sqrt{2}$ (c) $7 - \sqrt{5}$

Exam questions

1 (a) c^{12} (b) B^5 (c) d^6 (d) x (e) y (f) z^4
2 (a) 16 (b) 4 (c) $\frac{1}{9}$ (d) 1 (e) $\frac{2}{3}$ (f) 3 (g) 1 (h) $\frac{125}{27}$
3 (a) $5\sqrt{2} - 2$ (b) $79 - 30\sqrt{6}$ (c) $p - 1$
4 (a) $\frac{\sqrt{5}}{5}$ (b) $\frac{6 + 3\sqrt{2}}{2}$ (c) $-3\sqrt{2} - 5$ (d) $\frac{13 + 4\sqrt{3}}{121}$

Polynomials

Checkpoints

1 $f(-1) = 0 \quad f(3) = 0 \quad f(-4) = 0$
$\therefore x^3 + 2x^2 - 11x - 12 = (x + 4)(x + 1)(x - 3)$

Exam questions

1 (a) $x^3 + 6x^2 + 3x - 10$
(b) $2x^3 - 11x^2 + 12x + 9$
(c) $4x^4 - 20x^3 - 23x^2 + 120x + 144$
2 (a) $(x + 2)(x - 1)(x + 3)$
(b) $x(x + 4)(x + 1)(x - 5)$
(c) $(x - 2)(x - 2)(x + 1) \equiv (x - 2)^2(x + 1)$
(d) $(x + 2)(x - 2)(x + 3)$
(e) $(2x - 1)(x + 2)(x + 3)(x - 1)$

Quadratic equations

Checkpoints

1 (a) $x^2 + 9x + 20$ (b) $x^2 + x - 30$ (c) $(x + 6)(x + 1)$
2 (a) $(x + 2)^2 - 4 - 12 = 0 \Rightarrow (x + 2)^2 = 16$
$x + 2 = \pm 4 \quad \therefore x = 2$ or -6
(b) $(x + 8)^2 - 64 - 80 = 0 \Rightarrow (x + 8)^2 = 144$
$x + 8 = \pm 12 \quad \therefore x = 4$ or -20
3 $x = \frac{-3 \pm \sqrt{3^2 - (4)(4)(-2)}}{2(4)} = \frac{-3 + \sqrt{41}}{8}$ or $\frac{-3 - \sqrt{41}}{8}$

Exam questions

1 (a) $(x + 2)(x - 7)$ (b) $(x + 4)(x + 5)$
(c) $(2x - 3)(x - 6)$
2 (a) $(x - 2)^2 - 4 - 21 = 0$
$(x - 2)^2 = 25 \quad \therefore x = 7$ or -3
(b) $(x + 4)^2 = 49 \quad \therefore x = 3$ or -11
3 (a) $x = 1.7$ or -4.7 (to 2 s.f.)
(b) $x = 8$ or -3
(c) $x = 2$ or -3.5
(d) $x = 1.5$ or -3.9 (to 2 s.f.)
(e) Not possible
(f) $x = 1.1$ or -0.45 (to 2 s.f.)

Quadratic functions

Checkpoints

1

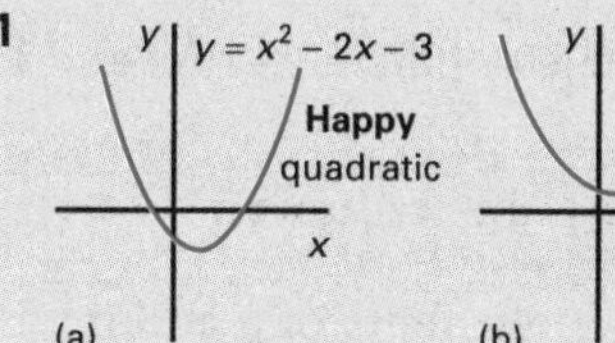

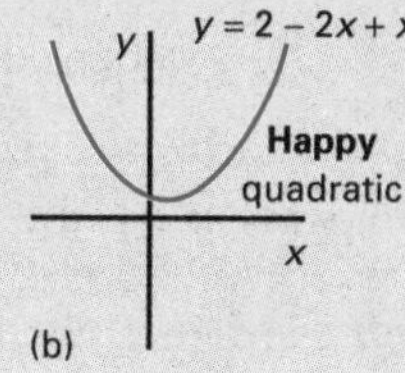

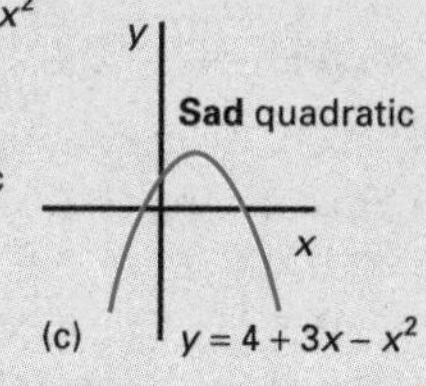

2

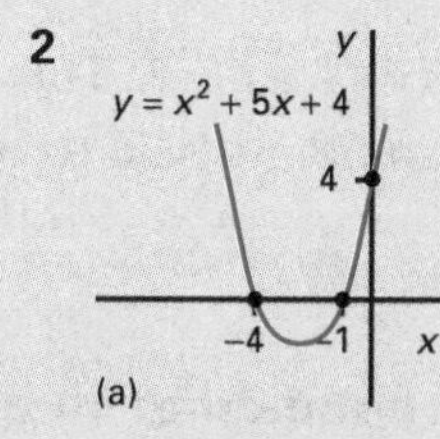

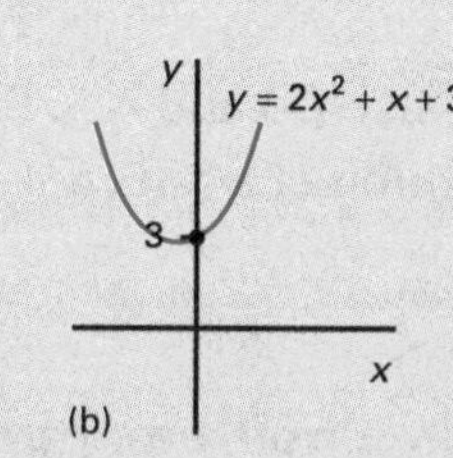

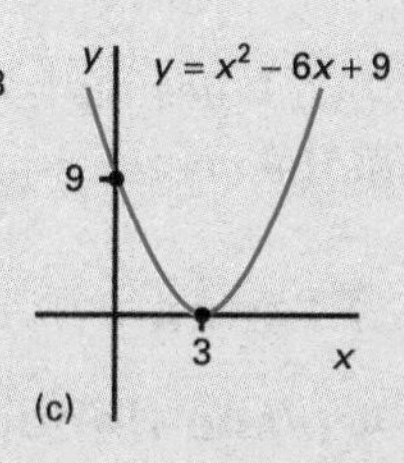

Exam questions

1 $y = x^2 - 7$

2

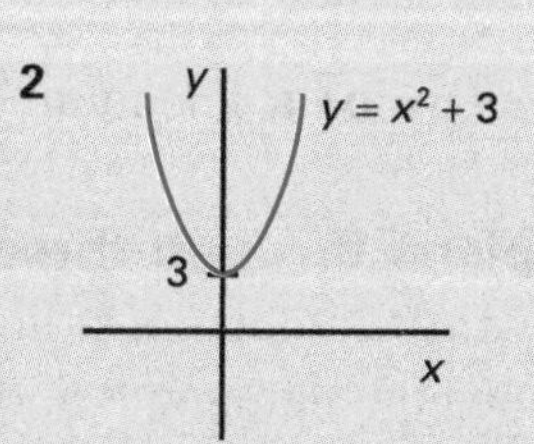

3

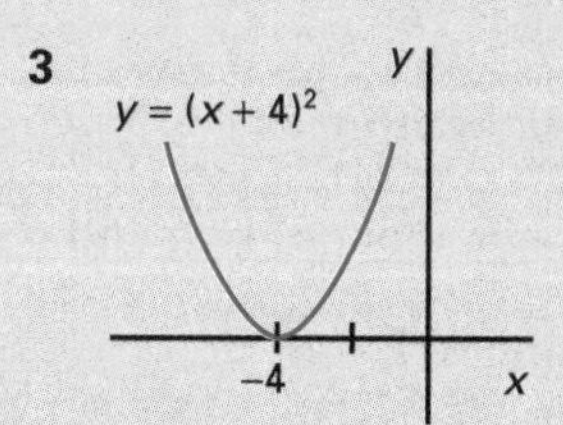

4

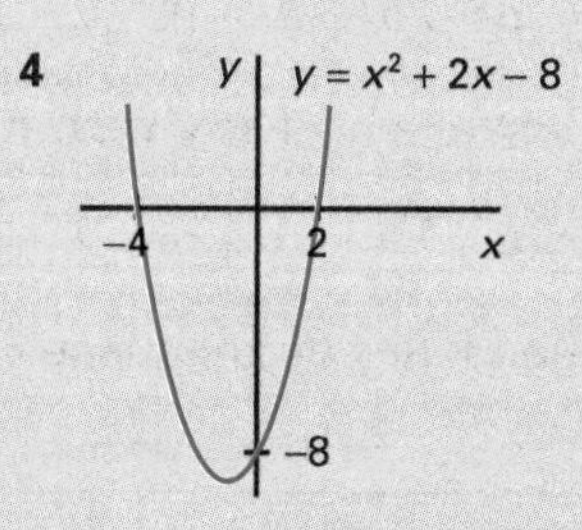

5

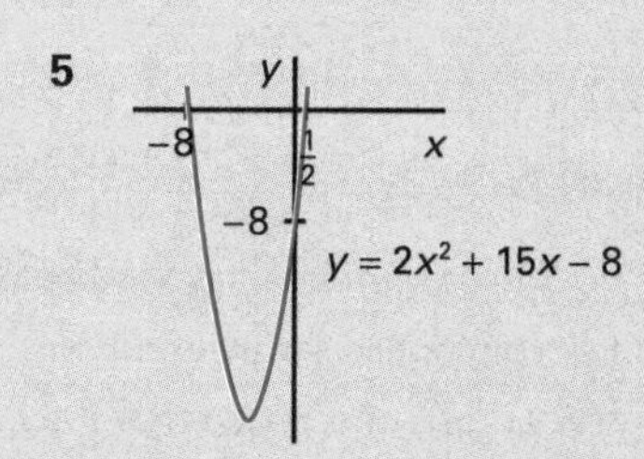

6

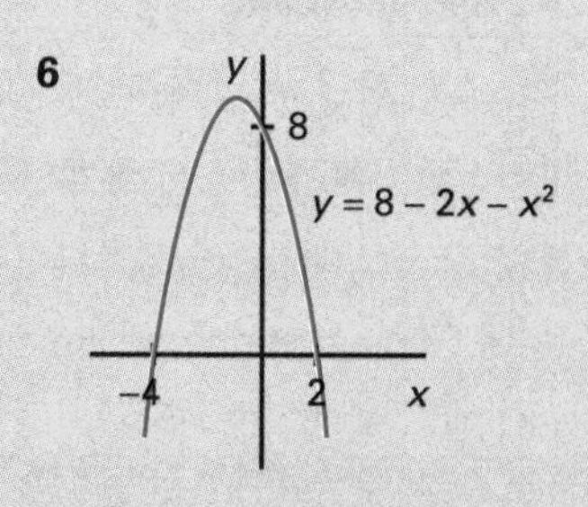

7

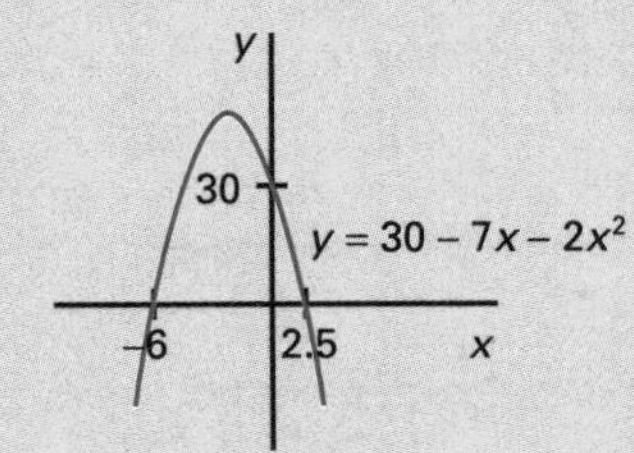

8

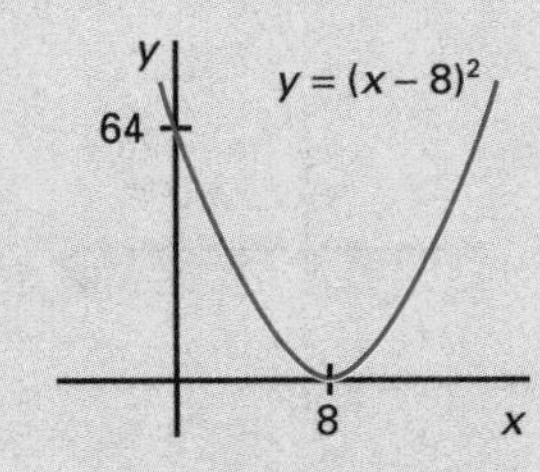

9

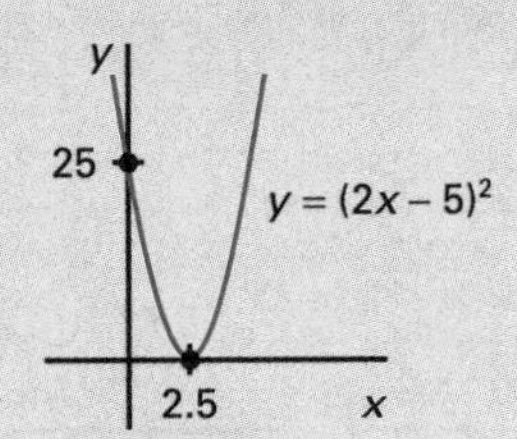

10

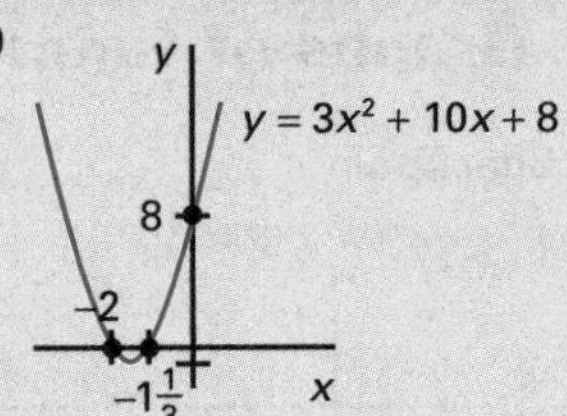

Simultaneous equations

Checkpoints

1 $5(16 - 2y) + 12y = 86 \quad 80 + 2y = 86 \quad \therefore y = 3$
$x + 2(3) = 16 \quad \therefore x = 10$

2 To 1 s.f.: $x = -4, y = 11$; $x = -2, y = -6$; $x = 4, y = 8$

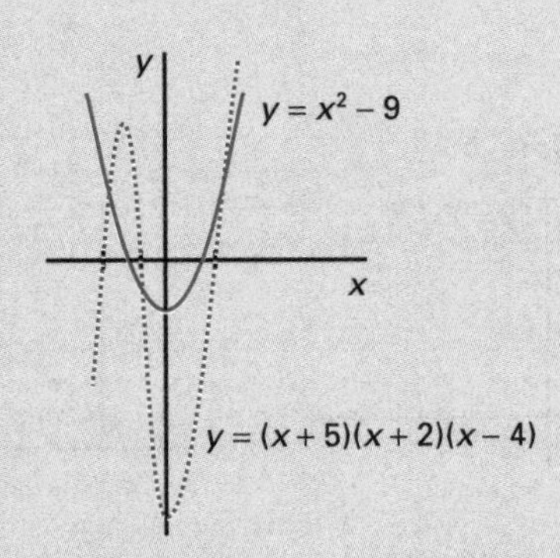

Exam questions

1 (a) $x = -1.5, y = 19$
(b) $x = 14, y = -8$
(c) Substitute $y = \dfrac{42 - 5x}{2}$ into $3x^2 + 4y^2 = 196$
$\Rightarrow 3x^2 + 1\,764 - 420x + 25x^2 = 196$
$\Rightarrow 28x^2 - 420x + 1\,568 = 0$
Solving gives: $x = 8$ or 7
Substituting each value into $5x + 2y = 42$ gives: $x = 8$, $y = 1$ and $x = 7$, $y = 3.5$

2 (a) From the graph below:
$x = -4.7, y = 11$; $x = 0.4, y = -5.4$; $x = 3.3, y = 8$

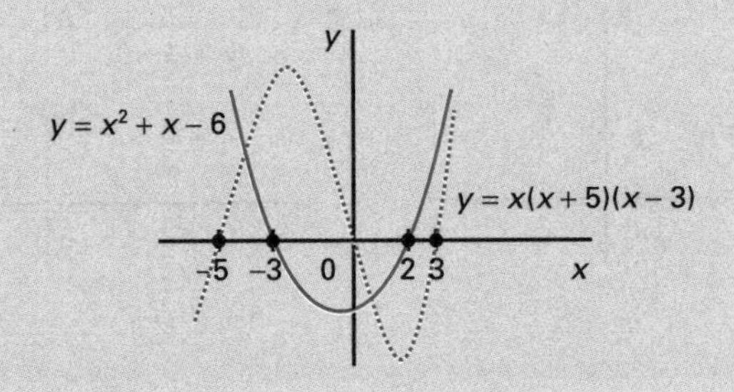

(b) From the graph below:
$x = -3.4, y = 10.8$; $x = -1.7, y = 2$, $x = 0.1, y = -1$; $x = 2, y = 3.2$

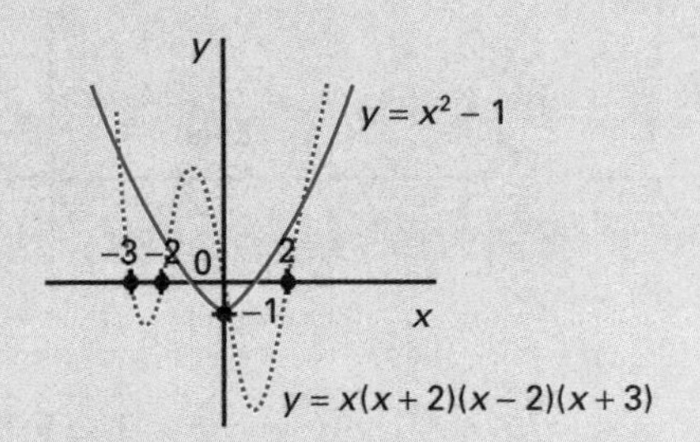

Inequalities

Checkpoints

1 (a) $13 > 4$ True (b) $5 > -4$ True (c) $20 > 2$ True
(d) $-20 > -2$ False

2 (a) $10 - 15 < 3x \quad \therefore -\frac{5}{3} < x$
(b) $4x - 3 < 3x - 6 \quad \therefore x < -3$
(c) $3x + 9 > 5x - 1 > 4x + 1$
$\therefore 5 > x$ and $x > 2$ i.e. $2 < x < 5$

3 $\therefore -3 < x - 2$

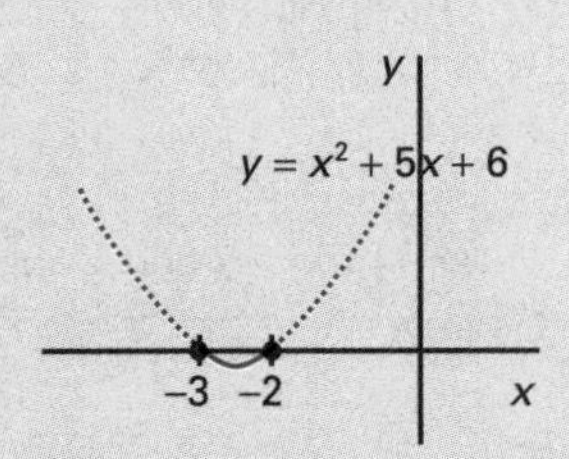

4 $\therefore x \le -1$ and $x \ge 5$

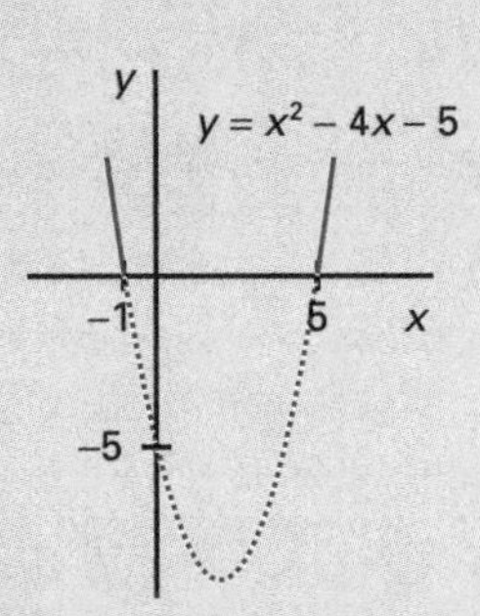

Exam questions

1 $x > 2$

2 $15 < 5x \Rightarrow 3 < x$
$2x < 23 \Rightarrow x < 11\frac{1}{2} \quad \therefore 3 < x < 11\frac{1}{2}$

3 $(x - 2)(x + 5) = 0 \quad \therefore x \ge 2$ and $x \le -5$

4 $(2x - 9)(x + 2) = 0 \quad \therefore -2 \le x \le \frac{9}{2}$

5 $5x^2 - 11x - 12 < 0$
$(5x + 4)(x - 3) < 0 \quad \therefore x > -\frac{4}{5}$ and $x < 3$

6 $4x^2 - 1 > 0 \Rightarrow (2x - 1)(2x + 1) > 0 \quad \therefore x > \frac{1}{2}$ and $x < -\frac{1}{2}$

7 $-5x - 14 \le 9x \quad \therefore x \ge -1$

8 $x(x + 4)(x - 3) \ge 0 \quad \therefore -4 \le x \le 0$ and $x \ge 3$

9 $x < -4$ and $x > 5$

10 $x < 19$ and $x > -7.5$

Functions

Checkpoints

1

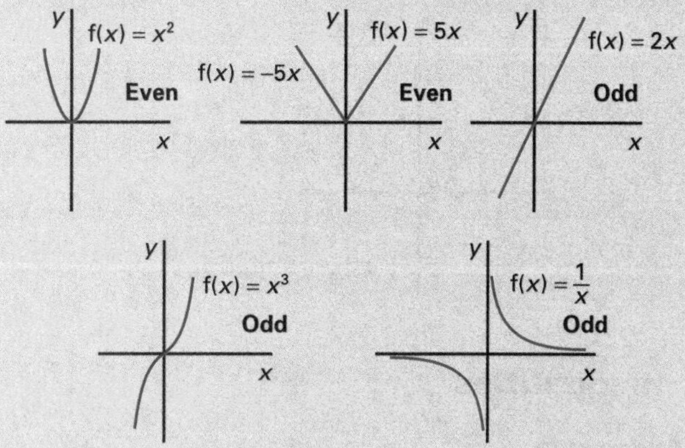

2 (a) $f^{-1}(x) = \frac{x}{4}$ Is a function

(b) $g^{-1}(x) = \frac{x-3}{5}$ Is a function

(c) $h^{-1}(x) = \sqrt{\frac{x}{4}}$ Is not a function as it is a one-to-many mapping.

3 $fg(x) = f(x^2 + 2x) = 4 - 3(x^2 + 2x) = 4 - 3x^2 - 6x$

$gf(x) = g(4 - 3x) = (4 - 3x)^2 + 2(4 - 3x)$

$= 24 - 30x + 9x^2$

4 For $f(x) = x^2 + 2$: range $f(x) \geq 2$ and $f^{-1}(x) = \sqrt{x-2}$

For $f(x) = 2 - x$: range $-3 < f(x) < 2$ and $f^{-1}(x) = 2 - x$

Exam questions

1 (a) $f(x)^3 = 7x^3 + 3$ (b) $g(7x + 3) = (7x + 3)^3$

(c) $y = 7x + 3 \Rightarrow \frac{y-3}{7} = x \quad \therefore f^{-1}(x) = \frac{x-3}{7}$

(d) $y = x^3 \Rightarrow \sqrt[3]{x} = y \quad \therefore g^{-1}(x) = \sqrt[3]{x}$

(e) $f^{-1}(\sqrt[3]{x}) = \frac{\sqrt[3]{x} - 3}{7}$

2 (a) Domain all x except $x = -3$; range all values except $f(x) = 0$

(b) Domain $x \geq -7$; range $g(x) > 0$

3 (a) odd (b) odd (c) even (d) odd (e) neither (f) even

(a)

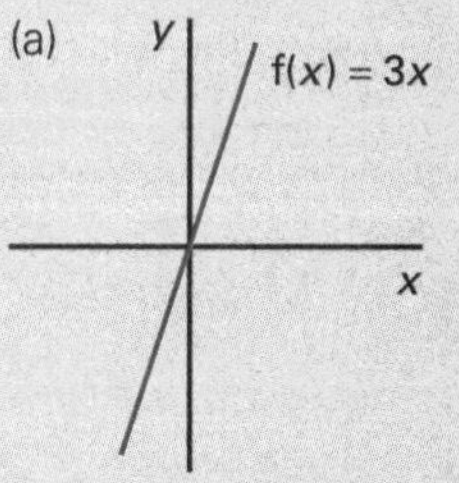

(b)

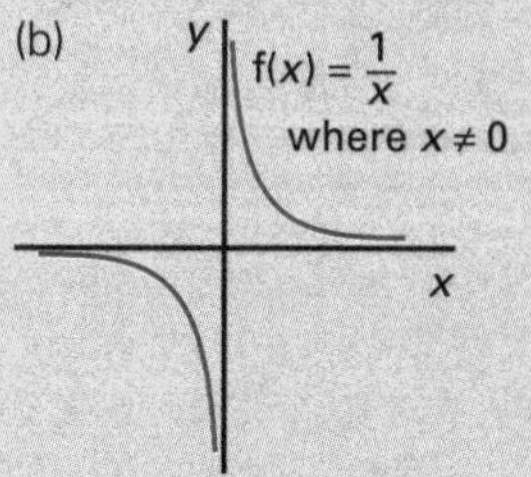

(c)

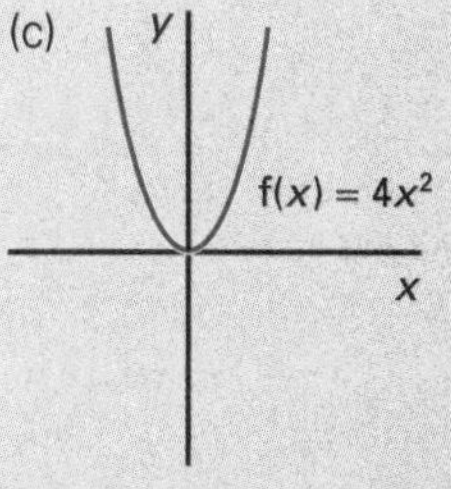

(d)

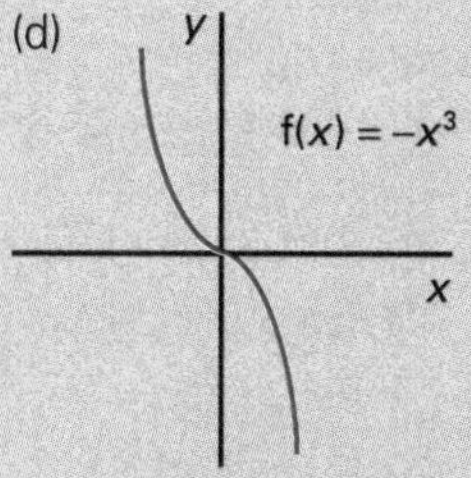

(e)

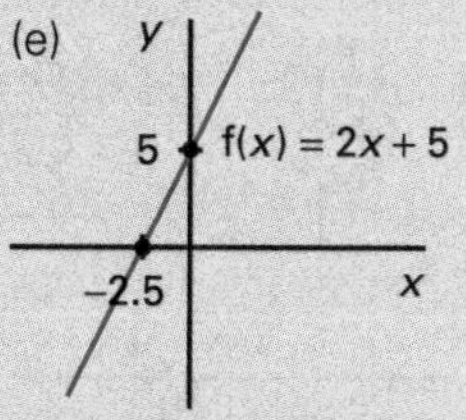

(f)

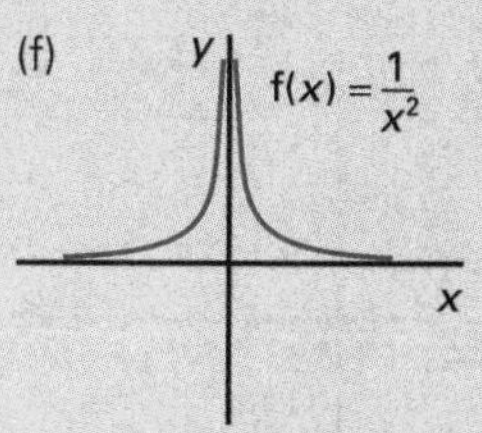

Graphs of functions

Checkpoints

1

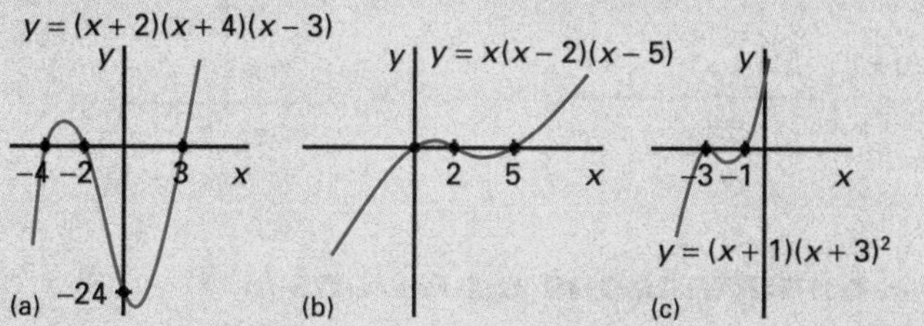

2

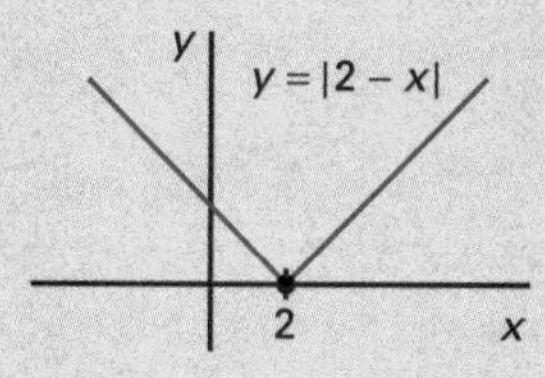

Exam questions

1

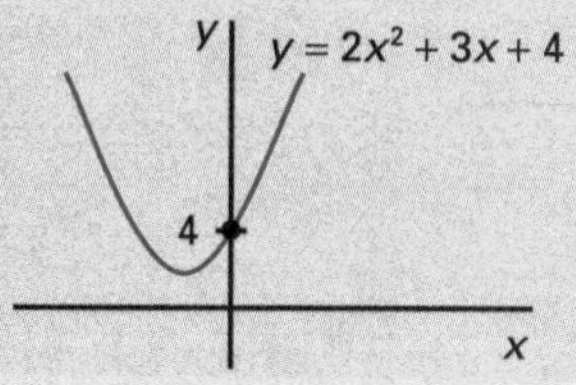

2

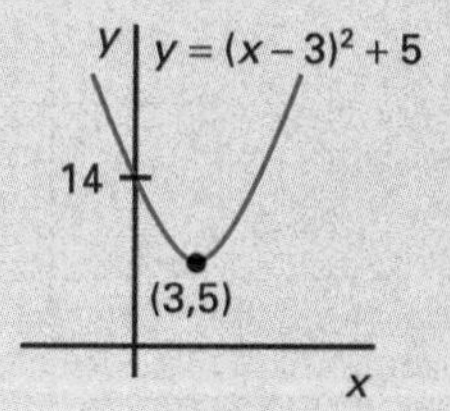

3

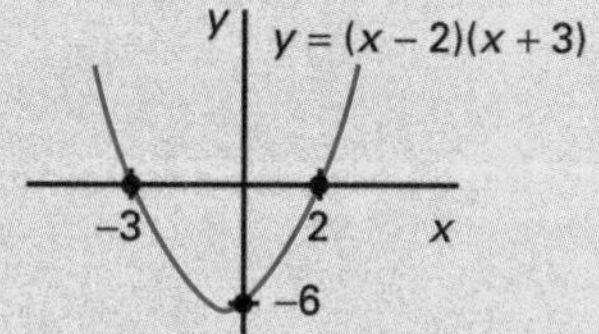

4

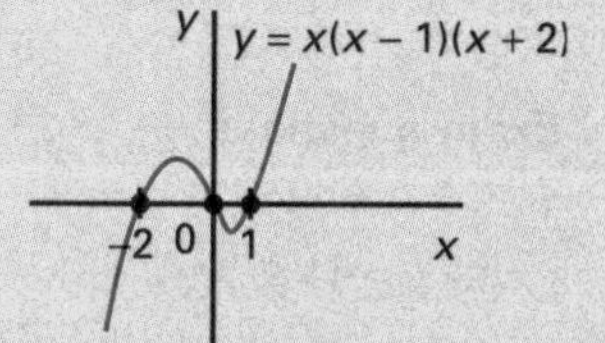

5

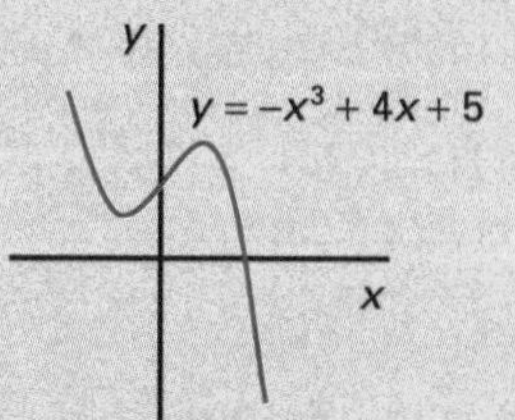

6

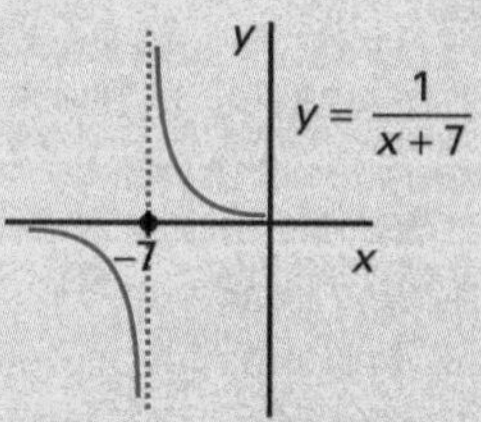

7

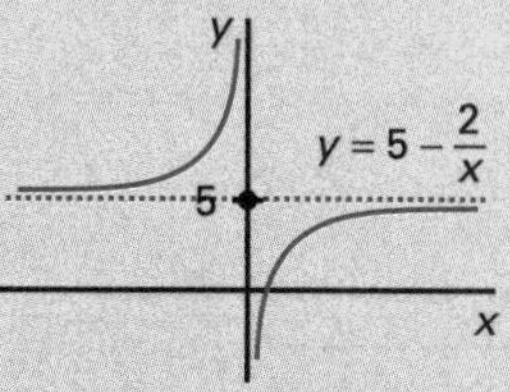

8

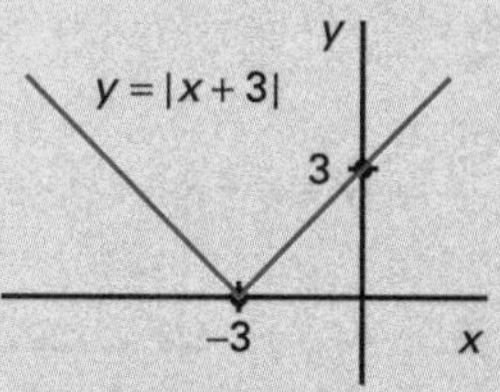

9

$y = |2x - 7|$

7, 3.5

10

$y = 2x^2$, $y = x$, Inverse function

11

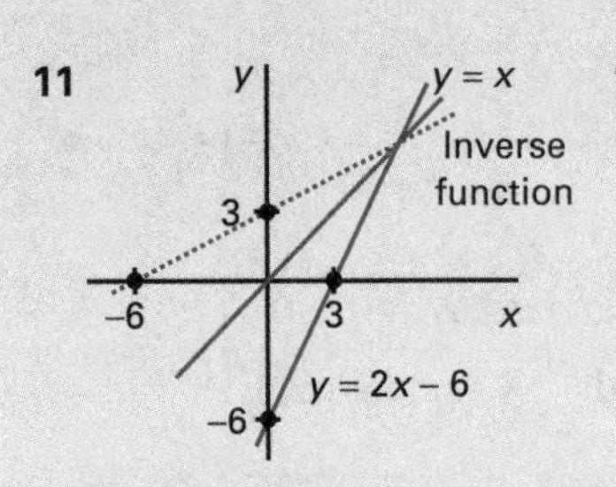

12

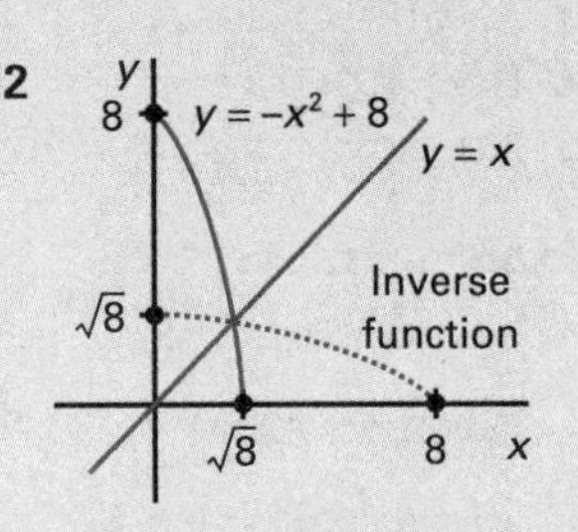

Transformations

Checkpoint

1

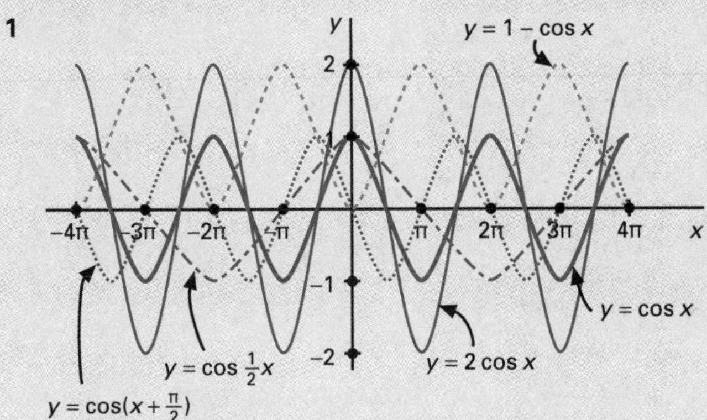

$y = 2\cos x$	amplitude is doubled
$y = \cos \frac{1}{2}x$	period is doubled
$y = \cos(x + \frac{\pi}{2})$	shifts by $\frac{\pi}{2}$ radians to the left
$y = 1 - \cos x$	inverts and shifts up by 1

Exam questions

1 (a) Translated up by 2
(b) Translated down by 8
(c) Translated to the right by 3
(d) Translated to the left by 5 and down by 2
(e) Translated to the left by 8 and up by 6
(f) Stretched by a factor of 2

2

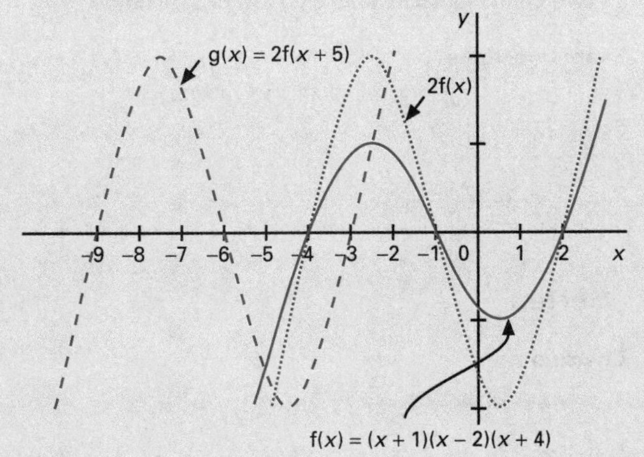

3

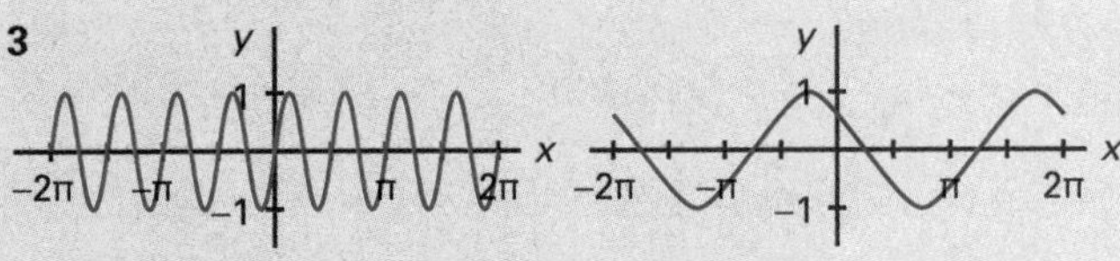

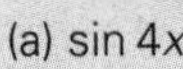

(a) $\sin 4x$ (b) $\cos(x + \frac{\pi}{4})$

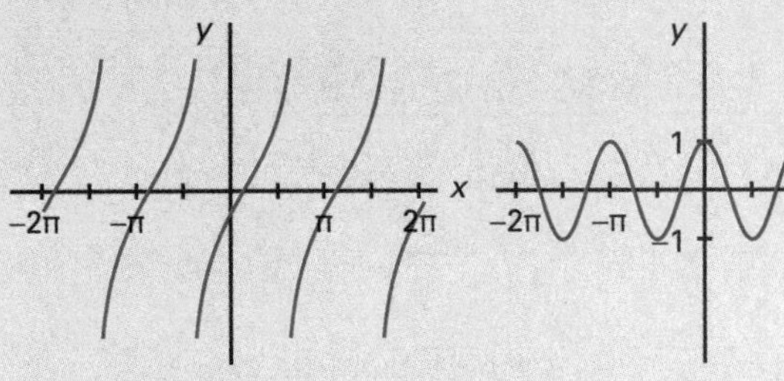

(c) $\tan(x - \frac{\pi}{6})$ (d) $\sin(2x + \frac{\pi}{2})$

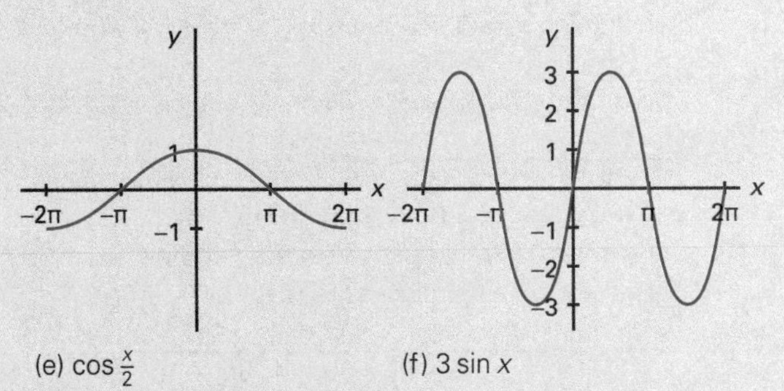

(e) $\cos \frac{x}{2}$ (f) $3 \sin x$

Rational functions

Checkpoints

1 (a) and (d).

2 (a) $\frac{4 \times 8 + 5 \times 7}{7 \times 8} = \frac{67}{56}$ (b) $\frac{8 \times 7}{9 \times 10} = \frac{56}{90}$ (c) $\frac{5 \times 3}{6 \times 2} = \frac{15}{12}$

3 (a) $5(x + 3)$
(b) $2x(y + 2x)$
(c) $5x - 20$ (d) $x^2 + 3x$

4 $2x + 7 = A(x + 5)^2 + B(x - 3)(x + 5) + C(x - 3)$
Substituting $x = -5$, $x = 3$ and looking at coefficients of x gives:

$A = \frac{13}{64}$, $B = -\frac{13}{64}$, $C = \frac{3}{8}$

5 $2x + 7 = A(x^2 + 5) + (Bx + C)(x - 3)$
Substituting $x = 3$ and looking at coefficients of x and x^2 gives:

$A = \frac{13}{14}$, $B = -\frac{13}{14}$, $C = -\frac{11}{14}$

Exam questions

1 (a) $\frac{2}{3x - 1} - \frac{1}{4x + 1} = \frac{2(4x + 1) - (3x - 1)}{(3x - 1)(4x + 1)} = \frac{5x + 3}{(3x - 1)(4x + 1)}$

(b) $\frac{3x + 18}{(x + 6)(x - 4)} - \frac{9}{x^2 - 5x + 4}$

$= \frac{3(x + 6)}{(x + 6)(x - 4)} - \frac{9}{(x - 1)(x - 4)}$

$= \frac{3(x - 1) - 9}{(x - 4)(x - 1)} = \frac{3}{x - 1}$

(c)
$$
\begin{array}{r}
x^2 + 6x + 8 \\
x-5 \,\overline{)\, x^3 + x^2 + 22x - 40} \\
\underline{x^3 - 5x^2} \qquad\qquad \\
6x^2 - 22x \qquad \\
\underline{6x^2 - 30x} \qquad \\
8x - 40 \\
\underline{8x - 40}
\end{array}
$$

2 (a) $\dfrac{4}{(x+1)(x-1)} = \dfrac{A(x-1) + B(x+1)}{(x+1)(x-1)}$

Substituting $x = 1$ and $x = -1$ gives:

$\dfrac{4}{(x-1)(x+1)} = \dfrac{-2}{x+1} + \dfrac{2}{x-1}$

(b) $\dfrac{2x-1}{(x+3)(x^2-1)} = \dfrac{A}{x+3} + \dfrac{Bx+C}{x^2-1}$

$2x - 1 = A(x^2 - 1) + (Bx + C)(x + 3)$

Substituting $x = -3$, $x = 1$ and $x = -1$ eventually gets:

$\dfrac{2x-1}{(x+3)(x^2-1)} = \dfrac{-7}{8(x+3)} + \dfrac{7x-5}{8(x^2-1)}$

(c) $\dfrac{x^2+1}{(x-2)(x+1)^2} = \dfrac{A}{x-2} + \dfrac{B}{x+1} + \dfrac{C}{(x+1)^2}$

$x^2 + 1 = A(x+1)^2 + B(x-2)(x+1) + C(x-2)$

Substituting $x = -1$, $x = 2$ and looking at the coefficients of x^2 produces the result:

$\dfrac{x^2+1}{(x-2)(x+1)^2} = \dfrac{5}{9(x-2)} + \dfrac{4}{9(x+1)}\ \dfrac{2}{3(x+1)^2}$

(d) $\dfrac{x+3}{x(x+5)^2} = \dfrac{A}{x} + \dfrac{B}{x+5} + \dfrac{C}{(x+5)^2}$

$x + 3 = A(x+5)^2 + Bx(x+5) + Cx$

Substituting $x = -5$, comparing coefficients of x^2 and substituting $x = 0$ gives:

$\dfrac{x+3}{x(x+5)^2} = \dfrac{3}{25x} - \dfrac{3}{25(x+5)} + \dfrac{2}{5(x+5)^2}$

Exponentials and logarithms

Checkpoints

1

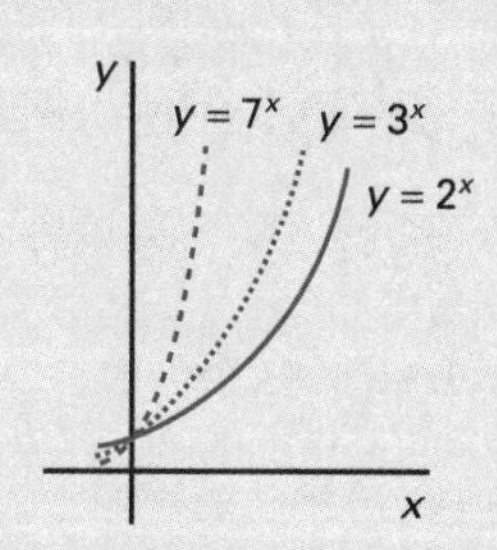

2 Press e^x button and then 1. It is irrational.

3 $\log_3 9 = 2$

Exam questions

1 (a) log 10 (b) $\log 4^2 \times 3 = \log 48$ (c) $\log \dfrac{12}{4} = \log 3$

(d) 3 (e) $\log 36^{\frac{1}{2}} - \log 5^2 = \log \dfrac{6}{25}$

(f) $\ln 25^{\frac{1}{2}} + \ln 3^2 - \ln 5 = \ln \dfrac{5 \times 9}{5} = \ln 9$

2 (a) 2 (b) −1

(c) $4 \times 4 \times 2 = 32 \quad \therefore 4^{2.5} = 32$. The answer is 2.5

(d) 2.64 (e) $\ln 28 = 3.33$

Applications

Checkpoints

1 If the λ is positive it is a growth, if the λ is negative it is a decay.

2 Newton

3 $e^0 = 1$

4 Base 10

5 1.07 represents 7% rate of interest that is added each year.

Exam questions

1 n = number in colony, t = time in days

When $n = 200$, $\dfrac{dn}{dt} = 20$ and $\dfrac{dn}{dt} = \lambda n$, therefore,

$20 = \lambda \times 200 \Rightarrow \lambda = 0.1$

Also, $n = Ae^{\lambda t}$. When $t = 0$ and $n = 200$

$200 = Ae^0 \quad \therefore A = 200.$

$n = 200e^{0.1t}$. When $t = 30$, $n = 200e^{0.1 \times 30} \quad \therefore n = 4\,017$

2 $\dfrac{dy}{dt} = \lambda y \Rightarrow y = Ae^{\lambda t}$ where y = temperature *difference* and t = time in minutes.

When $t = 0$, $y = 80$. $\therefore A = 80$

When $t = 3$, $y = 74$. $\therefore 74 = 80e^{3\lambda} \Rightarrow \log_e \dfrac{74}{80} = 3\lambda$

$\therefore \lambda = -0.025\,987.$

When $y = 60$, $60 = 80e^{-0.025\,987t}$

$\ln \dfrac{60}{80} = -0.025\,987t \quad \therefore$ time = 11 minutes.

On phone 11 − 3 = 8 minutes.

3 $2\,000(1.065)^n = 5\,000 \quad 1.065^n = 2.5$

$\therefore n = 14.6$ years

4 (a) $\dfrac{\log 24}{\log 7} = x \quad \therefore x = 1.63$ (to 3 s.f.)

(b) $5^{2x} = 12 \Rightarrow \dfrac{\log 12}{\log 5} = 2x \quad \therefore x = 0.772$ (to 3 s.f.)

Sequences

Checkpoints

1 $8n - 2$

2 When n is an odd number the (−1) will remain negative and so the odd terms will be negative. $(-1)^{n+1}$ would be used for sequences where the even terms are negative.

3 $9(4)^n$

4 The convergent sequence is heading towards zero. The divergent sequence is tending towards infinity.

Exam questions

1 (a) $3n + 4$ (b) $2(3)^n$ (c) $n^2 + 2$ (d) $2(-1)^{n+1}(2)^n$

(e) $128(-1)^{n+1}(\frac{1}{2})^n$

2 3, 9, 15, 21, 27

3 2, 3, 7, 13, 27

4 (a), (b), (c) and (d) are divergent; (e) is convergent.

Series

Checkpoints

1 $a + (n-1)d = 137a = 4d = 7 \quad \therefore n = 20$

$S_n = \frac{20}{2}[2 \times 4 + (20 - 1)7] = 1\,410$

2 $6 \rightarrow 12 = 7$ terms
3 5. Yes.
4 $a = 4$ $r = 2.$ $ar^{n-1} = 390.625$
$4(2.5)^{n-1} = 390.625$
$(2.5)^{n-1} = 97.656\ 25$
Taking logs of both sides
$\log 2.5^{n-1} = \log 97.656\ 25$
$n - 1 = 5$ $\therefore n = 6$
$S_6 = \frac{4(2.5^6 - 1)}{2.5 - 1} = 648.375$

Exam questions

1 (a) $a = 14$ $d = 4$ $a + (n - 1)d = 86$ $\therefore n = 19$
$S_{19} = \frac{19}{2}[2 \times 14 + (19 - 1)4] = 950$
(b) $r = 1.8$ $a = 3$ $3 \times 1.8^{n-1} = 102.036\ 672$
$(n - 1)\log 1.8 = \log 34.012\ 224$ $\therefore n = 7$
$S_7 = \frac{3(1.8^7 - 1)}{1.8 - 1} = 225.832\ 512$
(c) $a = 60$ $r = 0.4$ $S_\infty = \frac{60}{1 - 0.4} = 100$
2 $S_8 = 124$ $S_{15} = 390$
$124 = \frac{8}{2}(2a + 7d)$ ①
and $390 = \frac{15}{2}(2a + 14d)$ ②
Solving the simultaneous equations ① and ② gives $d = 3$ and $a = 5$
$S_{25} = \frac{25}{2}(2 \times 5 + (25 - 1)3) = 1\ 025$
3 $a + ar = 180$ ① and $182\frac{6}{7} = \frac{a}{1 - r}$ ②
Solving ① and ② simultaneously gives $r = \frac{1}{8}$ and $a = 160$
4 (a) divergent
(b) convergent; sum = 96
(c) convergent; sum = 64

Binomials

Checkpoints

1 Add the previous two terms from above.

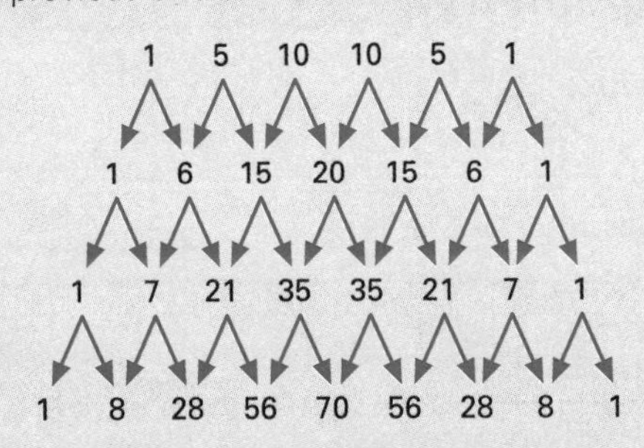

2 $^8C_3 = 56$ $^8C_5 = 56$ Looking at the 8th row of Pascal's Triangle you can see it is symmetrical about 70 and that the 3rd and 5th terms are the same.
3 $-1 < 4x < 1$, i.e. $-\frac{1}{4} < x < \frac{1}{4}$ are the values for which the expression is valid.

Exam questions

1 (a) $p^{10} + 10p^9q + 45p^8q^2 + 120p^7q^3$
(b) $(\frac{1}{5})^7 + {}^7C_1(\frac{1}{5})^6(2x) + {}^7C_2(\frac{1}{5})^5(2x)^2 + {}^7C_3(\frac{1}{5})^4(2x)^3$
$= (\frac{1}{5})^7 + 14(\frac{1}{5})^6x + 84(\frac{1}{5})^5x^2 + 280(\frac{1}{5})^4x^3$
(c) $4^8 + 8(4)^7(-\frac{3}{2}x) + {}^8C_2(4)^6(-\frac{3}{2}x)^2 + {}^8C_3(4)^5(-\frac{3}{2}a)^3$
$= 65\ 536 - 196\ 608x + 258\ 048x^2 - 193\ 536x^3$
2 (a) $^4C_2(3)^2(5x)^2 = 1\ 350x^2$ (b) $^9C_5(x)^4(-2y)^5 = -4\ 032x^4y^5$
(c) $^6C_1(2c)^5(3d) = 576c^5d$ (d) $^4C_2(6f)^2(-2g)^2 = 864f^2g^2$
3 (a) $\frac{1}{(1 + 2x)^5} = (1 + 2x)^{-5}$
Using binomial series, $2x$ replaces 'x' and n is -5.
$1 + (-5)(2x) + \frac{(-5)(-5 - 1)(2x)^2}{2!} + \frac{(-5)(-5 - 1)(-5 - 2)(2x)^3}{3!}$
$= 1 - 10x + 60x^2 - 280x^3$
(b) $\sqrt[3]{1 + 2x} = (1 + 2x)^{\frac{1}{3}}$
Using binomial series, $2x$ replaces 'x' and n is $\frac{1}{3}$.
$1 + \left(\frac{1}{3}\right)(2x) + \frac{(\frac{1}{3})(\frac{1}{3} - 1)}{2!}(2x)^2 + \frac{(\frac{1}{3})(\frac{1}{3} - 1)(\frac{1}{3} - 2)}{3!}(2x)^3$
$= 1 + \frac{2}{3}x - \frac{4}{9}x^2 + \frac{40}{81}x^3$
(c) $\frac{1}{\sqrt{1 - 2x}} = (1 - 2x)^{-\frac{1}{2}}$
Using binomial series, $-2x$ replaces x and n is $-\frac{1}{2}$.
$1 + \left(-\frac{1}{2}\right)(-2x) + \frac{(-\frac{1}{2})(-\frac{1}{2} - 1)}{2!}(-2x)^2 + \frac{(-\frac{1}{2})(-\frac{1}{2} - 1)(-\frac{1}{2} - 2)}{3!}$
$= 1 + x + \frac{2}{3}x^2 + \frac{5}{2}x^3$
4 $(1 + 2x)^{15} = 1 + 15(2x) + \frac{15(14)}{2!}(2x)^2 + \frac{15(14)(13)}{3!}(2x)^3$
$(1.02)^{15} = (1 + 0.02)^{15} = (1 + 2 \times 0.01)^{15}$
Substituting $x = 0.01$ into the expansion of $(1 + 2x)^{15}$.
$= 1 + 0.3 + 0.042 + 0.003\ 64 = 1.346$ (to 3 d.p.)

Proof

Checkpoints

1 $p \Leftarrow q$
2 $p \not\Rightarrow q$. Not all parallelograms have 90° vertices.

Exam question

The sum of the n terms of a G.P. is:
$S_n = a + ar + ar^2 + \ldots + ar^{n-3} + ar^{n-2} + ar^{n-1}$ ①
Re-writing the sum out, after multiplying it by r, gives:
$rS_n = ar + ar^2 + ar^3 + \ldots + ar^{n-2} + ar^{n-1} + ar^n$ ②
Subtracting ② from ①: $S_n - rS_n = a - ar^n$
Factorising: $S_n(1 - r) = a(1 - r^n)$
Dividing by $(1 - r)$: $S_n = \frac{a(1 - r^n)}{1 - r}$

The geometry of straight lines

Checkpoints

1 $y = -1\frac{1}{2}x + 15$
2 (a) -4; (b) $2\frac{1}{2}$
3 (a) $y - 4 = -2(x - 3)$ or $y = -2x + 10$
(b) $\frac{y + 5}{8} = \frac{x - 2}{-3}$ or $8x + 3y - 1 = 0$
4 (1,0)
5 (a) $y = 4x - 3$ (b) $x + 2y - 3 = 0$

Exam question

1 (a) Gradient of AB is $\frac{3-(-5)}{-5-7} = \frac{8}{-12} = -\frac{2}{3}$

Equation is $y - 3 = -\frac{2}{3}(x - (-5))$ (You will often get full marks at this stage.)

$\therefore 3y - 9 = -2x - 10$

$\therefore 3y + 2x + 1 = 0$

(b) Coordinates of D are average of coordinates of A and B, i.e. D is (1,–1).

Gradient of CD is $\frac{5-(-1)}{5-1} = \frac{6}{4} = \frac{3}{2}$

From (a) gradient of AB is $-\frac{2}{3}$

Hence perpendicular to AB as $-\frac{2}{3} \times \frac{3}{2} = -1$.

Equation of CD is $y - 5 = \frac{3}{2}(x - 5)$

$2y - 10 = 3x - 15$

$2y = 3x - 5$

(c) x-axis crossed when $y = 0$, so $0 = 3x - 5$, i.e. $x = \frac{5}{3}$

Examiner's secrets

Take care to be accurate with negative signs!

The coordinate geometry of circles

Check points

1 Centre (2,–1), radius $\frac{1}{2}$.

2 (a) Normal $4y - 3x + 2 = 0$ (b) tangent $4x + 3y - 36 = 0$.

Exam questions

1 The centre is at the midpoint of the diameter, i.e. (1,2). The diameter's length is the distance between (–3,–1) and (5,5), i.e. $\sqrt{(-3-5)^2 + (-1-5)^2} = 10$, so the radius is 5. Hence the equation is $(x-1)^2 + (y-2)^2 = 25$.

2 (a) $x^2 + y^2 + 2x - 4y - 4 = 0$ Complete the square to give $(x+1)^2 + (y-2)^2 - 9 = 0$. Hence C is (–1,2).

(b) $r = 3$ from above equation. ($x^2 + y^2 = 3^2$)

(c) A lies on the circle since $\left(\frac{4}{5} + 1\right)^2 + \left(\frac{22}{5} - 2\right)^2 = 9$.

The normal at A is the line through A and C, i.e. $3y - 4x - 10 = 0$. Gradient $= \frac{4}{3}$. The tangent at A is a line perpendicular to this (i.e. with gradient $= -\frac{3}{4}$) and passing through A:

$y - y_1 = m(x - x_1)$

$y - \frac{22}{5} = -\frac{3}{4}\left(x - \frac{4}{5}\right)$

$3x + 4y - 20 = 0$

(d) B lies on the tangent since $3 \times 4 + 4 \times 2 - 20 = 0$.

(e)

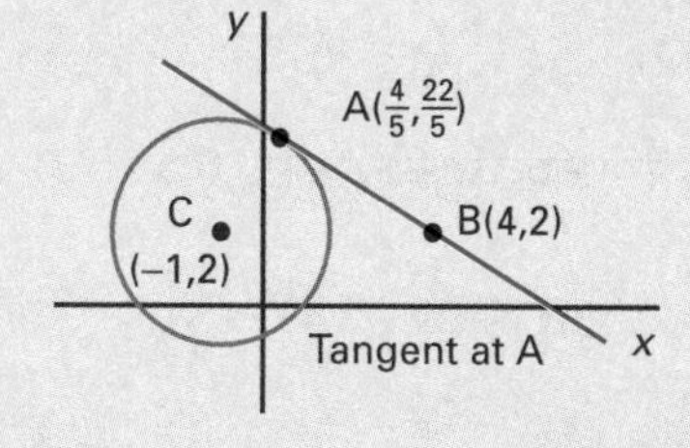

(f) 36.9°

Parameters

Checkpoints

1 At ($\pm a$,0) and (0,$\pm b$). When $a = b$, the axes are cut symmetrically i.e. a circle.

2 $x = 2at \quad y = at^2$

3 $y^2 = \left(\frac{x}{4}\right)^3$

4 (a) $x^3 + 2x^2 = y^2$ (b) $\left(\frac{x}{a}\right)^2 + \left(\frac{y}{b}\right)^2 = 1$

Exam questions

1 (a)

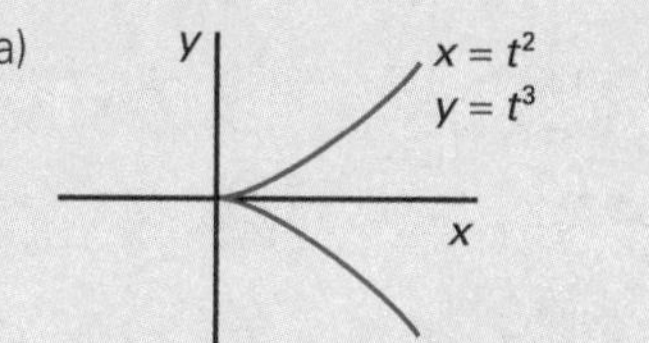

$t = \pm\sqrt{x}$

so $y = \pm(\sqrt{x})^3$

(b)

y
$x = 3t^2$
$y = 6t$
x

$t \pm \sqrt{\frac{x}{3}}$

so $y = \pm 6\sqrt{\frac{x}{3}}$

2 Standard ellipse equation: $x = 3 \sin t$, $y = 2 \cos t$.

Examiner's secrets

Note ± square roots. You will remember if you're on the lookout for symmetry in the curve. *Always label curve sketches!*

Trigonometry

Checkpoints

$\cos x = \frac{4}{5}$ and $\tan x = \frac{3}{4}$

Exam questions

1 angle ACB $= \tan^{-1}\frac{7}{8}$

angle ACB = 41.2° (to 3 s.f.)

2 $AB = \frac{10 \times \sin 32°}{\sin 78°}$

AB = 5.42 cm (to 3 s.f.)

3 $(BC)^2 = 8^2 + 11^2 - 2 \times 8 \times 11 \times \cos 47°$

BC = 8.06 cm (to 3 s.f.)

Radians and trigonometry

Checkpoints

1 57.3°

2 Arc length: 2π radians gives length $2\pi r$ so θ radians gives length $r\theta$. Area: 2π radians gives area πr^2 so θ radians gives area $\frac{\pi r^2\theta}{2\pi} = \frac{1}{2}r^2\theta$

Exam question

By Pythagoras, AC = BC = $\sqrt{125}$ cm.

$A\hat{C}B = 2\tan^{-1}\frac{1}{2} = 0.927\,3$ rads

so arc AXB = $r\theta$ = 10.37 cm.

Total perimeter = $2\sqrt{125} + 10.4 = 32.7$ cm to 1 d.p.

Area is $\frac{1}{2}r^2\theta = \frac{1}{2} \times 125 \times 0.927\,3 = 58.0\text{ cm}^2$ to 1 d.p.

Examiner's secrets

Do the maths without units, *but always answer in the units given in the question.*

Trigonometric relations

Checkpoints

1. cos, sin, sec and cosec: period 2π; tan and cot: period π.
2. sec θ: domain $\theta \neq (2n+1)\frac{\pi}{2}$ because here cos θ = 0 and reciprocal of 0 is undefined; similar reasons for restricted domains of cosec and cot.
   sec θ and cosec θ: must be ≤ −1 (reciprocal of cos and sin from −1 to 0) or ≥1 (reciprocal of cos and sin from 0 to 1)
   cot θ: can take any value (reciprocal of tan, which can take any value).
3. When the reciprocal is undefined, the function is zero.
4. $\cos^2\theta + \sin^2\theta \equiv 1$. Divide by $\cos^2\theta$:
   $$1 + \frac{\sin^2\theta}{\cos^2\theta} \equiv \frac{1}{\cos^2\theta}$$
   $\therefore\ 1 + \tan^2\theta \equiv \sec^2\theta$.
   Similarly with the other one.
5. One-to-many
6. No angle has a cos or sin > 1 or < −1 so there's no arccos or arcsin of numbers >1 or < −1.

Exam questions

1 sec θ(1 − sin θ) is
$$\frac{1-\sin\theta}{\cos\theta} = \frac{(1-\sin\theta)(1+\sin\theta)}{\cos\theta(1+\sin\theta)} = \frac{1-\sin^2\theta}{\cos\theta(1+\sin\theta)} = \frac{\cos^2\theta}{\cos\theta(1+\sin\theta)} = \frac{\cos\theta}{1+\sin\theta}$$

2 (a)

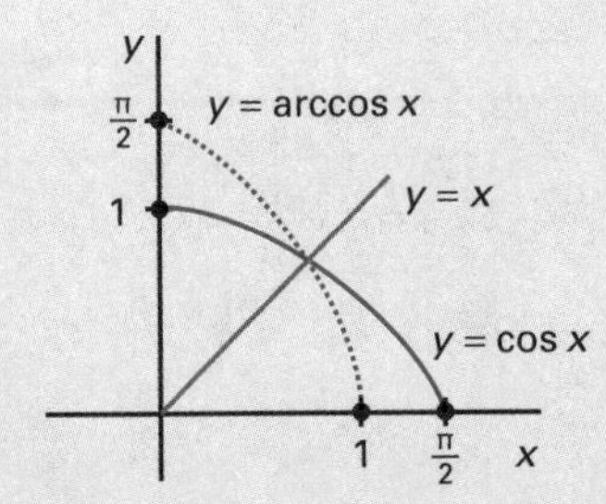

(b) 0.65 − cos 0.65 < 0; 0.75 − cos 0.75 > 0. Use $x_1 = 0.7$; then x_6 and x_7 are both equal to 0.74 to 2 d.p.

(c) x = arccos(arccos) x so cos x = arccos x. This happens when the graphs intersect. arccos x is the inverse of cos x so its graph is a reflection of cos x in $y = x$. Therefore when $y = \cos x$ and $y = \arccos x$ intersect they will also intersect with $y = x$ (see part a)). Hence when cos x = arccos x, cos x also = x. Hence solution is 0.74 to 2 d.p. as in part (b).

Further trigonometric relations

Checkpoints

1. $\sin 15° = \cos 75° = \dfrac{\sqrt{6}-\sqrt{2}}{4}$ $\quad\cos 15° = \sin 75° = \dfrac{\sqrt{6}+\sqrt{2}}{4}$
   $\tan 15° = 2 - \sqrt{3}$ $\quad\tan 75° = 2 + \sqrt{3}$
2. $\cos 2A = 1 - 2\sin^2 A \Rightarrow 2\sin^2 A = 1 - \cos 2A$
   $\Rightarrow 2\sin^2\dfrac{A}{2} = 1 - \cos A \Rightarrow \sin^2\dfrac{A}{2} = \dfrac{1}{2}(1-\cos A)$
   and similarly for the other one.
3. $R\cos(\theta - \alpha)$ is a translation of cos θ by +α along the x-axis followed by a stretch of factor R parallel to the y-axis.

Exam questions

1
$$\cos 2x + \cos x = 0$$
$$2\cos^2 x - 1 + \cos x = 0$$
$$(2\cos x - 1)(\cos x + 1) = 0$$
$$\cos x = \frac{1}{2} \text{ or } -1$$
$$x = \frac{\pi}{3}, \pi \text{ or } \frac{5\pi}{3}$$

2 (a)
$$8\sin x + 15\cos x = 10$$
Let $R\cos(x-\alpha) = 8\sin x + 15\cos x$
$$R(\cos x\cos\alpha + \sin x\sin\alpha) = 8\sin x + 15\cos x$$
$$R\cos\alpha = 15;\ R\sin\alpha = 8$$
$$R \text{ is } 17;\ \alpha = \arctan\frac{8}{15} \approx 28.1°$$
$$\text{So } 17\cos(x - 28.1) = 10$$
$$x = \arccos\frac{10}{17} + 28.1$$
$$= 82.0° \text{ or } 334.1° \text{ (to 1 d.p.)}$$

(b) (i)
$$\tan(\alpha+\beta) = \frac{\tan\alpha + \tan\beta}{1 - \tan\alpha\tan\beta}$$
$$-1 = \frac{\frac{4}{3} + \tan\beta}{1 - \frac{4}{3}\tan\beta}$$
$$\frac{4}{3}\tan\beta - 1 = \frac{4}{3} + \tan\beta$$
$$\frac{1}{3}\tan\beta = \frac{7}{3} \text{ so } \tan\beta = 7$$

(ii)

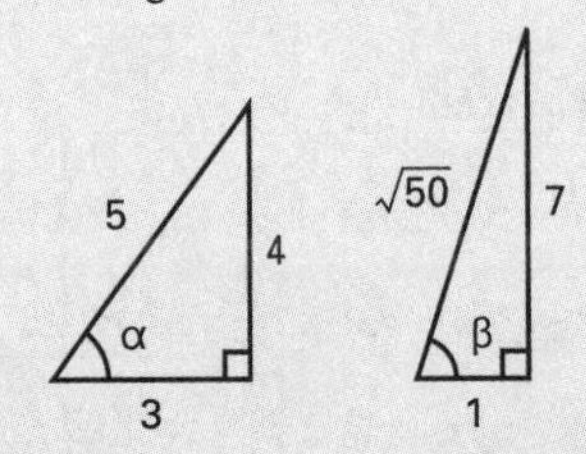

Draw triangles from knowledge of tan α and tan β. Solve triangles by Pythagoras to find sin α and sin β.

$\text{Sin }\alpha = \dfrac{4}{5}$, $\sin\beta = \dfrac{7}{\sqrt{50}}$

(iii) Put sin α, sin β, cos α and cos β from the above triangles into double-angle formulae for sin 2α and sin 2β. $\text{Sin }2\alpha = \dfrac{24}{25}$, $\sin 2\beta = \dfrac{7}{25}$

Examiner's secrets

Look out for clues in the questions as to which technique or identity to use. If you have them all at your fingertips, you won't get stuck!

Basic differentiation

Checkpoints

1 $y = c$ is a horizontal line.

2 (a) $3x^2$ (b) $-\frac{1}{p^2}$

3 $f'(x) = 15x^2 - 7$ $f''(x) = 30x$

4 $\frac{dh}{dt} = \frac{3}{4}$ cm per day

Exam questions

1 (a) $6x - 4$ (b) $\frac{1}{2}x^{-\frac{1}{2}} = \frac{1}{2\sqrt{x}}$ (c) $\frac{5}{2}x^{\frac{3}{2}} + \frac{2}{3}x^{-\frac{5}{3}}$

(d) $-12x^{-5} = -\frac{12}{x^5}$ (e) $2x + 6 = 2(x + 3)$

2 $2 + \frac{6}{x^4}$

3 (a) $\frac{dy}{dx} = 3x^2 - 12x + 10; -2$ (b) (1,3) and (3,1)

Applications of differentiation

Checkpoints

1 $y = 12x - 21$

2 $12y + x = 38$

Exam questions

1 $\frac{dy}{dx} = 24x^3 - 6x$ $\frac{dy}{dx} = 630$

$x = 3,\ y = 459$

$y - 459 = 630(x - 3)$

2 $y = x^2 - 4 + \frac{4}{x^2}$ $\frac{dy}{dx} = 2x - \frac{8}{x^3}$

$2x - \frac{8}{x^3} = 0$

$2x^4 - 8 = 0$

$(x^2 - 2)(x^2 + 2) = 0$

$x = \pm\sqrt{2},\ y = 0$

3 $\frac{dy}{dx} = 4x^3 - 6x^2 + 2x$

$x(2x - 1)(x - 1) = 0$

$x = 0$ or $\frac{1}{2}$ or 1

(0,0) min

$(\frac{1}{2}, \frac{1}{16})$ max

(1,0) min

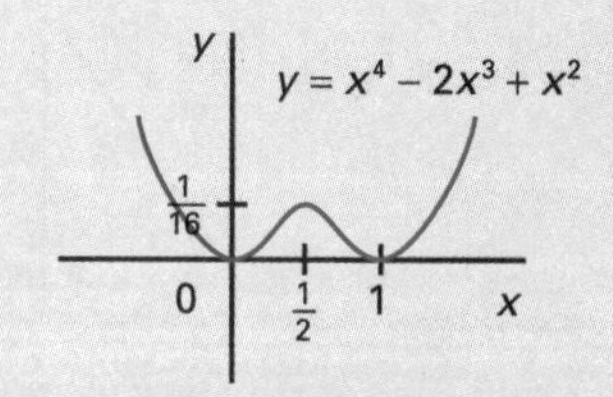

Rules of differentiation

Checkpoints

1 $4e^2$

2 20

3 (a) $28(7x + 2)^3$ (b) $\frac{x}{\sqrt{x^2 + 1}}$ (c) $\frac{5}{5x + 1}$ (d) $20e^{4x}$

4 $\frac{15}{\pi}$ ms^{-1}

Exam questions

1 (a) $\frac{4}{4x - 5}$ (b) $40x(5x^2 + 1)^3$

2 (a) $\frac{n}{x}$ (b) $n \ln x$; $\frac{n}{x}$

3 $y' = \frac{2}{\sqrt{4x - 3}}$; $5y - 2x = 11$

Further rules of differentiation

Checkpoints

1 (a) $4 \cos 4x$ (b) $-2x \sin(x^2)$ (c) $-8 \sec^2 2x$

2 $2y - x = \sqrt{3} - \frac{\pi}{3}$

3 (a) $1 + \ln x$ (b) $\frac{\sin x - x \cos x}{\sin^2 x}$ (c) $3e^{3x-4}$

4

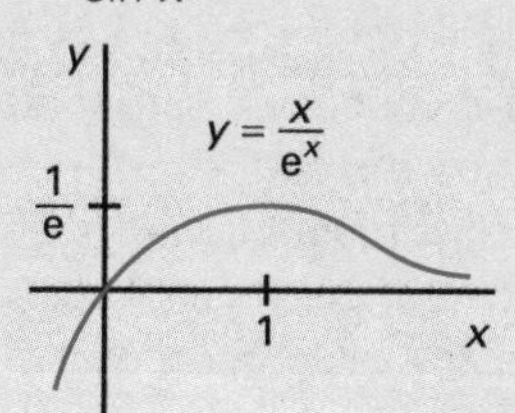

Exam questions

1 (a) $\frac{2x}{1 + x^2}$ (b) $\frac{-x \sin x - \cos x}{x^2}$ (c) $\frac{8 + x}{2\sqrt{x}} + \sqrt{x}$

2 $\frac{dy}{dx} = (2x - 1) \times 3 + 2 \times (3x + 4)$

$12x + 5 = 13$

$x = \frac{2}{3}$ $y = 2$

3 $\frac{dx}{dy} = \cos y$ $\frac{dy}{dx} = \frac{1}{\cos y} = \frac{1}{\sqrt{1 - x^2}}$

Implicit and parametric differentiation

Checkpoints

1 (a) $2y\frac{dy}{dx} = 8$ (b) $2x + 2y\frac{dy}{dx} = 3e^x$

(c) $y^2 + 2y\frac{dy}{dx} - 6x + \cos y\frac{dy}{dx} = 1$

2 $y - 2 = -\frac{7}{8}(x - 4)$

3 0.229 (3 s.f.)

Exam questions

1 $3x^2 + 3y^2\frac{dy}{dx} + 3x\frac{dy}{dx} + 3y = 0$

$4 + \frac{dy}{dx} + 2\frac{dy}{dx} - 1 = 0$ $\frac{dy}{dx} = -1$

2 $x' = 2t,\ y' = 12t^3 + 24t^2$ $\frac{dy}{dx} = 6t^2 + 12t$

$t = -1,\ \frac{dy}{dx} = -6,\ x = -3,\ y = -5$

$y - (-5) = -6(x - (-3))$ $y + 5 = -6(x + 3)$

Basic integration

Checkpoints

1 This would give $\frac{x^0}{0}$.

2 This is because $y = c$ differentiates to $\frac{dy}{dx} = 0$.

3 (a) $\frac{x^5}{5} - \frac{5x^4}{4} + c$

(b) Remember to multiply out the brackets first. $x^3 - \frac{1}{x} + c$

4 $y = x^2 + 3x - 2$

5 (a) $\int_1^4 3x^2 + 4\sqrt{x}\,dx$

$= \left[x^3 + \frac{8}{3}x^{\frac{3}{2}}\right]_1^4 = \left[\left(4^3 + \frac{8}{3} \times 4^{\frac{3}{2}}\right) - \left(1^3 + \frac{8}{3}\right)\right]$

$= \left[64 + \frac{64}{3} - 1 - \frac{8}{3}\right] = 44\frac{1}{3}$

(b) $\int_0^3 \frac{6}{2x+1}\,dx$

$= 3[\ln|2x+1|]_0^3$

$= 3[\ln 7 - \ln 1]$

$= 3 \ln 7$

Exam questions

1 (a) $\frac{x^3}{3} + \frac{1}{x} + c$ (b) $\frac{x^2}{8} + \frac{1}{2x} + c$

2 $y = x^4 - \frac{1}{x^2} + c$ $17 = 2^4 - \frac{1}{2^2} + c$

$c = 1\frac{1}{4}$ $y = x^4 - \frac{1}{x^2} + 1\frac{1}{4}$

3 $\left[2x^{\frac{1}{2}}\right]_1^4 = 4 - 2 = 2$

4 $\left[3e^{2x} + \frac{3}{4}\ln x\right]_1^3 = 3e^6 + \frac{3}{4}\ln 3 - 3e^2$

Rules of integration

Checkpoints

1 (a) $2 \sin x + 5 \cos x + c$ (b) $\frac{\tan 4x}{4} + \frac{\tan 5x}{5} + c$

2 $\frac{1}{2}\left[\frac{\sin 6x}{6} - x\right] + c$

3 $\frac{(1+5x)^{12}}{60} + c$

4 $\frac{\sin^4 x}{4} + c$

5 $\frac{1}{4}\ln 3$

Exam questions

1 $u = 1 + x^3$

$du = 3x^2\,dx$

$\frac{1}{3}\int \frac{du}{u} = \frac{1}{3}\ln u = \frac{1}{3}[\ln(1 - x^3)]_0^1 = \frac{1}{3}\ln 2$

2 $u = 1 + x^2$

$du = 2x dx$

$\int \sqrt{u}\,\frac{du}{2} = \frac{1}{2}\int u^{\frac{1}{2}}\,du = \frac{1}{3}\left[u^{\frac{3}{2}}\right] = \frac{1}{3}\sqrt{(1+x^2)^3} + c$

Further rules of integration

Checkpoints

1 $\int \ln x \cdot 1\,dx = \ln x \cdot x - \int x \cdot \frac{1}{x}\,dx = x \ln x - x + c$

2 $-x^2 \cos x + 2x \sin x + 2 \cos x + c$

3 $\frac{7}{5}\ln|x+2| - \frac{12}{5}\ln|3 - x| + c$

4 $\int 1 - \frac{25}{8(x+5)} + \frac{9}{8(x-3)}\,dx$

$= x - \frac{25}{8}\ln|x+5| + \frac{9}{8}\ln|x-3| + c$

Exam questions

1 (a) $-x \cos x + \sin x + c$

(b) $\left[\ln x \cdot \frac{x^3}{3}\right]_1^2 - \int_1^2 \frac{x^3}{3} \times \frac{1}{x}\,dx = \left[\frac{x^3}{3}\ln x - \frac{x^3}{9}\right]_1^2 = \frac{8}{3}\ln 2 - \frac{7}{9}$

2 $A = 3, B = 5, C = -1$

$\left[\frac{3}{2}\ln|1+2x| - \frac{5}{2(1+2x)} - \ln|1+x|\right]_0^1$

$= \frac{3}{2}\ln 3 - \frac{5}{6} - \ln 2 + \frac{5}{2} = \frac{5}{3} + \ln\left(\frac{\sqrt{27}}{2}\right)$ or $\frac{5}{3} + \ln\frac{3\sqrt{3}}{2}$

Applications of integration

Checkpoints

1 12 square units; area of a trapezium

2 192π

3 radius = 1; area as in example

Exam questions

1 $-\left[\frac{1}{x}\right]_1^2 = \left[-\frac{1}{2} - (-1)\right] = \frac{1}{2}$

1 (1,1) is the intersection

$A = \int_0^1 x^2\,dx + \frac{1}{2} \times 1 \times 1 = \left[\frac{x^3}{3}\right]_0^1 + \frac{1}{2} = \frac{5}{6}$

3 $V = \pi\int_0^4 4x\,dx$

(a) $= \pi[2x^2]_0^4 = 32\pi$

(b) $V = \pi\int_0^4 x^2\,dy = \pi\int_0^4 \frac{y^4}{16}\,dy = \pi\left[\frac{y^5}{80}\right]_0^4 = 12\frac{4}{5}\pi$

4 $A = \int_0^{\frac{\pi}{2}} 3 \sin t \times (-5) \sin t\,dt = -15\int \sin^2 t\,dt$

$= -\frac{15}{2}\int (1 - \cos 2t\,dt) = -\frac{15}{2}\left[t - \frac{\sin 2t}{2}\right]_0^{\frac{\pi}{2}} = -\frac{15\pi}{4}$

Differential equations

Checkpoints

1 $y = \frac{3x^2}{2} + 2x + 1$

2 $\frac{dy}{y} = \frac{dx}{x}$ leads to $\ln y = \ln x + \ln c$ and hence the answer.

Exam questions

1 $\dfrac{1}{\sec y}\dfrac{dy}{dx} = \dfrac{1}{x^2}$ $\quad \int \cos y\, dy = \int x^{-2}\, dx \quad \sin y = -\dfrac{1}{x} + c$

$c = \dfrac{\sqrt{2}+1}{\sqrt{2}} \quad y = 0, x = 2 - \sqrt{2}$

2 Let N be the number present.

$\therefore \dfrac{dN}{dt} = -kN \quad \int \dfrac{dN}{N} = -k\int dt \quad \ln N = -kt + \ln c$

$t = 0, N = P \quad t = 300, N = \dfrac{P}{2} \quad \therefore P = C, K = \dfrac{\ln 2}{300}$

$t = 697$ years when $N = 0.2P$

Numerical solution of equations

Checkpoints

1 $e^x - 3 = 0$ between $x = 1$ and $x = 1.1$

2 There is a solution between 0 and 1. Start with 1 and iterate to give 0.5, 0.8, 0.61, 0.73, 0.65, 0.70, 0.67, i.e. 0.7 to 1 d.p.

3 If the gradient of $f(x)$ is too steep, the staircase or cobweb diverges.

(a) positive gradient of $y = f(x) < 1$ at X, so 'staircase' approaches X.

(b) positive gradient of $y = f(x) > 1$ at X, so 'staircase' diverges from X.

(c) negative gradient of $y = f(x) > -1$ at X, so 'cobweb' approaches X.

(d) negative gradient of $y = f(x) < -1$ at X, so 'cobweb' diverges from X.

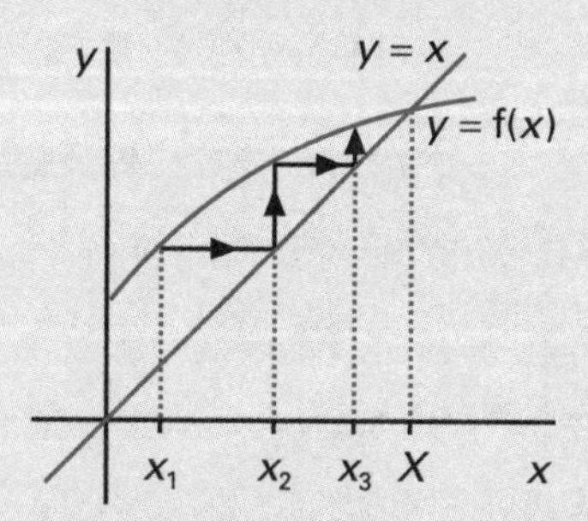

(a) Staircase converging to X

(b) Staircase diverging

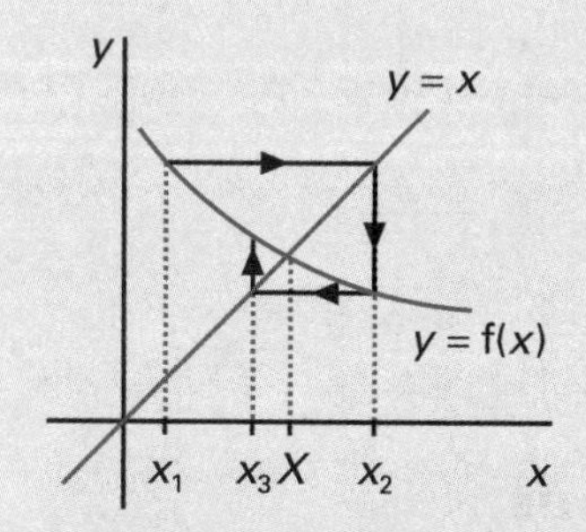

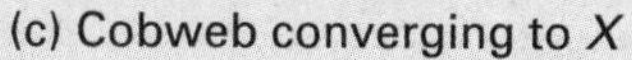

(c) Cobweb converging to X

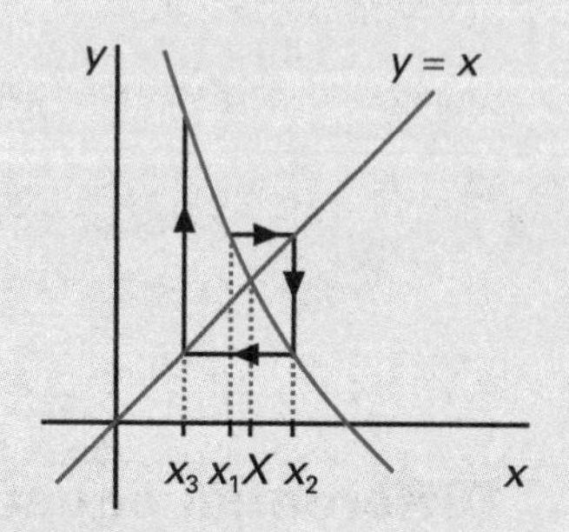

(d) Cobweb diverging

Exam questions

1 $1 - \sin x - x = 0 \Rightarrow 1 - \sin x = x$. This has a solution when $y = 1 - \sin x$ intersects $y = x$.

(a)

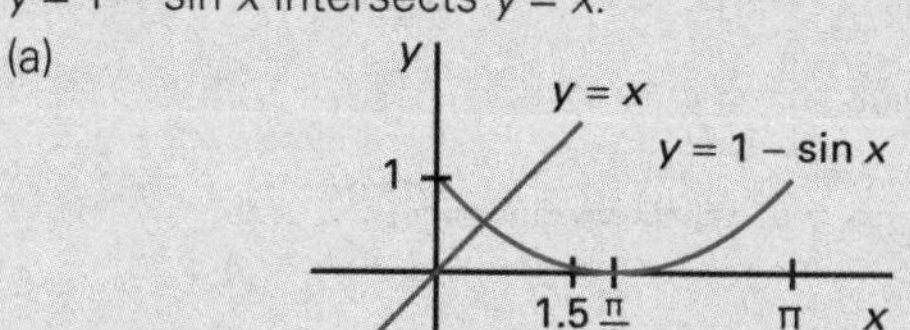

(b) $1 - \sin x - x = 0.02\ldots$ when $x = 0.5$ and $-0.16\ldots$ when $x = 0.6$. In other words, $1 - \sin x - x$ changes sign between $x = 0.5$ and $x = 0.6$. It is closer to zero for $x = 0.5$.

2

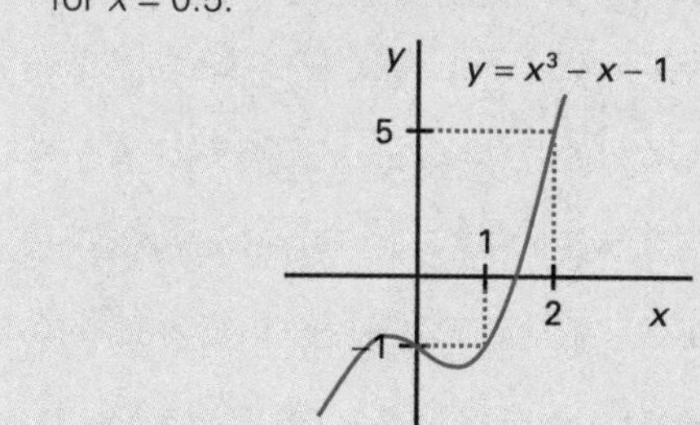

$x = (x+1)^{\frac{1}{3}}$ use $x_{n+1} = (x_n + 1)^{\frac{1}{3}} \quad \therefore x \sim 1.32$.

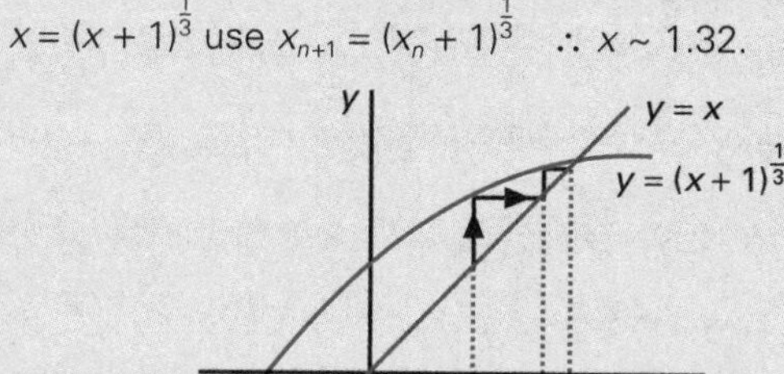

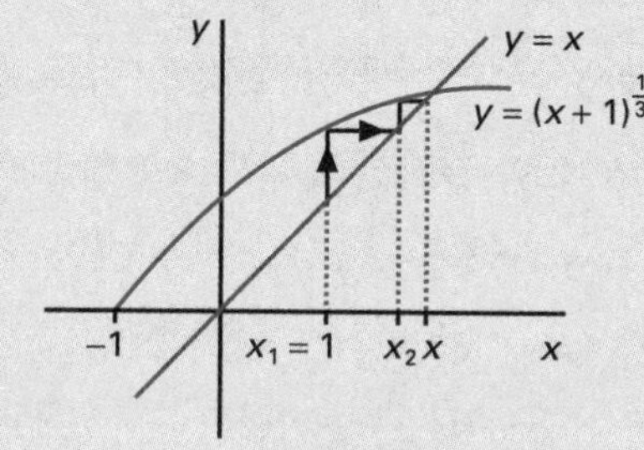

Numerical integration

Checkpoints

1 Because all the y-coordinates except the first and last appear in two strips.

2 Area $\approx \frac{1}{2}[3 + 4 + 2(8 + 17 + 12)] = 40.5$

3 Approx area is 679.233, so % error is 2.502 (to 4 s.f.).

Exam questions

1 Four strips give five y-coordinates. Strip width $h = 0.5$.

$$\int_1^3 \sqrt{1+x^2}\, dx \approx \frac{1}{4}\{\sqrt{1+1^2} + \sqrt{1+3^2} + 2(\sqrt{1+1.5^2} + \sqrt{1+2^2} + \sqrt{1+2.5^2})\} \approx 4.51$$

2 $\dfrac{0.25}{3}\{0 + 0.629\,960\,5 + 4(0.168\,776\,6 + 0.482\,744\,6) + 2 \times 0.329\,316\,8\}$

$= 0.324\,557$

Vectors

Checkpoints

1 10; 7 **2** $\begin{pmatrix}6\\0\\-3\end{pmatrix}$ **3** $\frac{1}{6}\begin{pmatrix}4\\2\\-4\end{pmatrix} = \begin{pmatrix}\frac{2}{3}\\\frac{1}{3}\\-\frac{2}{3}\end{pmatrix}$

4 $\mathbf{i} = \begin{pmatrix}1\\0\end{pmatrix} \mathbf{j} = \begin{pmatrix}0\\1\end{pmatrix} \begin{pmatrix}-3\\-1\end{pmatrix} = -3\mathbf{i} - \mathbf{j}$ **5** $\begin{pmatrix}-1\\3\\-1\end{pmatrix}$

Exam question

1 (a)

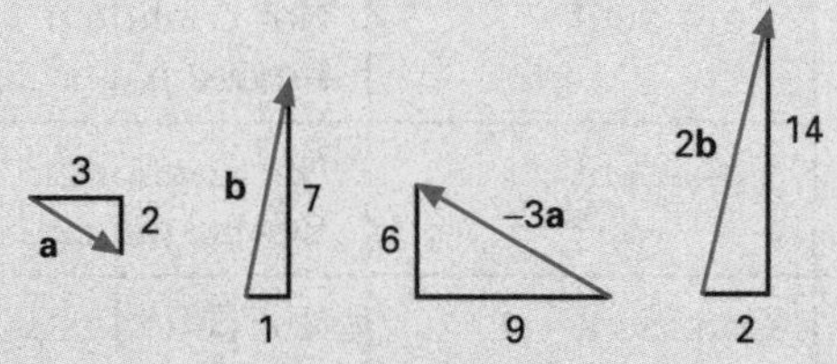

(b) $\mathbf{c} = 3\mathbf{i} - 2\mathbf{j} + 2(\mathbf{i} + 7\mathbf{j}) = 5\mathbf{i} + 12\mathbf{j}$

$|\mathbf{c}| = \sqrt{5^2 + 12^2} = 13$

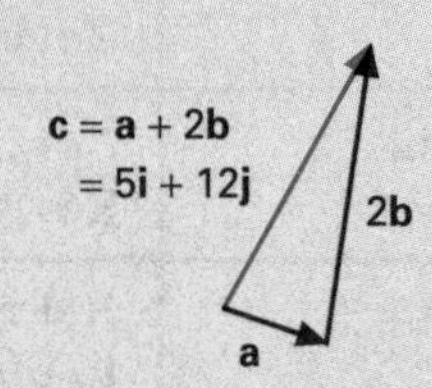

Examiner's secrets

Label your diagrams to make them make sense!

Position vectors and lines

Checkpoints

1 $-\mathbf{a}$ is the same length as $\mathbf{a}$ in the opposite direction.

2 (a) $\sqrt{6}$ (b) 5

3 $\mathbf{r} = \begin{pmatrix}0\\0\\2\end{pmatrix} + t\begin{pmatrix}-1\\3\\-2\end{pmatrix}$

4 (a) $\mathbf{r} = \begin{pmatrix}0\\0\\2\end{pmatrix} + t\begin{pmatrix}-1\\3\\-4\end{pmatrix}$

(b) Twice as far from A as B is but on the opposite side; at A; half way between A and B; at B; as far from B as A is but on the opposite side.

Exam questions

1 (a) $\mathbf{r} = \mathbf{a} + t(\mathbf{b} - \mathbf{a}) = \begin{pmatrix}1\\0\\9\end{pmatrix} + t\begin{pmatrix}-4\\4\\-7\end{pmatrix}$

(b) Unit vector in direction $\begin{pmatrix}-4\\4\\-7\end{pmatrix}$ is $\frac{1}{9}\begin{pmatrix}-4\\4\\-7\end{pmatrix}$

2 (a) Line through C parallel to $\overrightarrow{AB}$:

$$\mathbf{r} = \mathbf{c} + t(\mathbf{b} - \mathbf{a}) = \begin{pmatrix}3\\0\\4\end{pmatrix} + t\begin{pmatrix}1\\2\\1\end{pmatrix}$$

(b) When $t = -2$, you get the point D.

(c) $AD = |\mathbf{d} - \mathbf{a}| = |-2\mathbf{i} - 6\mathbf{j} + 3\mathbf{k}| = 7$

(d) $\overrightarrow{AE} = \overrightarrow{AC} + \overrightarrow{CE}$ (triangle rule) so

$\overrightarrow{AE} = (-2\mathbf{j} + 5\mathbf{k}) + (3\mathbf{i} + 8\mathbf{j} - 2\mathbf{k}) = 3\mathbf{i} + 6\mathbf{j} + 3\mathbf{k}$

The vector from A to E is three times that from A to B, so if you go from A to B and continue along the same line you get to E.

Applications of vectors

Checkpoints

1 -8 **2** 48.2° **3** (1,2,3)

Exam question

(a) (i) If A, B, C collinear, $\mathbf{c} = \lambda\mathbf{a} + (1 - \lambda)\mathbf{b}$.

Simultaneous equations from the coefficients of **i**, **j** and **k** are

$$\left.\begin{aligned}2\lambda + 2(1 - \lambda) &= 2\\ 3\lambda + 6(1 - \lambda) &= \mu\\ 6\lambda + 9(1 - \lambda) &= 8\end{aligned}\right\} \Rightarrow \mu = 5.$$

(ii) If $\mathbf{c}$, $\overrightarrow{AB}$ perpendicular,

$$\mathbf{c}(\mathbf{b} - \mathbf{a}) = 0 \Rightarrow \begin{pmatrix}2\\\mu\\8\end{pmatrix} \cdot \begin{pmatrix}0\\3\\3\end{pmatrix} = 0 \Rightarrow \mu = -8$$

(b) (i) $L_1 = \mathbf{b} + t\mathbf{a}$. Substitute vectors **a**, **b**.

(ii) $L_2 = \mathbf{a} + s(\mathbf{d} - \mathbf{a})$. Substitute vectors **a**, $(\mathbf{d} - \mathbf{a})$.

(iii) If the lines intersect, $\begin{pmatrix}2\\6\\9\end{pmatrix} + t\begin{pmatrix}2\\3\\6\end{pmatrix} = \begin{pmatrix}2\\3\\6\end{pmatrix} + s\begin{pmatrix}4\\-7\\-4\end{pmatrix}$

Solve the top and bottom rows as simultaneous equations: $s = -\frac{3}{16}$, $t = -\frac{3}{8}$. These do not work in the equation given by the middle row. There is no solution for s and t that works in all three equations, so no intersection, i.e. skew lines.

(c) Scalar product between directions of lines:

$$\cos\theta = \frac{1}{\sqrt{49} \times \sqrt{81}}\begin{pmatrix}2\\3\\6\end{pmatrix} \cdot \begin{pmatrix}4\\-7\\-4\end{pmatrix} = -\frac{37}{63} \Rightarrow \theta = 126°$$

so the acute angle between the lines is $180° - 126° = 54.0°$.

Examiner's secrets

You'll always have room to set out solutions more fully than they're given here. Every step you explain can be worth a mark even if the final answer's wrong!

Revision checklist

Pure mathematics

By the end of this chapter you should be able to:

1	Manipulate indices and surds, use the factor and remainder theorems.	Confident	Not confident. **Revise** pages 6–9
2	Solve quadratic equations and graph quadratic functions.	Confident	Not confident. **Revise** pages 10–13
3	Solve simultaneous equations, linear and quadratic inequalities.	Confident	Not confident. **Revise** pages 14–17
4	Understand functions and their graphs, including inverses.	Confident	Not confident. **Revise** pages 18–21
5	Describe and sketch the graphs of transformations.	Confident	Not confident. **Revise** pages 22–23
6	Work with rational functions including simplification into partial fractions.	Confident	Not confident. **Revise** pages 24–25
7	Understand the laws of logarithms plus the connection between exponentials and logarithms.	Confident	Not confident. **Revise** pages 26–29
8	Identify sequences and series, including APs, GPs and binomial.	Confident	Not confident. **Revise** pages 30–35
9	Know about the geometry of straight lines and circles including the equation of a line.	Confident	Not confident. **Revise** pages 38–41
10	Use the sine and cosine rules, arc length and area of a sector formulae.	Confident	Not confident. **Revise** pages 44–46
11	Understand graphs of trig functions, trig formulae and solve trig equations.	Confident	Not confident. **Revise** pages 46–51
12	Differentiate polynomial functions, and use the rules of differentiation.	Confident	Not confident. **Revise** pages 52–53 and 56–59
13	Use differentiation to find equations of tangents, and turning points.	Confident	Not confident. **Revise** pages 54–55
14	Integrate polynomials, and use the rules of integration.	Confident	Not confident. **Revise** pages 62–67
15	Use integration to find areas and volumes.	Confident	Not confident. **Revise** pages 68–69
16	Form and solve differential equations.	Confident	Not confident. **Revise** pages 70–71
17	Solve equations numerically.	Confident	Not confident. **Revise** pages 72–73
18	Use the trapezium rule to approximate areas.	Confident	Not confident. **Revise** pages 74–75
19	Manipulate vectors in two- and three-dimensions.	Confident	Not confident. **Revise** pages 76–79
20	Use the scalar product and establish if pairs of lines intersect.	Confident	Not confident. **Revise** pages 80–81

Discrete mathematics

Discrete mathematics has only been examined at A-level for about 10 years. (It is sometimes called decision mathematics.) This means that the subject is still developing, with the drawback for you that different books and awarding bodies use their own notation. It is important that you check with your own awarding body's specification.

Many of the topics in this chapter are concerned with algorithms. These are sets of rules which you must follow in order to solve problems. Drawing large, clear and well-labelled diagrams is vital in obtaining accurate solutions.

You will find many of the problems and techniques learnt here can be linked to Economics, Business Studies and Technology.

Exam themes

- Graph theory
- Network problems
- Linear programming

We only cover a selection of ideas here, so you will need to check for any gaps, particularly if you are taking more than one discrete module.

Topic checklist

	AQA	CCEA	EDEXCEL	OCR/A	OCR/MEI
Graph theory	D1	D1	D1	D1	D1
Spanning trees	D1	D1	D1	D1	D1
Shortest paths	D1	D1	D1	D1	D1/D2
Two classic problems	D1/D2	D2	D2	D1	D2
Critical path analysis	D1/D2	D1	D1	D2	D1
Linear programming	D1	D1	D1	D1	D1

Graph theory

In this section you will be learning about a different type of graph from those you met using coordinates or in Statistics.

The jargon

Make sure that you know the words that will be used on your exam paper.

Checkpoint 1

Draw a table for this graph.

F H L G J

Basic graphs

In this graph, A, B and C are *vertices* (or *nodes*). They are joined by *edges* (or *arcs*). You can also show this in a table.

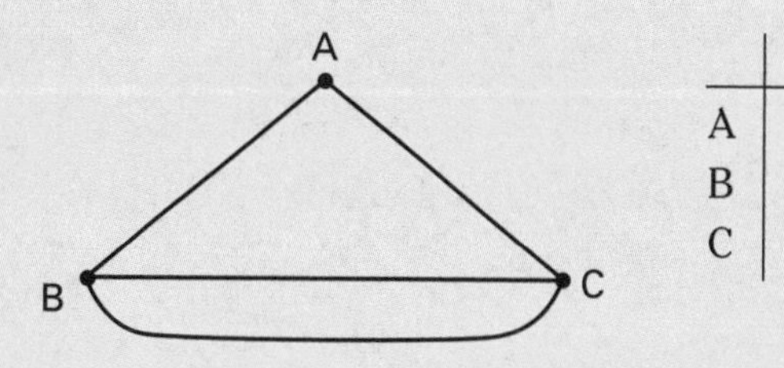

	A	B	C
A	0	1	1
B	1	0	2
C	1	2	0

What do you notice about the entries in the table? They are symmetrical about the leading diagonal.

What is the connection between the sum of the entries in the table and the number of edges on the graph? Sum = 2 × number of edges.

Directed graphs

The graph above was undirected. This means that you can travel from A to B and from B to A. In a directed graph this is not possible. In this graph, there are some directed edges; the direction is shown by the arrows. How does this change the table?

Checkpoint 2

Draw a graph from this table.

	A	B	C	D
A	1	2	1	0
B	1	2	0	2
C	2	0	0	1
D	0	2	1	0

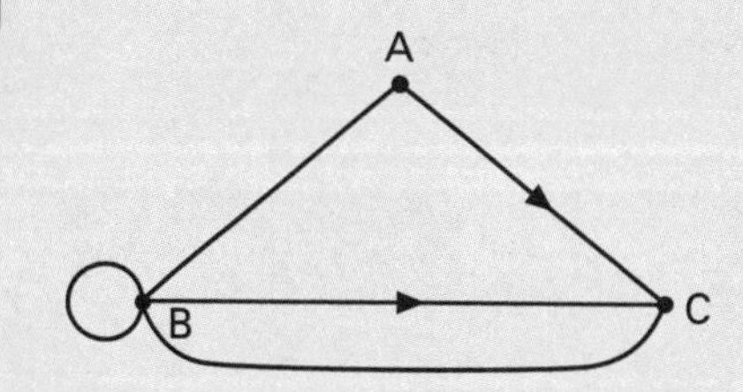

From/to	A	B	C
A	0	1	1
B	1	2	2
C	0	1	0

The entries are not symmetrical. The sum does not = 2 × number of edges. Notice that the *loop* at B is counted twice in the table.

Paths, connectedness, cycles and trees

The jargon

Again, check the words for your exam.

If you write down a list of vertices where each pair of successive vertices is joined by an edge and no vertex is *passed through* more than once, you have a *path*.

- A *cycle* (or *circuit*) is a closed path.
- A graph is *connected* if you can find a path between any two points on the graph.

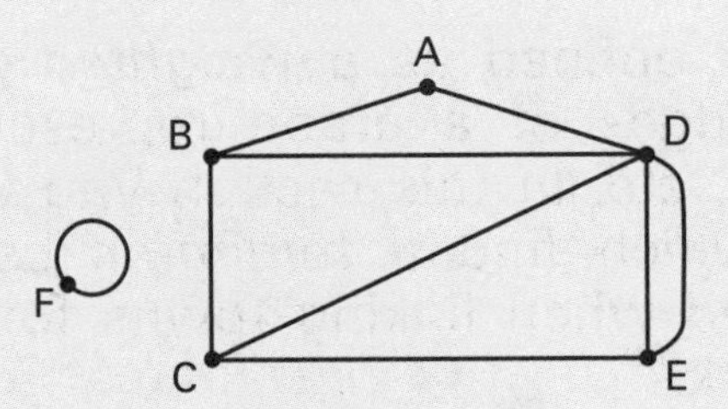

In this graph, ABD is a path, but ACE is not because A and C are not joined by an edge. DECD is a cycle. (DCED is the same cycle.) The graph is not connected because there is no path which contains F.

If you look back at the first figure on the opposite page, there are some cycles on a connected graph. If you remove some edges, such as the two between B and C, then the graph is still connected, but has no cycles. This is called a *tree*:

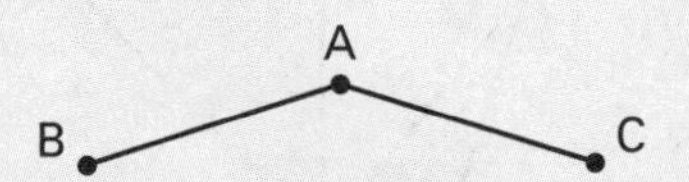

A graph such as this which connects all the vertices of the original graph is called a *spanning tree*. There are obviously many different ways in which you can remove edges and thus many different spanning trees for each graph.

Degrees and traversability

The number of edges which meet at a vertex is called the *degree* (or *order*) of the vertex. In the graph at the top of this page, the degrees of the vertices are A 2, B 3, C 3, D 5, E 3, F 2. A and F are called *even* vertices. B, C, D and E are called *odd* vertices.

How does the degree of the vertices influence this? Is there a rule? For an Eulerian trail all vertices must be even. For a semi-Eulerian trail the start and finish points are odd, the other vertices being even.

Exam questions answers: page 108

1 Draw a simple connected graph with seven vertices all having degree 2. (3 min)

2 A simple connected graph has seven vertices, all having the same degree *d*.
 (a) State the possible values of *d*.
 (b) For *each* of these values of *d*, state the number of edges of the graph. (7 min)

3 List the paths from A to B in this graph. (4 min)

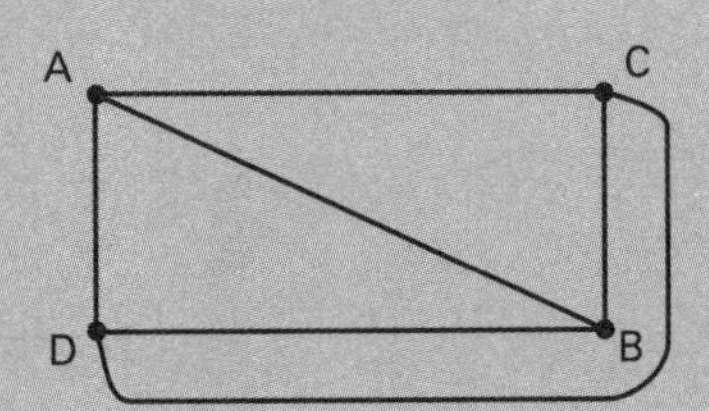

4 How many paths are there from P to Q in a complete graph joining P, Q, R, S and T? (Each vertex connects to every other one.) (5 min)

Checkpoint 3

Draw all the trees that link four vertices.

Examiner's secrets

Although trees may look different, they may be the same (isomorphic). Try looking from different angles, or look at the tables and compare them.

The jargon

If you can go over every edge of a graph once and return to the starting point, you have an *Eulerian trail*. If you cannot return to the start it is a *semi-Eulerian trail*.

Spanning trees

Links

Look back at trees, page 97.

A *network* is defined as a *weighted* graph. This is where the edges of a graph represent times, distances, costs, etc. In this section you will apply two algorithms which find a *minimum connector*. This could be used when linking towns for a cable television network.

Prim's algorithm

This network shows the approximate distance, in miles, between six towns in North West England. We need to find a minimum connector. This will be a *minimum spanning tree*, so will not necessarily be a path.

Watch out!

Prim's algorithm starts by choosing any *vertex* and will keep joining to the nearest *vertex*.

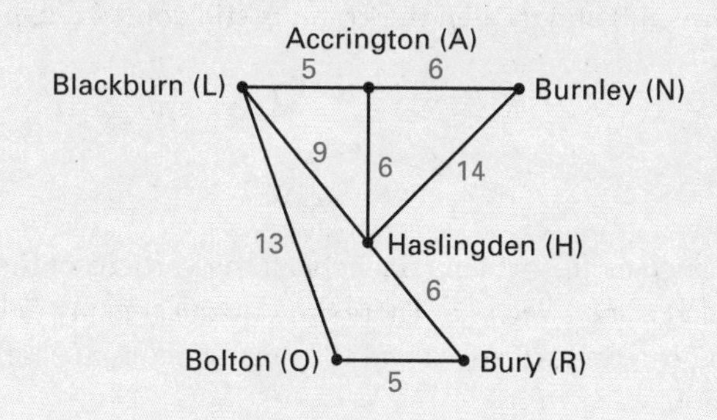

1 Find the shortest edge from any point. (If there are two, pick either.) Start at A. Pick LA = 5.

2 Find the shortest edge that joins either of these vertices (i.e. L or A). AH = 6, AN = 6. Pick either.

3 Repeat the process until all vertices have been included, *but* there must not be any cycles used. AN = 6, HR = 6. Pick one.

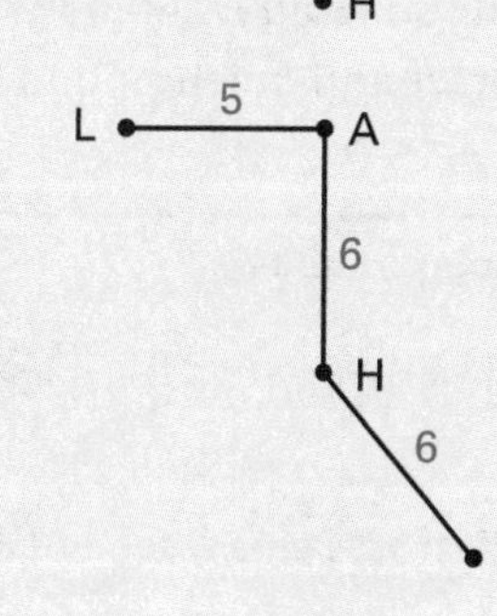

Action point

Complete the solution if you picked the other length.

RO = 5

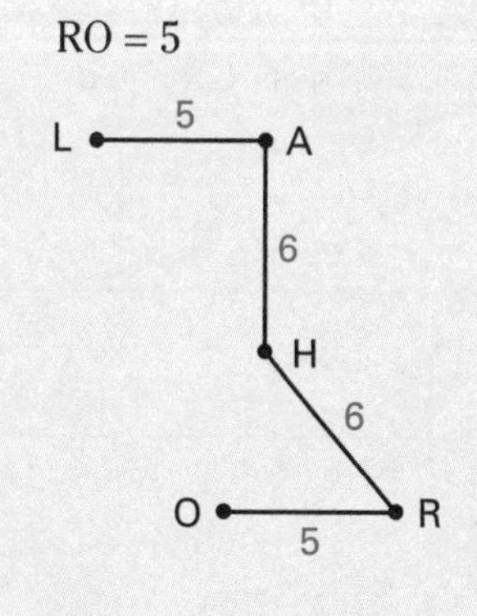

AN = 6

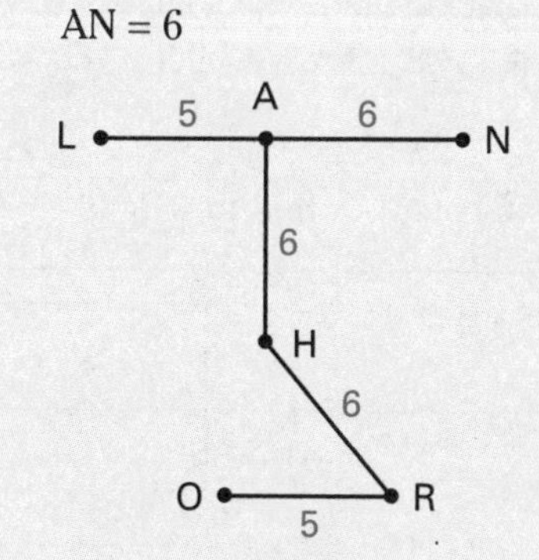

Examiner's secrets

There are many different ways of doing this. Make sure that you have shown your diagram at each stage so that the examiner can see that you have used the correct algorithm. The way that you have worked gains as many marks as the final solution.

Action point

Rewrite the information in a table and show that you can get the solution.

So, the minimum connector is 5 + 6 + 6 + 6 + 5 = 28 miles. Prim's algorithm is often used when the data are presented in a table.

Kruskal's algorithm

This differs from Prim because you have to consider the shortest edges, irrespective of their position. As with Prim, if there is a choice, select at random, and again you must *avoid cycles*.

This network shows the approximate distances, in miles, between five places on the Isle of Man.

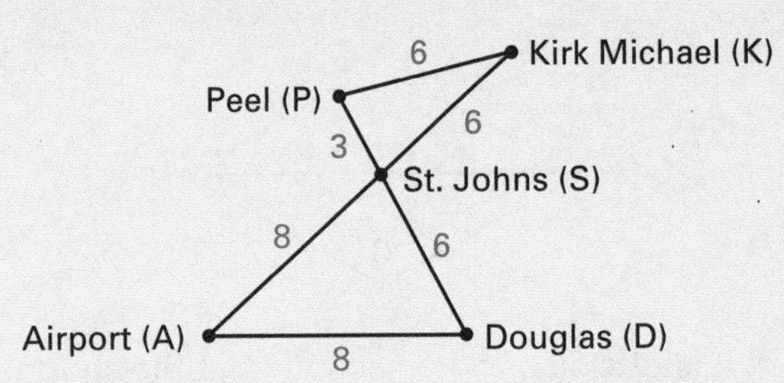

1 The shortest edge is PS = 3.

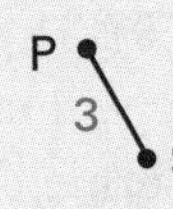

2 The next shortest edge is either PK, KS or SD, which are all 6. Pick any. SD

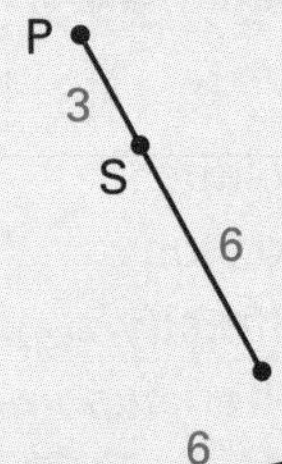

3 The next shortest is still 6, either PK or KS. Pick either. PK

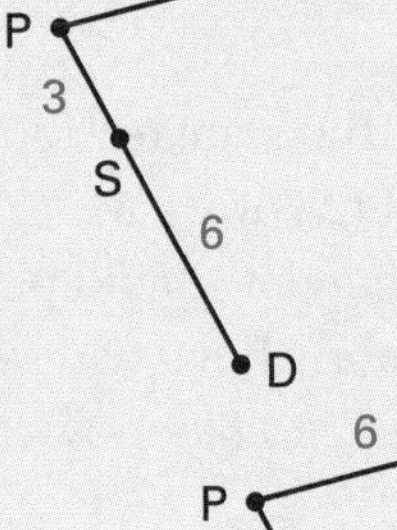

4 The next shortest is still 6, KS, *but* this completes a cycle so ignore.
The next shortest is 8, either SA or DA. Pick one. It doesn't matter which as it completes the solution.

So, the minimum connector is
6 + 3 + 8 + 6 = 23 miles.

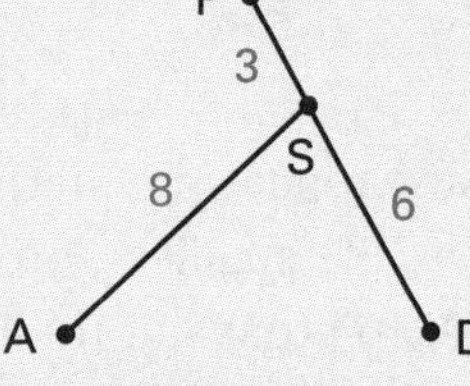

Watch out!

Kruskal's algorithm starts by choosing the *shortest edge* and then picks on successive *shortest edges*.

Checkpoint 1

Apply Kruskal's algorithm to the North West England example and then Prim's algorithm to the example shown here.

Checkpoint 2

Find two different minimum connectors. Verify that they are the same length.

Exam questions answers: page 108

1 Apply Kruskal's algorithm to find the minimum spanning tree for this network linking five Irish towns. (7 min)

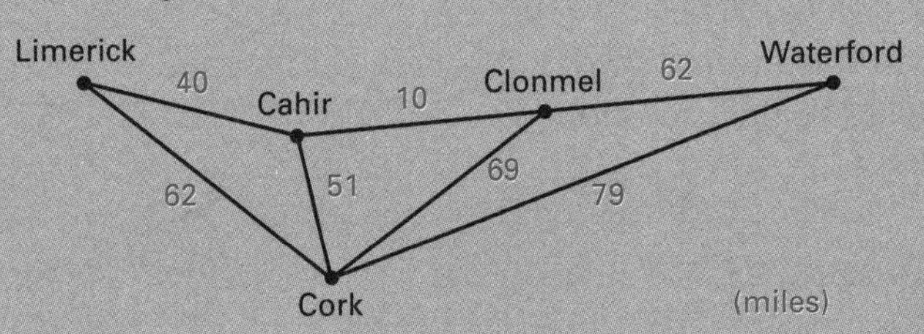

2 A graph has five vertices and six edges. Its minimum spanning tree has a length of 50 units. Three of the edges are of length 10 units. Draw a possible graph. (5 min)

Shortest paths

In this section you will be looking at problems involving shortest paths, quickest routes and paths which give a minimum cost. You will be expected to use a labelling technique to identify the shortest path.

Dijkstra's algorithm 

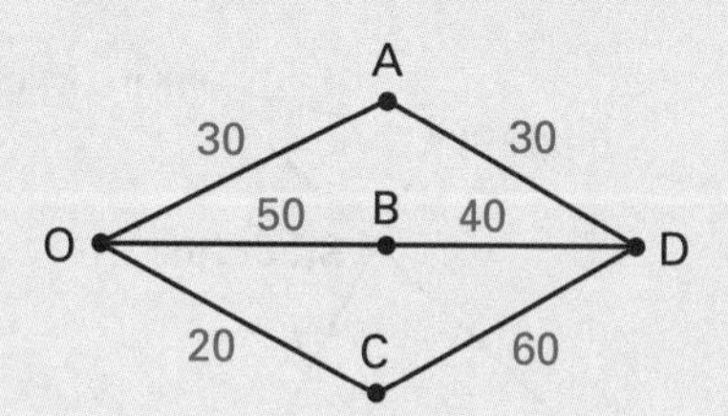

In this network it is obvious, by inspection, that the shortest path from O to D is OAD (= 60). This can be used to illustrate Dijkstra's algorithm. The technique you will use involves building up distances from O using temporary, then permanent, labels.

1 Start at O. Label this [0]. This is a *permanent* label.

2 From O, go along each edge and label the directly connected vertices with *temporary* labels.

3 Find the smallest temporary label (20 at C), Make this permanent [20]. Repeat the method above with all edges *from C*, so label D with 80.

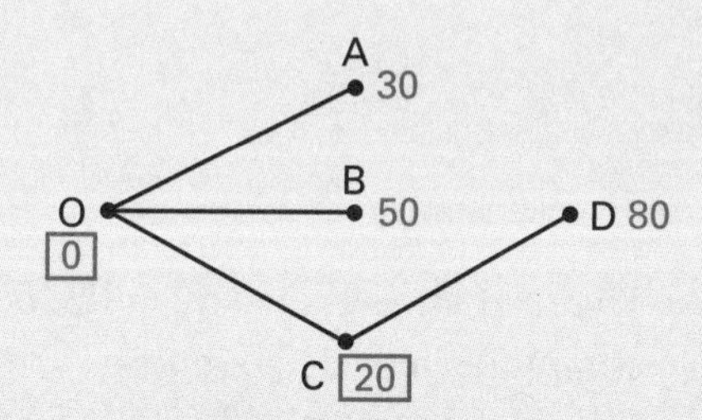

4 Now repeat the method in 3. 30 at A is the smallest, box this. AD is 30, so cross out the 80 at D and replace it with 60.

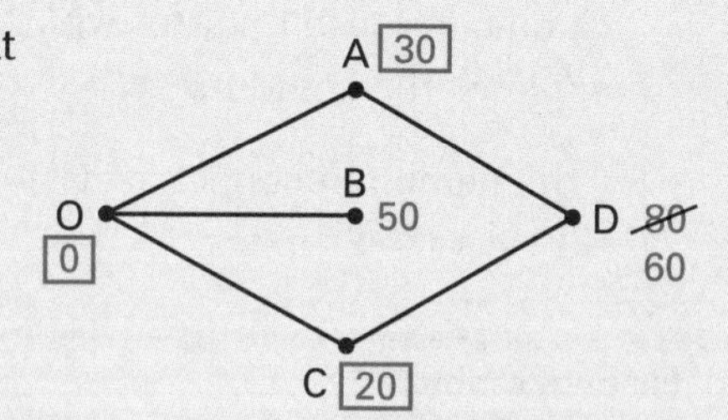

5 50 at B is the smallest temporary label left. Box it. BD is 40, so the distance to D is not reduced, which gives the final solution, 60 at D.

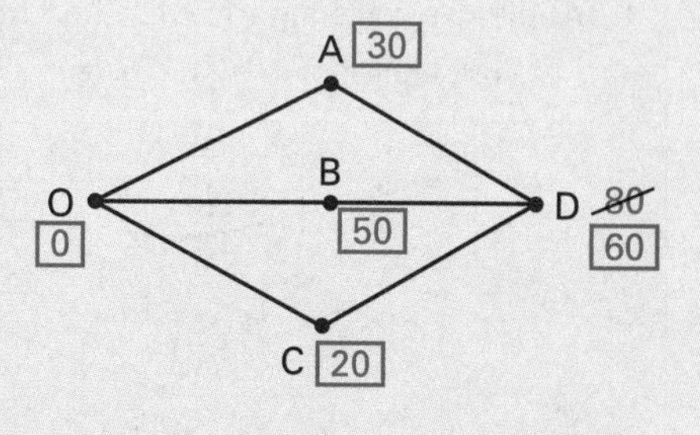

With more complicated networks, you then have to work backwards along the edges to identify the shortest path.

The jargon

Box ⇒ permanent
No box ⇒ temporary

Watch out!

In some more complicated networks, not all the arrows are going in the same direction!

Examiner's secrets

Notice that all the vertices have a permanent box once the algorithm has been completed. You should scan your network when you've finished to make sure you've considered every vertex.

Example

The network shows the quickest times that an athlete can complete various sections of a race. Use Dijkstra's algorithm to find the shortest path from O to F and fastest total time. If at the end of each intermediate section, there is a 20-second delay, how does this affect the solution?

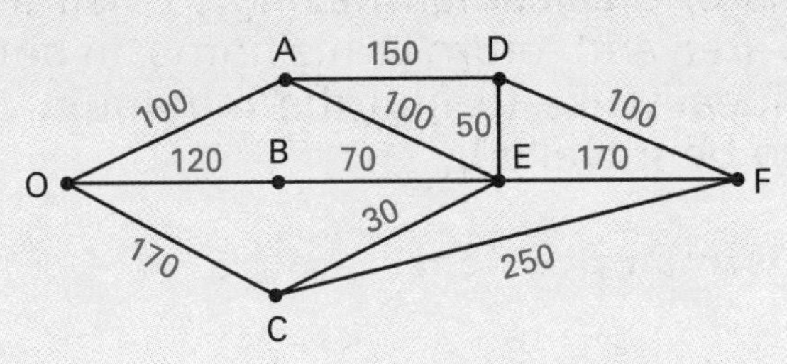

If you trace the algorithm, you will get this solution. Note the reducing of the temporary labels at D, E and F.

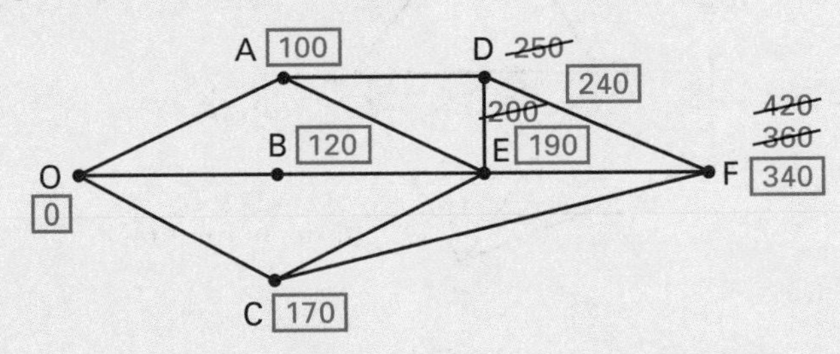

The path (working backwards) is O B E D F. The time is 340 seconds.

If you add 20 seconds at the end of each intermediate section then this route takes 340 + 60 = 400 seconds. Is there a quicker route?

Trace the algorithm again, adding 20 seconds each 'turn'.

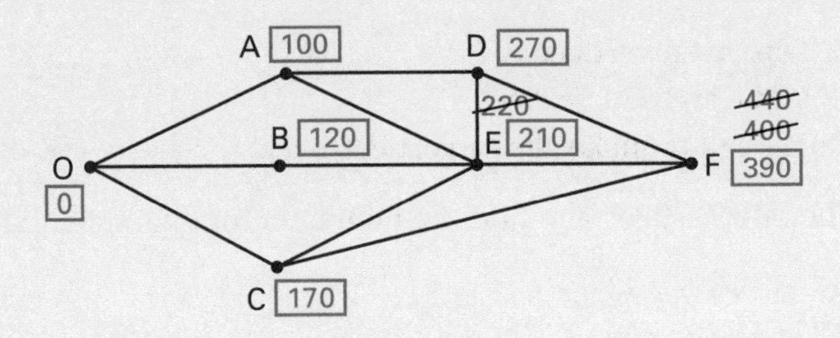

The path is O A D F with a time of 390 seconds.
(When tracing you should have found a value at D of 280 when going from E, when the temporary label was 270. This is quite common, just ignore it.)

Action point

Work through the solution yourself, as you would in the exam. Make sure you are clear about temporary and permanent labels.

Examiner's secrets

The values, which have a line through because a smaller value was found, have still been left visible. Don't erase these crossed-out numbers as they all form part of the working.

Examiner's secrets

You will only gain full marks if it is clear that you have used the algorithm correctly.

Exam question answer: page 108

Use Dijkstra's algorithm to find the shortest distance from K to P in this network. State the path and the distance. (10 min)

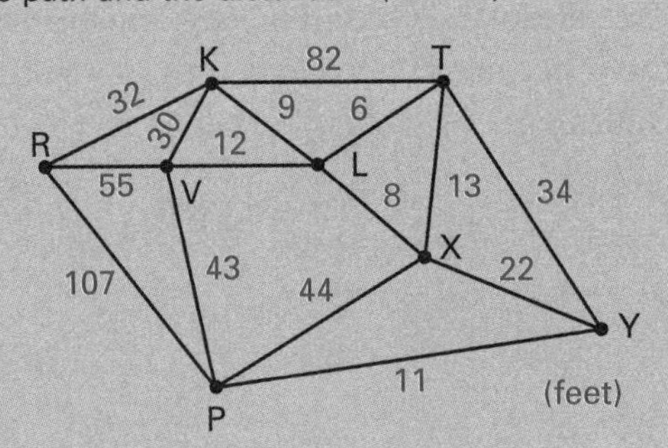

Two classic problems

There are two classic problems that you will be asked to solve. One is to cover *all* the *edges* in a network. The most common application is a postman having to deliver to all roads along a route. The other problem is to visit *all* the *vertices* in a network. The most common application is of a salesman having to visit all of the shops in his area and return to his home. In both cases you would ideally like to find the minimum distance that needs to be travelled.

The jargon

The Chinese postman problem is sometimes called the route inspection problem. With this problem you need to cover *all* the edges.

Links

Look back at 'Degrees and traversability', page 97.

Links

Remind yourself about Eulerian trails (page 97).

Watch out!

The shortest distance will not always be direct.

Checkpoint 1

Write out some other routes when the same extra edges are added.

Chinese postman problem

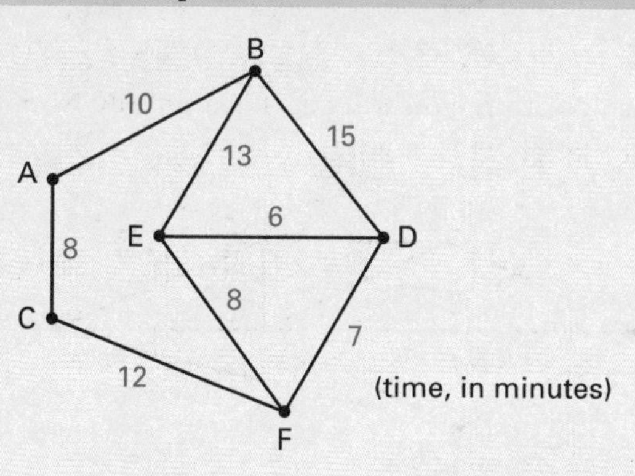

In this network, the depot is at D. The postman needs to walk along every edge, ideally once, and return to D. If this is not possible, extra edges will need to be added. You need to find the minimum of these extra edges.

You need to look at the degree of each vertex:

- A and C are even vertices.
- B, D, E and F are odd vertices.
- You then need to make all the vertices even.

Pair off the odd vertices and find the minimum total for each pair.

BD + EF = 15 + 8 = 23
BE + DF = 13 + 7 = 20
BF + DE = (BE + EF) + DE = (13 + 8) + 6 = 27

So, the extra edges should be BE and DF. The total time taken will be

10 + 8 + 12 + 8 + 13 + 15 + 7 + 6 + 20 = 79 + 20 = 99 minutes.

A suitable route *could* be D B A C F E B E D F D.

Travelling salesman problem

In this network the salesman is based in, say, Amsterdam. He needs to visit each of the other five cities and return to Amsterdam. (All distances are in miles.)

	Amsterdam	*Athens*	*Barcelona*	*Berlin*	*Brussels*	*Cologne*
Amsterdam	–	1 340	770	365	105	128
Athens	1 340	–	1 160	1 112	1 292	1 200
Barcelona	770	1 160	–	925	658	692
Berlin	365	1 112	925	–	401	300
Brussels	105	1 292	658	401	–	110
Cologne	128	1 200	692	300	110	–

The method used is called the *nearest neighbour* algorithm. It is a *greedy* algorithm because you always take the easiest ('greediest') route at each stage, which does not necessarily give the shortest route.

1 From Amsterdam, the nearest neighbour is 105 miles, Brussels.
2 From Brussels, the nearest neighbour is 110 miles, Cologne.
3 From Cologne, the nearest neighbour is 128 miles, *but* this is Amsterdam, already visited, so the next nearest is 300 miles, Berlin.
4 From Berlin, the nearest neighbour not yet visited is 925 miles, Barcelona.
5 From Barcelona you must visit Athens before returning to Amsterdam.

The distance travelled is 105 + 110 + 300 + 925 + 1 160 + 1 340 = 3 940 miles. This is an *upper bound* which means that the minimum distance will not be greater than this, and will probably be shorter.

Watch out!

Some tables are not symmetrical, so look for 'from/to' appearing at the corner of the table.

Examiner's secrets

You must show some working or explanation within your answer, particularly where you had to make a choice.

Checkpoint 2

Trace the algorithm starting at Athens. Is the distance reduced?

Checkpoint 3

Which starting point would give the minimum distance using this algorithm?

Exam questions

answers: page 108

1 A council worker has to paint the lines on a sports pitch. The lengths are in metres.

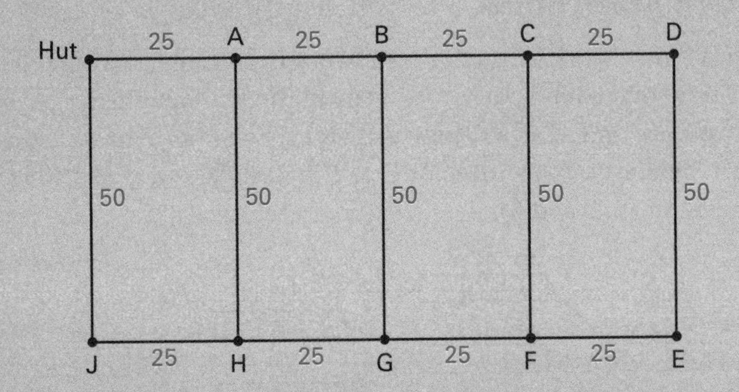

Find the shortest distance that he needs to walk after starting at the hut. State a route that he could take. (12 min)

2 The distances between six cities, in miles, are given.

	Bradford	*Carlisle*	*Leeds*	*Manchester*	*Newcastle*	*Sheffield*
Bradford	–	110	9	34	97	37
Carlisle	110	–	119	119	59	152
Leeds	9	119	–	40	92	33
Manchester	34	119	40	–	132	38
Newcastle	97	59	92	132	–	125
Sheffield	37	152	33	38	125	–

Use the nearest neighbour algorithm starting at Bradford to find a travelling salesman's route. State the distance. (12 min)

Critical path analysis

In business and production it is often necessary to schedule activities so that time and resources are not wasted.

Activities

This table gives a list of tasks that need completing, together with the time that each takes and any restrictions on their order.

Activity	*Time (days)*	*Predecessors*
A	3	–
B	2	A
C	4	A
D	3	A
E	3	B, C
F	2	C, D
G	1	D, E, F

The jargon

The activity network is often called the *precedence diagram*. Check what your specification uses.

First, you have to construct a clearly labelled activity network, with boxes ready to use in the next stage. One method of labelling is shown. In the centre box you should write the time for each activity.

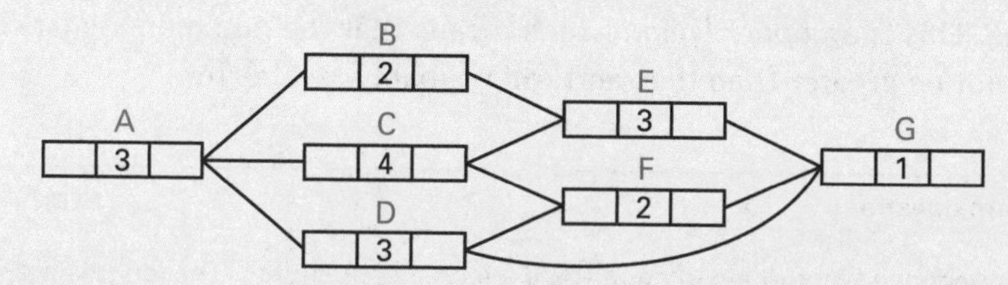

Examiner's secrets

You must label the diagram clearly. There are many different ways to do this, *but make sure that yours is clear and complete.*

Earliest start times

The jargon

This is also called a *forward pass* through the network.

You should now work through the network from A to G, writing the earliest time that each activity can start in the left-hand boxes. A, B, C and D are easy. At E, 7 is written, as B and C must both finish before E can start. Similarly, 7 is written at F, as it follows C and D and 10 at G (the largest of 10, 9 and 6).

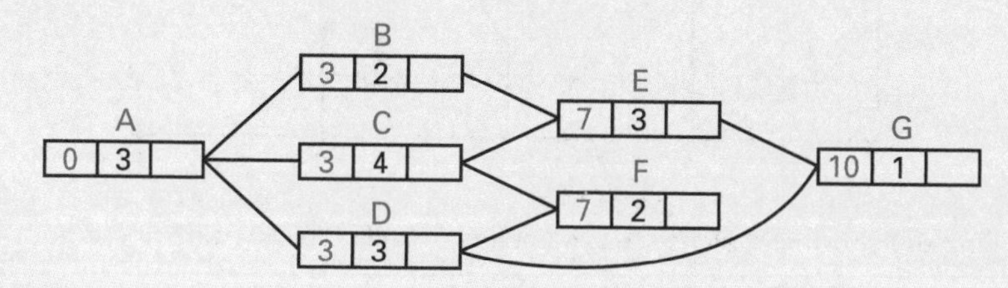

Latest finish times

The jargon

This is also called a *backward pass* or *reverse pass* through the network.

You now work backwards from G to A, writing the latest finish times in the right-hand boxes. The first decision is again at C, the choices being 10 – 3 = 7 or 10 – 2 = 8. The smaller answer 7 is written. Similarly 8 is written at D and 3 at A. The activity network is now complete.

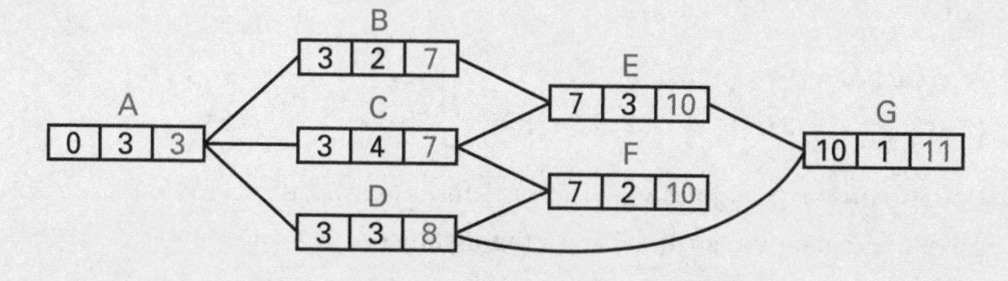

Critical path

You will see that, with some boxes the first two numbers add to make the third. These are *critical activities* and lie on a *critical path*. In this case there is only one critical path, which is A C E G. The minimum time is 11 days.

Float

In all the other boxes, the difference between the sum of the first two numbers and the third is called the *float time*, i.e. B = 2, D = 2, F = 1. This is the time by which the activity could be moved, or allowed to take longer, without affecting the 11-day completion time.

Resourcing

A *cascade chart*, or *Gantt chart*, is a bar chart where each activity is represented by a bar showing when it is scheduled. This can be used to construct a *resource histogram* when information is available about the number of people available at any time. The float can then be utilised in rescheduling the times of the activities.

In the case looked at above, suppose that all activities need two workers, apart from B, which needs three, and there is an extra cost if more than five are needed on any day. The diagrams show that D should move to start on day 4 so that only one day needs more than five workers.

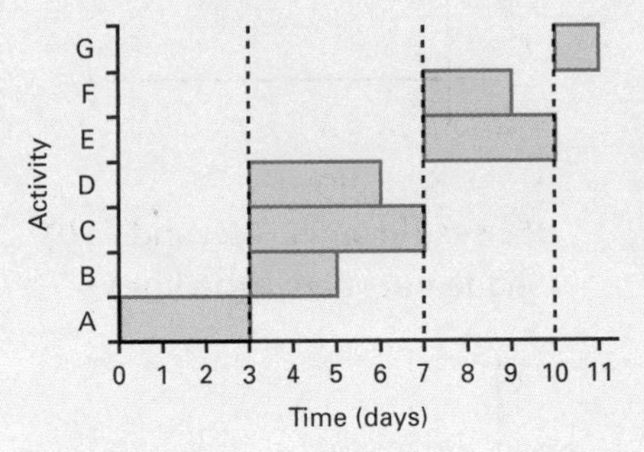

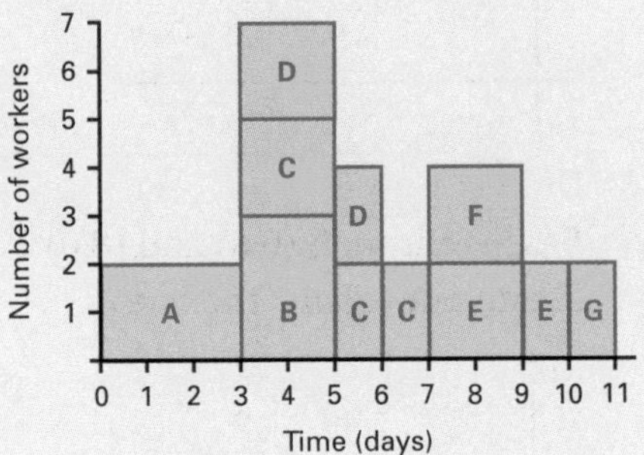

Action point

Make a copy of the table at the top of the previous page. Close this book now and do: (a) the activity network; (b) fill in the earliest start times and the latest finish times; (c) write down the critical path and the minimum time; (d) open the book and make sure you got them all correct.

Examiner's secrets

You will often have to choose which of the activities with possible float to use, as they may be on the same path.

Examiner's secrets

Note the dotted vertical lines on the cascade chart and the labelling of the letters on the resource histogram.

Exam question answer: page 109

The table shows a list of activities, the time that each takes and any immediate predecessors.

Activity	A	B	C	D	E	F	G	H	I
Days	6	16	15	24	9	13	10	7	6
Predecessors	–	A	A	–	D	B, C, E	F	E	G, H

(a) Construct an activity network. Find the critical path and state the time.

(b) Construct a cascade diagram and state the activities that could begin later without increasing the total time.

(c) Draw a resource histogram, given that each activity needs two people. If one person has to take three days off, when is the best time for this to happen? (35 min)

Linear programming

Links

See inequalities, page 16.

Examiner's secrets

It is better to shade out as this leaves the feasible region free for you to work out the solution more easily.

Watch out!

< and > signs mean that your lines should be broken (or dotted), otherwise use a solid line.

Checkpoint 1

How many integer solutions are there if the line had been $x + y = 4$?

Examiner's secrets

It is often better to draw this line on the graph so that the examiner can see that you know what to do.

This topic follows from work on inequalities at GCSE.

What to do

You will usually be given a series of statements which have to be translated into inequalities. A manufacturing context is very common. You will then have to show each of the inequalities on a graph. When you have shaded the correct sides of the lines, a feasible region will appear. The answers to the problems you are asked to solve will be found in this region, often on the boundaries at the vertices. You will often be asked for an optimal solution, or to find a maximum profit. Drawing a further, general line is needed at this stage.

Example

The region R is bounded by the inequalities $x \geq 0$, $x \leq 7$, $y \leq 5$, $x + y \leq 8$, $2x + 5y \geq 10$. Show the inequalities on a graph and state the coordinates of the points within R such that $x + y = 3$, where x and y are integers.

1 Draw $x = 0$ and $x = 7$
Shade out (to the left of $x = 0$ and to the right of $x = 7$)

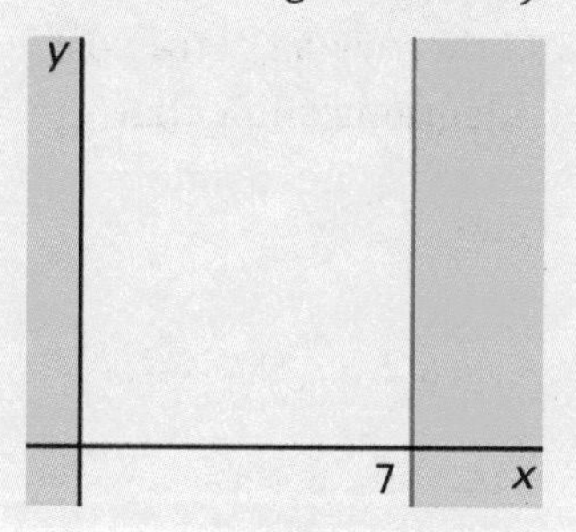

2 Draw $y = 5$
Shade out (above $y = 5$)

y
5
x

3 Draw $x + y = 8$ ($y = 8 - x$)
Passes through (0,8) and (8,0)
Shade out (*above* the line)

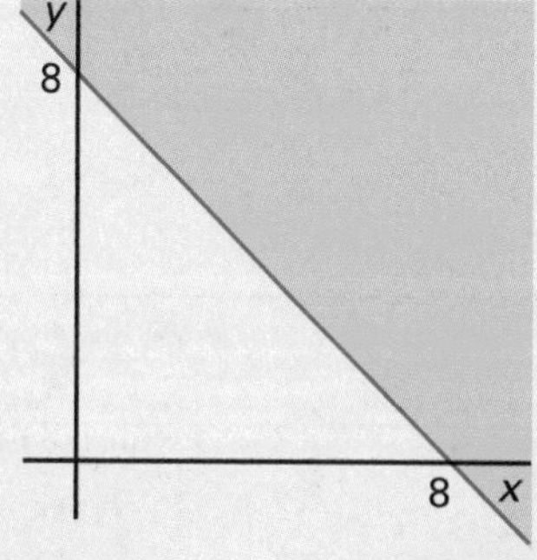

4 Draw $2x + 5y = 10$
Passes through (0,2) and (5,0)
Shade out (*below* the line)

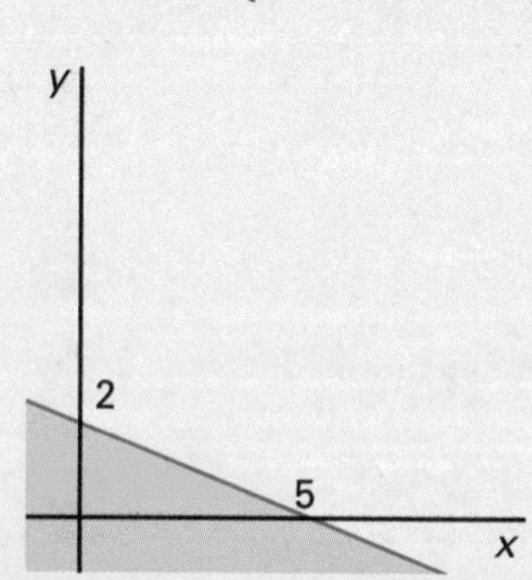

5 Label R.

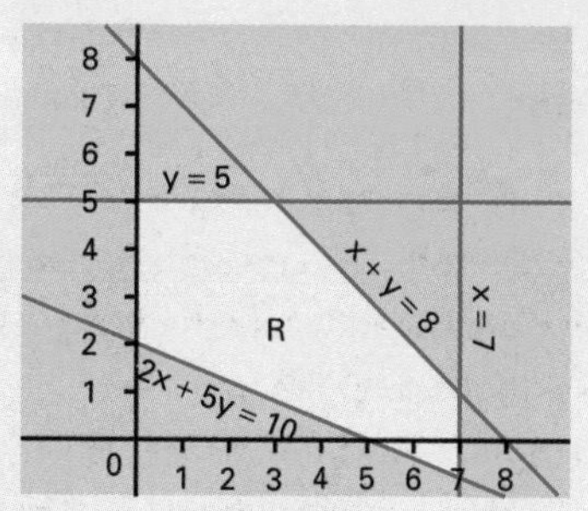

6 Look for points where $x + y = 3$. These are (0,3) and (1,2).

Example

The annual subscription for a sports club is £60 for adults and £15 for juniors. The club needs to raise £2 100 from subscriptions to cover its costs. The number of members is to be limited to 60. There must be as many adult members as juniors but not more than twice as many.

Represent this situation graphically. Find the number of adult and junior members which will raise the most money and the largest total membership which will just cover the costs.

You first need to set up some inequalities.

1 Define the variables. Let A be the number of adult members. Let J be the number of junior members.
2 Total of subscriptions is $60A + 15J$. This must be at least £2 100. $\therefore 60A + 15J \geq 2\ 100$ which simplifies to $4A + J \geq 140$.
3 Number of members is limited to 60. $\therefore A + J \leq 60$.
4 Limit on adult and junior members. $\therefore J \leq A \leq 2J$.
5 Draw a graph showing these inequalities.

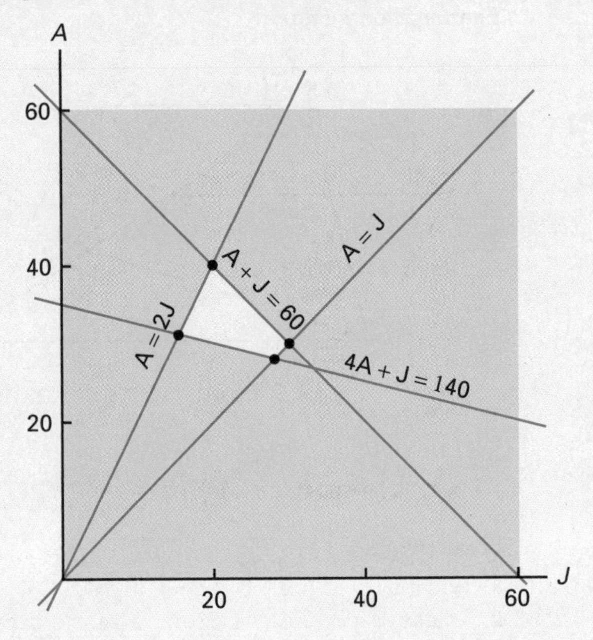

(20,40) gives $20 \times 15 + 40 \times 60 =$ £2 700 which is the most money. Look at the other vertices: (30,30), (28,28) and (15.5,31). The last is not possible, so look at the nearest point inside R. (16,31) gives 47 members.

Examiner's secrets

It is easier to simplify before you draw the graph.

Examiner's secrets

This can be written as two separate inequalities.

Watch out!

Check that your inequality signs are the correct was round and that they will give the correct type of line on the graph.

Checkpoint 2

If there were 20 junior members, how many adults could there be and what is the range of the amounts of money that would be raised?

Exam question answer: page 109

A company has a choice of two types of machine. Type J costs £50 a day to hire, needs one person to operate it and makes 30 items a day. Type K costs £20 per day to hire, needs four people and makes 70 items a day. The owner can spend up to £1 000 a day, has 64 people and has the space for a maximum of 25 machines.

(a) Set up some inequalities and show them graphically.
(b) Find the maximum number of items that can be manufactured each day.

(25 min)

Answers
Discrete mathematics

Graph theory

Checkpoints

1

	F	G	H	J	L
F	0	1	1	0	0
G	1	0	0	1	0
H	1	0	0	1	1
J	0	1	1	0	2
L	0	0	1	2	0

2

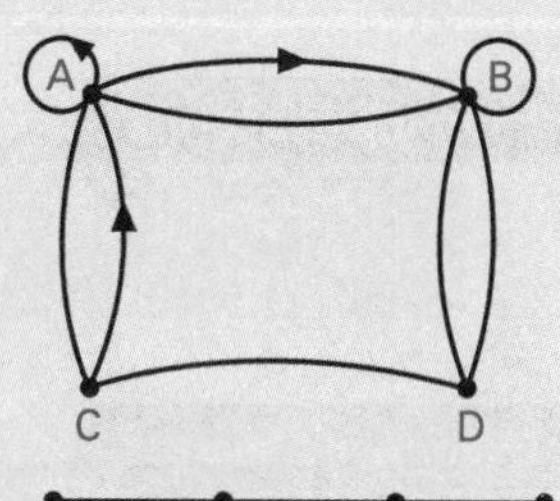

3

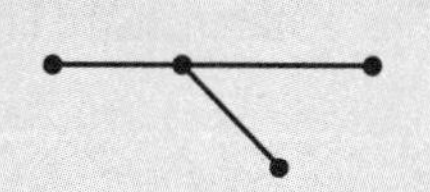

Examination questions

1

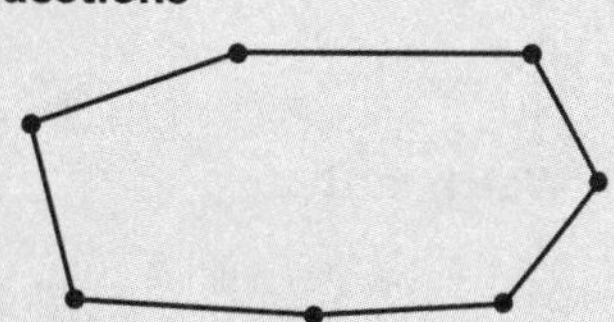

2 (a) 2, 4, 6
(b) 7, 14, 21

3 AB, ACB, ADB, ACDB, ADCB.

4 $1 + 3 + 6 + 6 = 16$

Spanning trees

Checkpoints

1

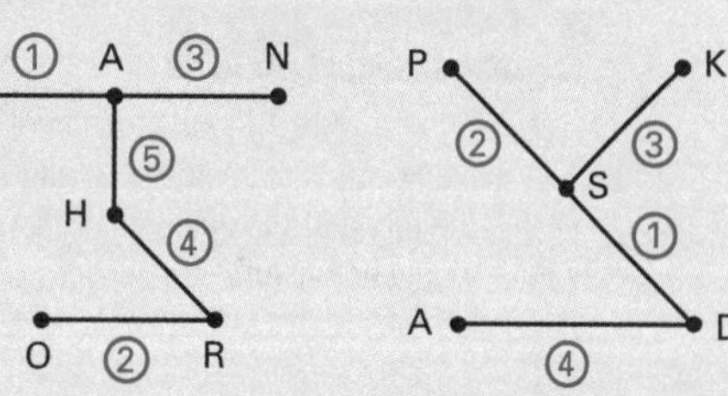

The circled numbers show the order in which the edges were added.

2

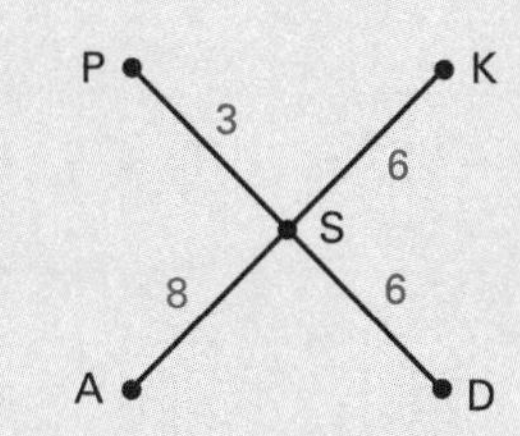

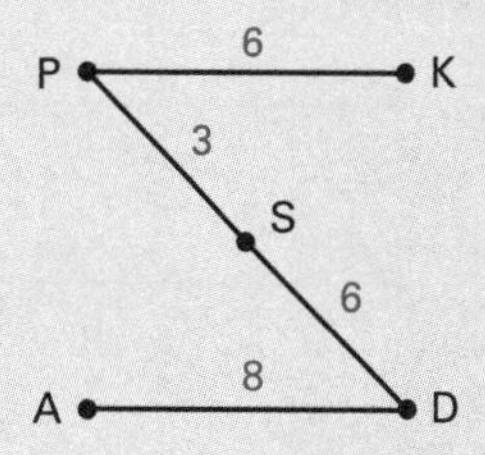

Both 23 miles.

Examination questions

1

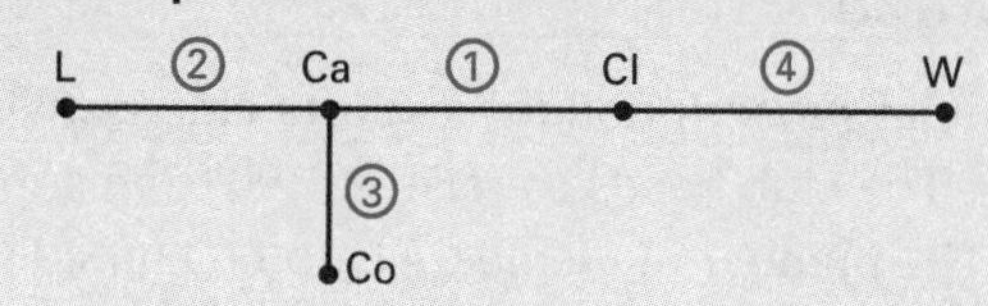

The order is shown by the circled numbers. Note that the 62 from Limerick to Cork is not used as this would complete a cycle.

2

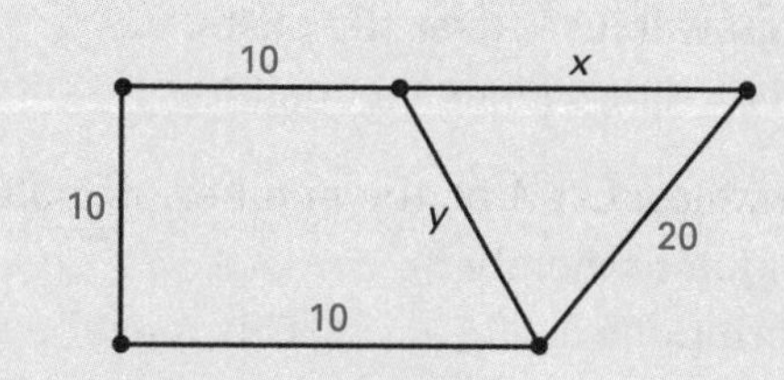

This is one possibility. Here x and y can take any values.

Shortest paths

Examination question

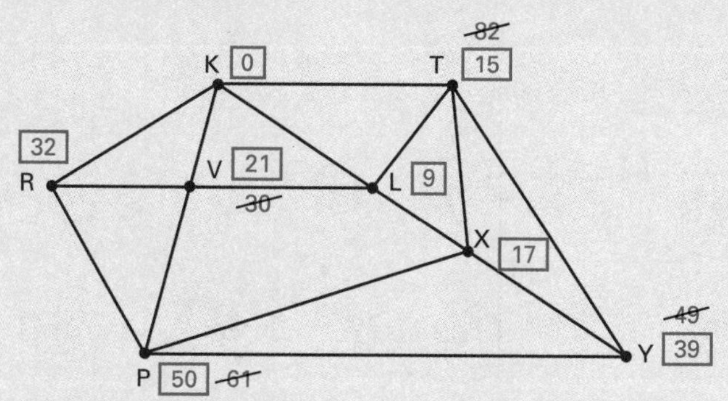

The shortest path is K L X Y P. Distance is 50 feet.

Two classic problems

Checkpoints

1 Two are D F C A B E D B E F D and D E B D F E B A C F D.

2 Athens–Berlin–Cologne–Brussels–Amsterdam–Barcelona–Athens. 3 557 miles.

3 Barcelona 3 463 miles

Examination questions

1 $8 \times 25 + 5 \times 50 = 450$ for the lines. Extra edges are $25 + 25 + 50 = 100$. $\therefore 450 + 100 = 550$ metres.
Hut–A B C D E F G B C F G H A H J–Hut

2 Bradford–Leeds–Sheffield–Manchester–Carlisle–Newcastle–Bradford. 355 miles.

Critical path analysis

Examination question

(a)

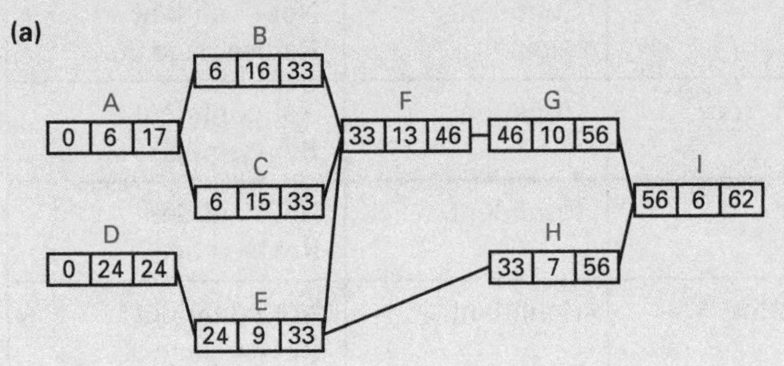

D E F G I; 62 days

(b)

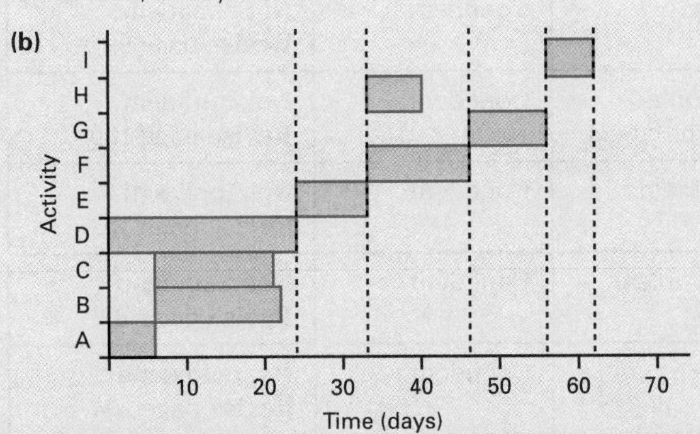

A, B, C, H

(c)

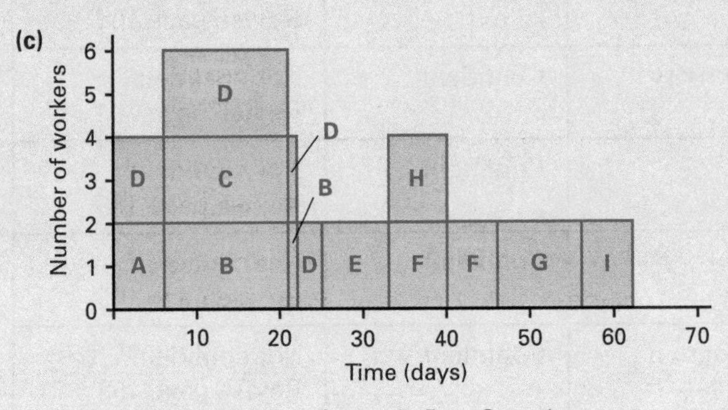

Between days 21 to 25 for an A, B or C worker.
Between days 40 to 46 for an H worker.

Linear programming

Checkpoints

1 4

2 30 to 40; £2 100 to £2 700

Examination question

(a) $J + K \leq 25$

$J + 4K \leq 64$

$50J + 20K \leq 1\,000$ (same as $5J + 2K \leq 100$)

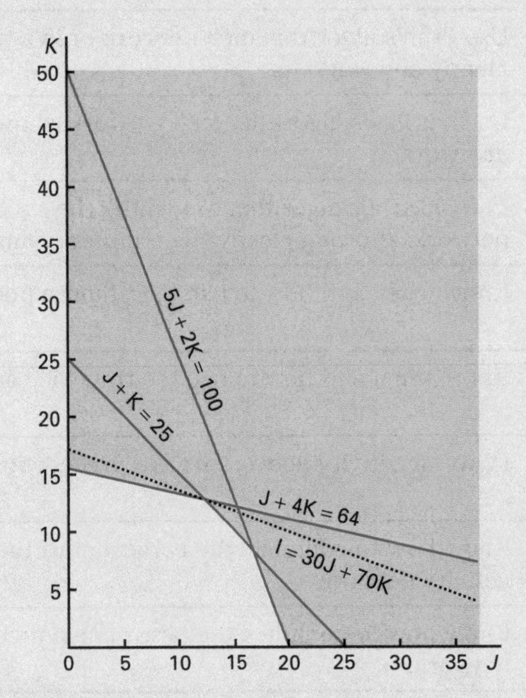

(b) 1 270 (note use of dotted line $I = 30J + 70K$)

Revision checklist
Discrete mathematics

By the end of this chapter you should be able to:

1	Distinguish between basic and directed graphs.	Confident	Not confident. **Revise** page 96
2	Understand the difference between paths, cycles and trees.	Confident	Not confident. **Revise** page 96
3	Know the difference between an Eulerian and semi-Eulerian trail.	Confident	Not confident. **Revise** page 97
4	Use Prim's algorithm on a network or in a table, showing clearly any working.	Confident	Not confident. **Revise** page 98
5	Use Kruskal's algorithm on a network, showing clearly any working.	Confident	Not confident. **Revise** page 99
6	Use Dijkstra's algorithm to identify the shortest path on a network, showing clearly the temporary and permanent labels.	Confident	Not confident. **Revise** page 100
7	Know when and how to use the Chinese postman problem.	Confident	Not confident. **Revise** page 102
8	Know when and how to use the travelling salesman problem.	Confident	Not confident. **Revise** page 102
9	Draw an activity network from a precedence table.	Confident	Not confident. **Revise** page 104
10	Know how to complete the earliest start times in an activity network.	Confident	Not confident. **Revise** page 104
11	Know how to complete the latest finish times in an activity network.	Confident	Not confident. **Revise** page 104
12	Work out the critical path from an activity network.	Confident	Not confident. **Revise** page 105
13	Calculate the float time on activities.	Confident	Not confident. **Revise** page 105
14	Draw and interpret a Gantt chart and a resource histogram.	Confident	Not confident. **Revise** page 105
15	Convert statements into inequalities, and draw these inequalities onto graphs.	Confident	Not confident. **Revise** page 106

Statistics

If you are thinking of going on to study topics like marketing, demography, sociology, psychology, biology, and many more besides, then the statistics you use at A level are going to be a big help. Without statistical methods we would not be able to do things like describe a set of data in terms of centre and spread, or to measure the strength or direction of a relationship between two variables. All this and more is looked at in this section. The table below shows what has been included, and roughly where in each specification it falls. As a part of the Applied Maths specification, statistics involves many skills from pure maths. In particular, you will need to be able to manipulate and use equations and formulae, differentiate and integrate functions of a single variable (usually x), and to sketch graphs. Do take time to check with your specification, however, as this section may cover things you don't need. Some of the earlier material will not be tested directly, but may be needed to draw inferences from data sets.

Exam themes

- Data description – the use of averages and measures of spread as well as diagrams
- Probability – applying concepts to situations involving uncertainty, or counting problems
- Modelling – finding and using appropriate theoretical models; regression and correlation
- Estimation – making inferences about a population, given sample values

Topic checklist

	AQA	CCEA	EDEXCEL	OCR/A	OCR/MEI	WJEC
The mode	S1	S1	S1	S1	S1	
The median	S1	S1	S1	S1	S1	
Cumulative frequency	S1	S1	S1	S1	S1	
The mean and standard deviation	S1	S1	S1	S1	S1	
Statistical diagrams	S1	S1	S1	S1	S1	
Probability	S1	S1	S1	S1	S1	S1
Probability trees	S1	S1	S1	S1	S1	S1
Permutations and combinations				S1	S1	S1
Discrete random variables	S2	S1	S1	S1	S1	S1
The binomial distribution	S1	S1	S2	S1	S1	S1
The Poisson distribution	S2/S3	S1	S2	S2	S2	S1
Continuous random variables	S2	S1	S2	S2/S3	S3	S1
The normal distribution	S1	S1	S1	S2	S2	S2
Normal approximations	S3		S2	S2	S2	S2
Hypothesis testing	S2/S3	S2	S2/S3	S2	S2/S3	S2
Sampling, estimates and confidence	S1	S2	S3	S2	S3	S2/S3
Independent normal distributions	S3	S2	S3	S3	S3	S2
Goodness of fit tests	S4		S3	S3	S3	
Correlation	S1	S2	S1	S1	S2	
Regression	S1	S2	S1	S1	S2	S3
Rank correlation			S3	S1	S2	
Testing correlation coefficients	S3	S2	S3		S2	

The mode

The mode – or modal value – is simply the most frequently occurring item of data. Where information is given in the form of a grouped frequency table, we can only give the modal class. To estimate a mode from grouped data, use either a histogram or cumulative frequency curve but – remember – this is only an estimate and as such should be treated very carefully.

The jargon

Frequency tables are essentially tally charts. *Grouped frequency tables* put data into categories, like 5 to 9 minutes, 10 to 14 minutes, and so on, for convenience.

Finding the mode

To find the mode from raw data you need to make a **frequency table** and look for the highest frequency. There may be more than one mode, or in fact no mode at all. If data are in the form of a **grouped frequency** table, then the modal class should be quoted.

Example: The following frequency table gives the hours spent watching TV per week by a year 12 class.

Time (hours)	$0 \le t < 1$	$1 \le t < 2$	$2 \le t < 3$	$3 \le t < 4$
Frequency	2	4	8	3

The modal class for number of hours of TV watched is 2 to 3 hours, with frequency 8.

Action point

Make your own table to remember the advantages and disadvantages of the various averages as you go.

Advantages of the mode

Example: A shop sells shoes in a town where the mean shoe size is 9.78, but they sell more size 8 shoes than any other. Therefore 8 is the modal size. Advantages:

- The mode is not affected by extreme values: people with size 13 feet are possible, but in this case not profitable.
- It will always take a value that appears in the data set: 9.78 is not a very useful number for the shopkeeper.

The jargon

Cumulative frequency is just a running total of the individual frequencies.

Graphical estimation of the mode

Estimation of the mode from the cumulative frequency curve

Here we see a cumulative frequency curve for data on the amount of time spent watching TV per week by a year 12 class.

The mode is the item or class with the highest frequency. Therefore, the cumulative frequency will make the biggest jump around the mode. This will correspond with the steepest part of the cumulative frequency curve and will be a point of inflexion. Draw a tangent at the steepest part, and read the mode from the x-axis. Here, our estimate for the modal length of time is 2.6 hours.

Links

See 'Cumulative frequency', page 116.

Checkpoint 1

Draw a cumulative frequency curve for this data, which shows the marks of 120 students in an exam, and use it to estimate the mode.

Marks	31–40	41–50	51–60	61–70
Frequency	15	35	50	20

Estimation of the mode from a histogram

Example: Here is a histogram to represent the same data on the amount of time spent watching TV per week by a year 12 class.

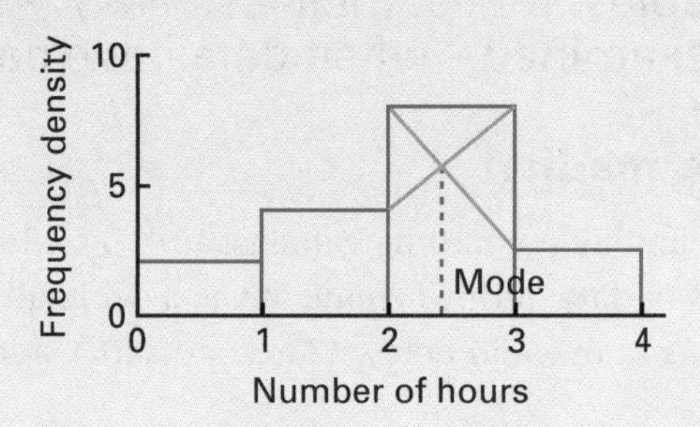

This way of estimating the mode uses the idea that the original distribution can be superimposed on the histogram. We find the tallest bar and join its corners to the adjacent bars to get a rough idea of its shape. The mode is read off of the horizontal axis at 2.4 hours.

Important point: This technique works best when we have equal class widths. With histograms of unequal class widths, we can really only give the modal class, which will be the 'bar' containing the largest area.

Checkpoint 2

Find an estimate of the mode of the data in checkpoint 1 by drawing a histogram.

Classifying distributions using the mode

In frequency distribution diagrams, the mode corresponds to the hump in the curve. Whether the hump is to the left or right tells us whether we have a negatively – or positively – skewed distribution.

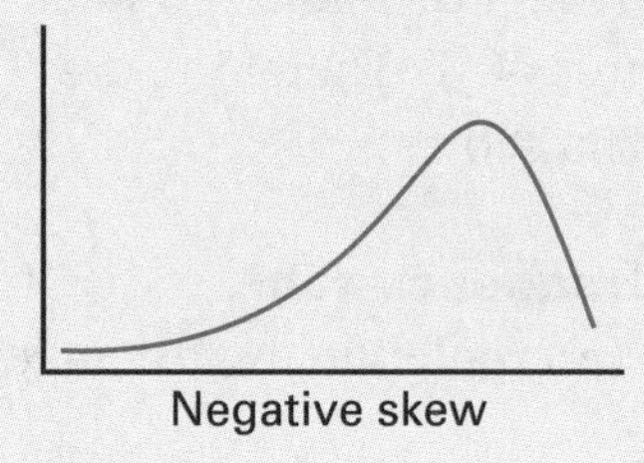
Negative skew

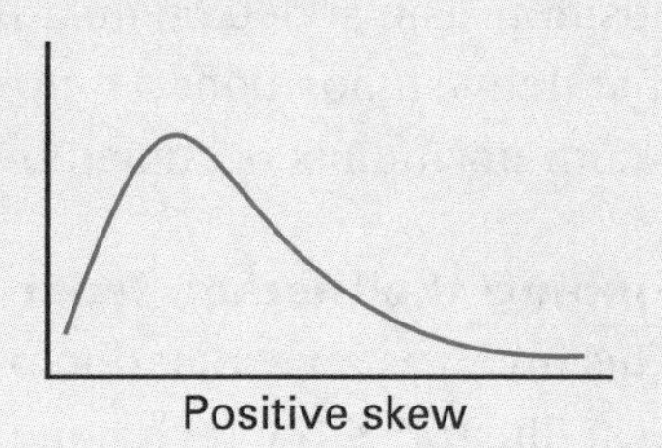
Positive skew

Distributions can also be bimodal, i.e. have two humps. The modes do not necessarily have to be equal. They may simply have no mode, in which case they are known as '**uniform distributions**'.

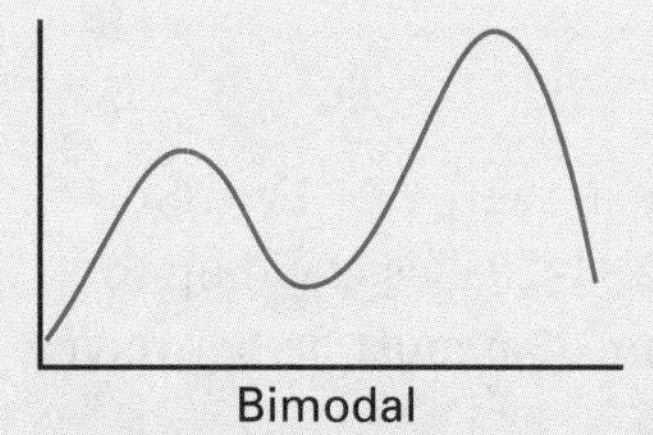
Bimodal

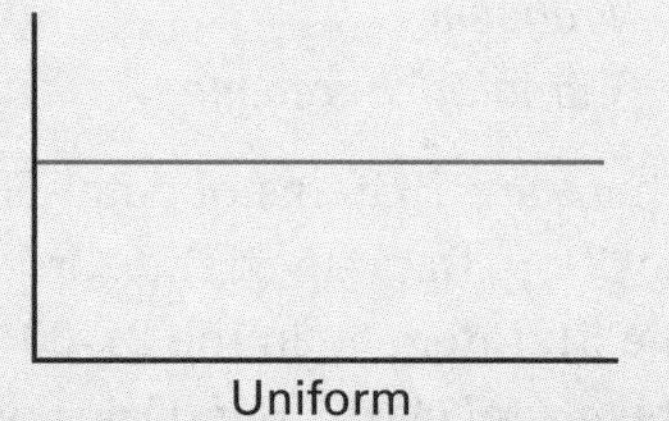
Uniform

The jargon

Skew is another term we can use to describe the shape of a distribution without resorting to diagrams. For example, saying that a distribution has negative skew will tell the reader that most of the data lie to the left of (i.e. below) the mode.

Test yourself

Practise drawing and labelling these four different diagrams.

Action point

For each type of skew, try to think of two example distributions that would fit: for example, for positive skew you could be looking at the marks achieved by students sitting a very hard test.

Exam question answer: page 156

The table below shows the number of women in various age groups who own a computer that is connected to the internet in the village of Milligurough. Find an estimate of the mode (a) by drawing a cumulative frequency curve, and (b) by drawing a histogram. (25 min)

Age of women (years)	9–15	16–24	25–35	36–49	50–80
Frequency	12	23	22	8	2

The median

The median of a set of data is given by the middle value, when those data are arranged in order of size. This is easy with raw data or when using ungrouped frequency tables. It becomes less easy – and so more likely to be examined – when data have been grouped.

Examiner's secrets

The most common mistake made with medians is that candidates forget to put their data in order first.

Finding the median

The formula for finding the middle value is: $\frac{1}{2}(n + 1)$, where n is the number of items of data. If this formula gives a decimal, like 23.5, then you find the items above and below (23rd and 24th), add them together and divide by 2.

Example: The median of the numbers 2, 4, 6, 8, is $(4 + 6) \div 2 = 5$.

Checkpoint 1

Find the median of these numbers: 1, 5, 7, 4, 2. Try to guess what the median would be if you doubled all the numbers. What about if you added 6 to them all?

Finding the median from an ungrouped frequency table

You will need to copy part of the table, in order to add a cumulative frequency row. This tells us which position is occupied by the first item in each category. We then work out where our item lies.

Example: The scores on a five-sided spinner are given in the following frequency table. Find the median score.

Score	1	2	3	4	5
Frequency	4	6	7	3	4
Cumulative frequency	4	10	17	20	24

There were 24 spins, so the median spin occupies $\frac{1}{2}(24 + 1) = 12.5$th position. The added cumulative frequency row of the table shows us that items in positions 11, up to 17, are scores of '3'. Therefore, the 12.5th position is occupied by a 3, the median score.

Examiner's secrets

Quickly copying out the table with a cumulative frequency row makes calculations so much easier. The versions shown here are a guide to how *your answer* would look.

Finding the median from a grouped frequency table

This involves an estimation process called **interpolation**. An example best illustrates the technique.

The jargon

Interpolation is what you do when you estimate a value between two others. For example, suppose a small jar of sweets contains 50 sweets, and a large jar contains 100. We interpolate roughly between these two values to get the medium jar containing 75 sweets. As you will see, the proper method is much more accurate!

Example: Calculate an estimate of the median weight, shown to the nearest gram, in the following grouped frequency table.

Weight (g)	1–10	11–20	21–30	31–40	41–50
Frequency	10	13	28	15	9
Cumulative frequency	10	23	51	66	75

There are 75 items of data. The median lies in the $\frac{1}{2}(75 + 1) = 38$th position. Data up to the 23rd item lie in the 11–20 class, and up to the 51st item lie in the 21–30 class. So the median must lie between the *actual* class boundaries of 20.5 and 30.5.

The following diagram is very useful under examination conditions:

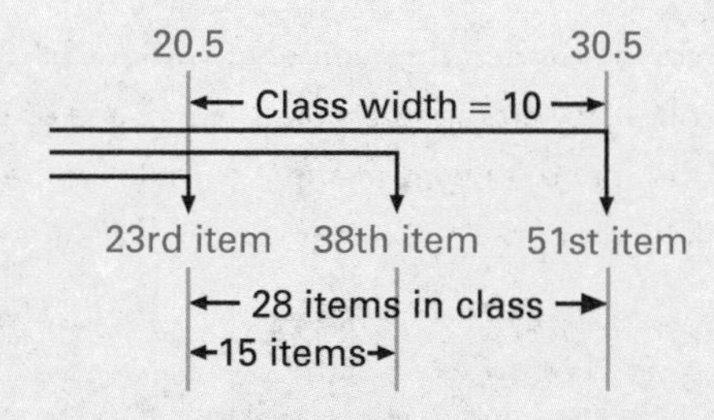

The assumption that data are evenly spread over each class interval is very important. It means that we need fifteen twenty-eighths of the interval, which is ten units wide. This must start from the actual lower class boundary, 20.5. We get the following formula and calculation:

$$\text{lower class boundary} + \frac{\text{number of items up to median}}{\text{number of items in the class}} \times \text{class width}$$

This becomes: $20.5 + (\frac{15}{28})10 = 25.857\ 14\ldots = 25.9$ (1 d.p.).

Estimating the median from a histogram

This is a graphical version of interpolation. Here is a histogram for the data on weights.

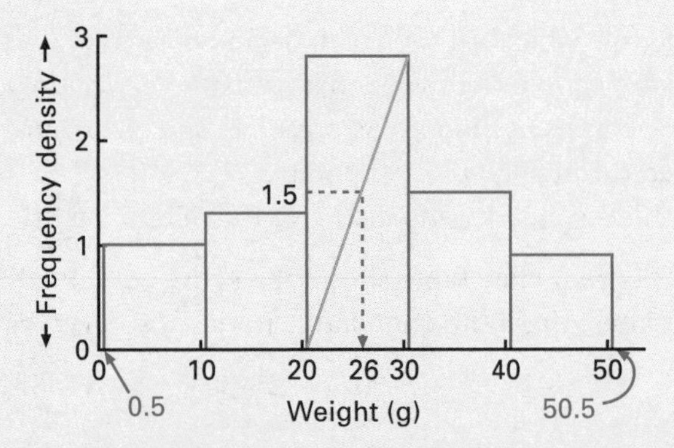

The median divides the area into two halves. We need the 15th item in the third bar to get exactly 38 items on each side, and so go to a height of $\frac{15}{10} = 1.5$. (We divide by the classwidth to give our answer in frequency density terms.) We go to the diagonal and down, and read the median off the x-axis, at 26 g.

Advantages of the median

Example: The salaries paid to workers at a factory are: £8 000, £8 500, £9 000, £13 000, £25 000. The median salary is £9 000.

- The median is not affected by extreme values, i.e. the boss's high salary, more accurately reflecting actual wages paid.
- As long as data are in order of size, or in a table, calculations involve very few numbers, cutting down time.

The jargon

All *class boundaries* used are the *actual class boundaries*, in this case 20.5, since this is the smallest value that would be rounded up to 21 – the bottom of the interval. The class boundaries then become 20.5 to 30.5, giving a class width of 10.

Checkpoint 2

Find the median of these data by drawing a histogram.

Weight (kg)	20–24	25–29	30–34	35–39	40–44
Frequency	8	15	21	17	2

Examiner's secrets

When you are using very long numbers, especially in the middle of calculations, then you only need to write the first three or four digits down as long as you follow with the '. . .'. This shows that you are taking the right steps with the correct numbers, without the need to write down a ten-digit display every time!

Action point

Turn over the page and try to make a list of the advantages of the median and the mode. What about the mean?

Exam question

answer: page 156

The times taken for 55 pupils to eat their lunch, to the nearest minute, are given below. Work out the median time taken (a) by drawing the histogram, (b) by calculation. Which method do you consider to be the most accurate, and why? (20 min)

Time (min)	3–4	5–9	10–19	20–29	30–44	45–60
Frequency	2	7	16	21	9	0

Cumulative frequency

Cumulative frequency is, in effect, a running total of all the frequencies up to that point. Cumulative frequency curves, or 'ogives' as they are sometimes called, are a graphical technique which allow estimates of medians, upper and lower quartiles, and percentiles to be carried out easily.

The jargon

Actual class boundaries are tricky; the actual upper class boundary is usually the smallest value a number in the *next class up* could take without being moved down.

Construction of cumulative frequency curves

This is best illustrated by example. However, here are a few important points:

- Only ever plot cumulative frequency against *actual upper class boundaries*.
- Choose a scale which allows a high degree of accuracy – these diagrams are simple to draw, and so must be *perfect* to get full marks.
- Make your curves as smooth as possible, and make sure they actually go through all plotted points.
- Label both axes clearly, and also give the diagram a title.

Example: The frequency table shows the scores obtained by a group of test candidates. Draw the cumulative frequency curve for this data.

Score	1–10	11–20	21–30	31–40	41–50
Frequency	5	17	28	25	15

To draw the curve, we need to make up a table, showing the actual upper class boundary (UCB) and associated cumulative frequency. Notice that the data are discrete, but the actual upper class boundary for the first class must still be 10.5. The table should look like this:

UCB	10.5	20.5	30.5	40.5	50.5
Cumulative frequency	5	5 + 17 = 22	22 + 28 = 50	50 + 25 = 75	75 + 15 = 90

Checkpoint 1

Copy and complete the following table:

Heights (cm)	1–4	5–8	9–12	13–16	17–20
Frequency	2	8	12	9	3
Upper class boundary					
Cumulative frequency					

Be careful about the size of axes. The x-axis will have to show the whole way from 0 to 50.5, with enough detail to be able to accurately plot at 10.5; similarly for the y-axis, from 0 to 90.

Here is a detailed diagram of the first two points plotted:

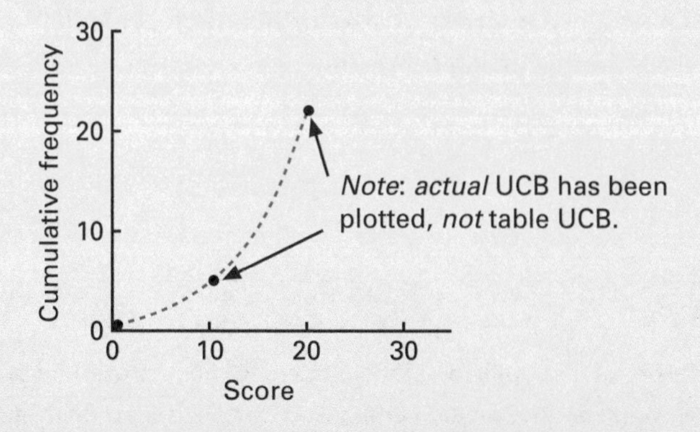

Examiner's secrets

It sometimes pays to copy out the table into your answer booklet, or at least the rows containing any class boundaries and any calculated values. Don't write them on the paper itself – you only hand in your answer sheet.

Checkpoint 2

Draw the cumulative frequency curve from the data in the table in checkpoint 1.

Cumulative frequency curve applications – quartiles and percentiles

The next graph shows the completed cumulative frequency curve for the previous example.

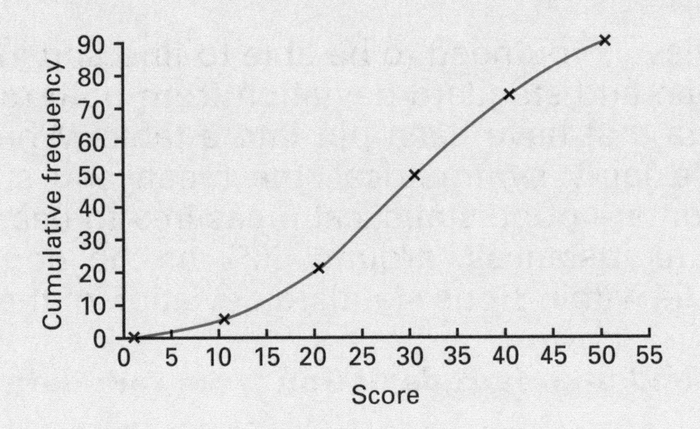

Cumulative frequency curve for test scores

We will use this diagram to illustrate the processes behind obtaining estimates of the median, upper and lower quartiles, and various percentiles, all from a cumulative frequency curve.

The *median* divides the distribution into two, and the *quartiles* divide the distribution into four equal pieces.

Lower quartile or Q_1: $\frac{1}{4}(n + 1)$
Median or Q_2: $\frac{1}{2}(n + 1)$
Upper quartile or Q_3: $\frac{3}{4}(n + 1)$
Interquartile range = $Q_3 - Q_1$.

The *percentiles* divide the distribution into 100 equal pieces.

90th percentile or P_{90}: $\frac{90}{100}(n + 1)$
10th percentile or P_{10}: $\frac{10}{100}(n + 1)$

To calculate the interquartile range (IQR) of the test scores: first find $Q_3 \approx$ 68th value, about 38 from the graph. Then find $Q_1 \approx$ 23rd value, about 21 from the graph. Then, the IQR = $Q_3 - Q_1 = 38 - 21 = 17$ marks. If 30% of all candidates pass, then to calculate the pass mark: 30% of 90 people ≈ 27 people. Go up to 27 people, along to the curve, and down to get ≈ 22.5 marks. To find the percentage of candidates who fail, if the cutoff point for failure is, say, 16 marks: find 16 marks, go up to the curve, and along to ≈ 12 people. The percentage who fail: $\frac{12}{90} \times 100 = 13\%$.

Exam question answer: pages 156–7

In a particular year, the heights of the tallest 250 police recruits are shown in the table below.

Height (cm)	170–	175–	180–	185–	190–	195–200
Number of police recruits	19	36	70	64	39	22

Plot a cumulative frequency curve to represent these data. Use your curve to estimate the median and interquartile range for these data. The shortest 34% of recruits are to form a special squad. Find the median and interquartile range of this squad. (20 min)

Examiner's secrets

The most common mistake with drawing a cumulative frequency curve is forgetting to plot the cumulative frequency against the upper class boundary and a cumulative frequency of zero against the lower class boundary of the first class.

Checkpoint 3

Work out estimates for the median and interquartile range of the data represented by the cumulative frequency curve on page 112.

The jargon

The *interquartile range (IQR)* is a *measure of spread* – it tells us the endpoints of the central 50% of the data. The larger the IQR, the more spread out the data.

Checkpoint 4

How do you think the median and interquartile range are affected by very high or low values? Is this a good, or bad thing?

The mean and standard deviation

At this level, you need to be able to find and interpret the mean and standard deviation from both raw data and data that have been put into a table. When your data are fairly symmetrical, the mean and standard deviation are good statistical measures to use. Under ideal circumstances, around 66% of the population should lie within ± one standard deviation of the mean.

The jargon

Σ, the Greek capital S for sum, is mathematical shorthand. It means 'add up everything that looks like this'. In our case, $\Sigma x = x_1 + x_2 + x_3 + \ldots + x_n$.

Mean and standard deviation from raw data

The mean is easy to find, but first of all we need some notation. The numbers in any set of data are referred to as $x_1, x_2, x_3, \ldots x_n$. We work out the mean, $\bar{x}$, by summing our values of x, and dividing by the number of values, n. Formally, this gives $\bar{x} = \frac{\sum x_i}{n}$.

The standard deviation of any set of numbers gives an idea of how spread out they are. It is the square root of the variance of the set of numbers. We use one of two formulae for the variance, taking the square root of the answer at the end to make calculations easier.

$$S^2 = \frac{\sum (x_i - \bar{x})^2}{n} \quad \text{or} \quad S^2 = \frac{\sum x_i^2}{n} - \bar{x}^2$$

Examiner's secrets

Write out enough working to show how you arrived at your answer, but don't be afraid to use your calculator functions to save time pushing buttons!

Checkpoint 1

The mean of a set of ten numbers was found to be 22. Can you find the new mean when a value of 30 was added?

Using ungrouped frequency tables

The table below shows the scores on a five-sided spinner.

Score	1	2	3	4	5
Frequency	7	8	6	10	9

We can see that there were seven 1s, eight 2s, six 3s, and so on. The sum of the scores is now $\sum f_i x_i$, where f_i is the frequency associated with x_i, i.e. f_1 is 7, f_2 is 8, and so on. The total number of spins is now the total of the frequencies.

The formula for the mean becomes $\bar{x} = \frac{\sum f_i x_i}{\sum f_i}$.

To calculate the mean:

$$\bar{x} = \frac{7 \times 1 + 8 \times 2 + 6 \times 3 + 10 \times 4 + 9 \times 5}{7 + 8 + 6 + 10 + 9} = \frac{126}{40} = 3.15$$

Examiner's secrets

With frequency tables, the most common mistake is to divide by the number of classes, rather than the total frequency.

The formulae for the variance become

$$S^2 = \frac{\sum f_i (x_i - \bar{x})^2}{\sum f_i} \quad \text{or} \quad S^2 = \frac{\sum f_i x_i^2}{\sum f_i} - \bar{x}^2$$

To calculate the variance:

$$S_x^2 = \frac{7 \times 1^2 + 8 \times 2^2 + 6 \times 3^2 + 10 \times 4^2 + 9 \times 5^2}{7 + 8 + 6 + 10 + 9} - 3.15^2$$

$$= \frac{478}{40} - 9.922\,5 = 2.027\,5$$

So the standard deviation of x, S_x, is $\sqrt{2.027\,5} = 1.424$ (3 d.p.).

Using grouped frequency tables

When data have been grouped, we do not have exact values for our calculations any more. We estimate our x_i values by using the midpoints of the actual class boundaries. The effect on the formulae is to replace x_i with m_i, where m_i is the midpoint of class i.

The formula for the mean becomes

$$\bar{x} = \frac{\sum f_i m_i}{\sum f_i}$$

The formulae for the variance become

$$S^2 = \frac{\sum f_i(m_i - \bar{x})^2}{\sum f_i} \quad \text{or} \quad S^2 = \frac{\sum f_i m_i^2}{\sum f_i} - \bar{x}^2$$

Example: The following table shows the attendance at a club.

Age (years)	0–9	10–19	20–29	30–39	40–49
Frequency	11	23	45	67	54

We need the actual class boundaries, and look at the type of data, to make our decision. Were this to be something like speed or height then the first upper class boundary would be 9.5. However, this is in years; with age you don't say you are ten until you are actually past your tenth birthday. So the first upper class boundary is actually 10.

Class boundaries	0–10	10–20	20–30	30–40	40–50
Midpoint, m_i	5	15	25	35	45

$$\bar{x} = \frac{\sum f_i m_i}{\sum f_i} = \frac{6\,300}{200} = 31.5$$

$$S^2 = \frac{\sum f_i m_i^2}{\sum f_i} - \bar{x}^2 = \frac{225\,000}{200} - 31.5^2 = 132.75$$

So an estimate for $\bar{x}$ is 31.2 years and for S is $\sqrt{132.75} = 11.52$.

Combining two sets of results

Suppose we had a set of test scores, mean $\bar{X}$, standard deviation S_x. Another set of test scores are obtained, having mean of $\bar{Y}$ and standard deviation S_y. What are the mean and standard deviation of the whole lot combined? Let's call the combined set Z. We use these two equations:

$$\bar{Z} = \frac{\sum X + \sum Y}{n_X + n_Y} \quad \text{and} \quad S_Z = \sqrt{\frac{\sum X^2 + \sum Y^2}{n_X + n_Y} - \bar{Z}^2}$$

Now, we need $\sum X$ and $\sum Y$, which we get from the equations for $\bar{X}$ and $\bar{Y}$. We also need $\sum X^2$ and $\sum Y^2$, which are easy enough to obtain from the equations for S_x and S_y.

Exam question answer: page 157

The table below shows the pH levels in soil sample from fields on a particular farm. Estimate the mean pH levels and standard deviation. Another farm had a further 40 samples taken, and the mean was found to be 6.24, whereas the standard deviation was 0.25. Find the mean and variance of pH levels on both farms. (20 min)

Height	4.0–4.7	4.8–5.5	5.6–6.3	6.4–7.1	7.2–7.9	8.0–
Frequency	1	4	8	28	8	1

The jargon

Anything that is further than 2 standard deviations away from the mean can be considered an *outlier* – an item which is completely extreme – which should maybe be ignored, as it is *not typical* and can *distort* the measures.

Action point

Write out the formulae for means and variances on a sheet of paper: start with raw data versions, then the version for ungrouped and, finally, grouped frequency tables.

Checkpoint 2

Estimate the mean and standard deviation of this set of data:

Interval	1–4	5–8	9–12	13–16
Frequency	5	4	8	3

Examiner's secrets

As with so many A-level exam questions, when you attack them and get stuck, try asking yourself, 'What *can* I do next?' rather than, 'What *should* I do next?' Try listing what you know, like information from the question, or equations that feel relevant, and an idea will usually present itself!

Statistical diagrams

Representing information in a graph or chart can be quite powerful – especially when comparing two distributions. You should be aware of all of the methods of displaying data, and pay special attention to *class boundaries* when drawing histograms, as many exam marks can be lost here.

Frequency distributions

Any data to be processed will normally be given in the form of a frequency distribution. This is just a table, telling you what numbers the data can take (x values), and how many of each (**frequency**).

Example: A frequency distribution showing test scores.

Score	1	2	3	4	5
Frequency	3	7	10	6	4

Watch out!

The values that the variable can take – in this example the 'score' row – can *look* quite weird, depending on what the actual data is about. Here it's easy, but it might not look so straightforward! More later . . .

Bar line graphs

Line graphs are best suited to *discrete* data. They are like a bar chart, only with lines. Here is a bar line graph for the test score data in the table above.

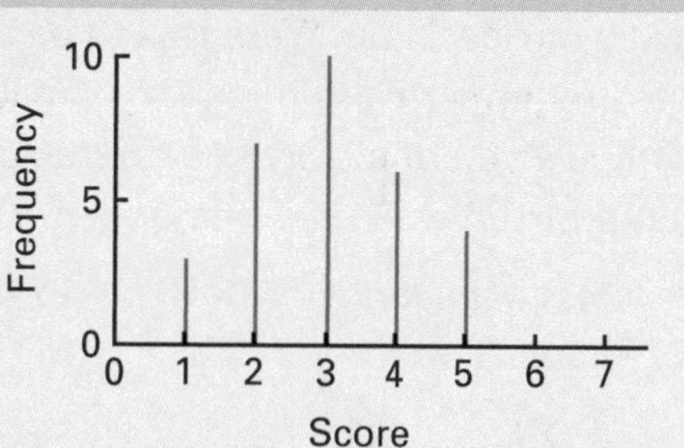

The jargon

Discrete data comes in nice whole-number packages. There are no decimals or fractions, just *counting* type numbers. Good examples are the number of chairs in a room, or the number of cars passing a particular point.

Stem and leaf diagrams

These are about as simple as a diagram can get, once you've got the hang of them. You decide on how big a step you want to increase by each time, normally tens, and these numbers form the 'stem'. The leaves are then the numbers that make up the 'units' column. Here are two quick examples. The second has been grouped in steps of five.

Example 1

0	1	This is just 01, or 1
10	4 7	
20	**3** 5	This '3' means '23'
30	1 4 6 7	
40	3 3 5 9	
50	**0** 3	This '0' means just '50'
60	1	

Example 2

0	3	
5	**1** 3	The '1' means '5 + 1' or 6
10	3 3 5	
15	**0** 2 2 4	The '0' means '15 + 0' or 15
20	4 5	
25	0 1	
30	0	

Watch out!

A very common mistake in diagrams where the steps are five is to think that a number like 26 is shown as a 6 next to the 20. It should be a 1 next to the 25, as 25 + 1 = 26. Also, try to keep all leaf numbers in columns, to show off the longer or shorter rows.

This gives rise to a back-to-back diagram that is very good at showing the different shapes of two distributions. You would have one central stem with leaves coming out from both sides. These types of diagram quickly and easily give the *modes* or *modal classes*, and show whether the distribution is skewed in any way.

Histograms

Histograms are best suited to representing *continuous* data. Unlike a bar chart, the height of the bar gives the *frequency* density. To get the frequency density, divide the frequency by the actual class width.

To help decide on the actual class boundaries, pick a class, and ask yourself: 'What is the smallest value an item in this class can take, before it is moved down?' Whatever your answer is, that becomes the *upper class boundary of the class below*. Work from right to left through classes.

Example: The table shows the speed of cars passing a checkpoint.

Speed (kph)	0–9	10–19	20–24	25–29	30–40
Frequency, F	1	4	9	15	3
Actual class boundaries	0 to 9.5	9.5 to 19.5	19.5 to 24.5	24.5 to 29.5	29.5 to 40.5
Class width, C	9.5	10	5	5	11
Frequency density (F ÷ C)	1 ÷ 9.5 = 0.11	4 ÷ 10 = 0.4	9 ÷ 5 = 1.8	15 ÷ 5 = 3	3 ÷ 11 = 0.27

We added the last three rows for calculations. The histogram is now plotted, with frequency density on the y-axis, speed on the x-axis, and bars which meet at actual class boundaries.

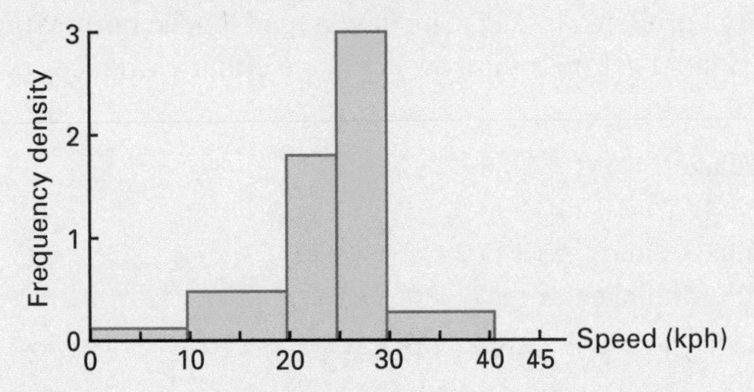

Box and whisker diagrams

When you know the median (Q_2) and upper (Q_3) and lower (Q_1) quartiles of two distributions then *box and whisker plots* are very useful. The thing to remember is to keep the scales the same, so that you can directly compare values.

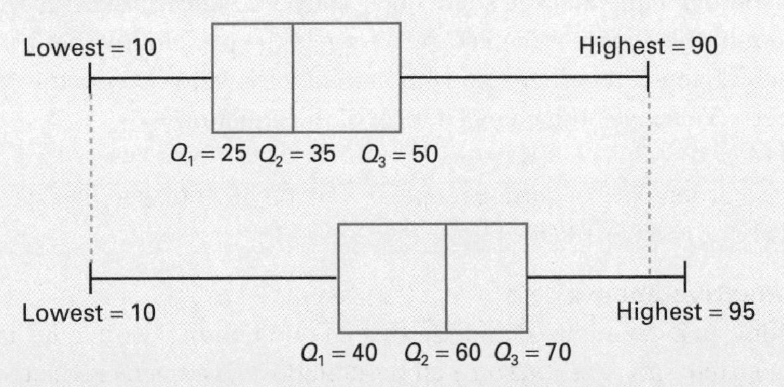

Exam question answer: page 157

A company took a survey of the ages of its employees. The results are shown in the frequency table below.

Age (years)	18–	20–	30–	40–	50–
Frequency	8	22	15	7	2

Construct a histogram to represent these data, stating fully any assumptions you make. (15 min)

The jargon

Continuous data are not counted; they are measured. Distance, time and weight are examples. Continuous data take values that have only been measured to a certain degree of accuracy – the nearest cm, mm, kg or mph.

Examiner's secrets

As you can see, the actual class boundaries might not be the same as those given. This is to take any rounding into account. Here speeds have been rounded to the nearest kph; so the largest value that could fall into the first category is 9.5 kph. If you're getting some very strange actual class widths, it's a good idea to check that you've carried out this step correctly!

Checkpoint

Construct a box and whisker diagram for the data represented by the cumulative frequency curve on page 117. Try to get a feel for how the curve would look if you shifted the 'box' to the left or right.

Action point

Try to imagine how the frequency distribution diagrams for these two box and whisker plots would look. The first will have positive skew, the second negative. Can you see why? It's to do with the bulk of the distribution being dragged one way or the other.

Probability

There is one basic law for probability, which gets changed as the nature of the events changes. That means that you have one formula to *learn* and *understand* and then you can do just about any probability question there is!

The law and its parts

Action point

Make a list to learn of the different forms of the equation as you meet them, including why they change – it'll save time in the exam.

The one equation you *must* learn is this:

$$P(A \cup B) = P(A) + P(B) - P(A \cap B)$$

Where:

$P(A)$ means 'The probability event A happens'
$P(B)$ means 'The probability event B happens'
$P(A \cap B)$ means *both* A and B happen
$P(A \cup B)$ means A *or* B (or both) happen

Examiner's secrets

Fitting the statements you are given into the equation makes seeing what you have and what you need so much easier. In many cases, that's the hard bit done!

Example: The probability I eat cornflakes for breakfast is 0.6 and the probability I drink tea is 0.25. One day in ten I'll have both. What is the probability I get up tomorrow and have either cornflakes or tea?

Answer:

$P(\text{cornflakes}) = P(C) = 0.6$
$P(\text{tea}) = P(T) = 0.25$
$P(\text{cornflakes and tea}) = P(C \cap T) = 0.1$

We need $P(\text{cornflakes or tea}) = P(C \cup T)$.

We use $P(C \cup T) = P(C) + P(T) - P(C \cap T)$
$= 0.6 + 0.25 - 0.1 = 0.75.$

We are now in a position to explore what happens when A and B take on certain characteristics.

Checkpoint 1

I watch *Eastenders* with probability 0.8. The probability I do the ironing is 0.3. The probability I do the ironing or watch *Eastenders* is 0.9. What is the probability I do both?

Mutually exclusive events

When a die is rolled, the two events 'Getting a one' and 'Getting a six' cannot *both* happen at the same time. They are *mutually exclusive events*. In the equation the $P(A \cap B)$ term is the probability that both A and B happen together – so with mutually exclusive events this must be zero. This gives the changed form of the equation:

$$P(A \cup B) = P(A) + P(B)$$

So, the probability of getting a one or a six on an unbiased die is:

$$P(1 \cup 6) = P(1) + P(6) = \tfrac{1}{6} + \tfrac{1}{6} = \tfrac{2}{6}$$

Checkpoint 2

Make a list of three pairs of mutually exhaustive events, either to do with rolling a die, or drawing one card from a pack. Hint: diamonds or clubs?

Exhaustive events

'Getting an even number' and 'Getting an odd number' with a die are called *exhaustive* events, since all possibilities have been exhausted. Therefore, for the exhaustive events A and B, our equation becomes

$$P(A \cup B) = P(A) + P(B) - P(A \cap B) = 1$$

Checkpoint 3

Make a list of three pairs of exhaustive events. Think of drawing a card from a pack, or spinning an eight-sided spinner. Hint: red card or black?

Events might be exhaustive, mutually exclusive, both or neither. Examples involving drawing two cards from a pack:

- → Drawing a card under 10, drawing a 2 or more (exhaustive only).
- → Drawing an ace, drawing a king (mutually exclusive only).
- → Drawing a red card, drawing a black (both).
- → Drawing an even card, drawing a heart (neither).

Conditional probability

A card is drawn from a well-shuffled pack. If you know it's a heart, what is the probability it's the ace? There are 13 hearts, so our probability is $\frac{1}{13}$. This situation translates into the statement: 'The probability of getting an ace, given that we have drawn a heart, is $\frac{1}{13}$.' We write it like this: $P(A \mid B)$ (probability of A given B). There's a neat formula for this:

$$P(A \mid B) = \frac{P(A \cap B)}{P(B)}$$

Example: A group of prisoners consists of seven robbers and five fraudsters. Two prisoners are to be chosen to run an errand.

(a) What is the probability that two robbers are chosen?

(b) What is the probability that a robber is chosen, given that one robber has already been picked?

Answer:

(a) We want a robber and a robber, or $P(R \cap R) = (\frac{7}{12})(\frac{6}{11}) = \frac{42}{132}$.

(b) A robber has been chosen, so there are 11 prisoners left, out of which six are robbers, so the probability of picking a robber is $\frac{6}{11}$.

Alternatively, using the formula and with obvious notation,

$$P(R \mid R) = \frac{P(R \cap R)}{P(R)} = -\frac{42}{132} \Big/ \frac{7}{12} = \frac{6}{11}$$

Notice the difference between $P(R \cap R)$ and $P(R \mid R)$. The first says: 'We have chosen two things, what is the probability we have chosen these two in particular?' The second says: 'We have chosen one, and now, *before the next one is chosen*, what is the probability that it is this one particular outcome?'

Independent events

Two events are independent when one has no influence over the occurrence of the other. An example could be my score when I roll a die, and whether or not it is raining outside. In this case, we get $P(A \mid B) = P(A)$, since B happening means nothing to A.

$$\text{Also, } P(A \mid B) = P(A)$$

$$P(A) = \frac{P(A \cap B)}{P(B)}$$

$$P(A \cap B) = P(A)\,P(B)$$

Exam questions answers: pages 157–8

1 In a group of 200 people, 131 own a computer, 157 own a mobile phone, and 104 own both. Find the probability that a person chosen at random (a) owns a computer or a mobile phone, (b) owns a computer or a mobile phone but not both, (c) owns a computer, given that they own a mobile phone, (d) does not own a mobile phone, given that they own a computer. (10 min)

2 Events A and B are such that $P(A) = 0.7$ and $P(B) = 0.2$. If $P(A \mid B) = 0.1$, find the probability that

(a) exactly one of the events occurs,

(b) at least one of the events occurs,

(c) both of the events occur,

(d) B occurs, given that A has already occurred. (10 min)

Examiner's secrets

Conditional probability is a thing some people find hard. It's really down to confidence and practice, but a good way to approach any question is to write down this formula and simply try to put the information in its right place.

Checkpoint 4

I have a bag containing 12 red and 4 green wine gums. What is the probability that the second sweet I eat is a red one, if my first was a green? What is the probability I eat two greens?

Examiner's secrets

Always leave fractions in their original form, unless asked to simplify. This will make adding or subtracting much easier!

Action point

Write out the formulae for independent and conditional probabilities. Try to think up some different experiments and fit the data into the formulae. Examples are a die being rolled twice; a card being picked at random from a deck and then picked again – with replacement first, and then without replacement. Find the probabilities of different outcomes, enter the data and see what you get!

Probability trees

Links

See 'Probability', page 122.

If ever there was an easy way to work out questions, this is it! It can give exhaustive, mutually exclusive, independent *and* conditional probabilities. All this from following a few simple rules.

Constructing a tree diagram

Tree diagrams consist of points, one for each event, and branches, one for each outcome.

Here are the basic rules:

- → Always keep event points in strict columns, according to order.
- → Each possible outcome of an event point has its own branch which must be clearly labelled.
- → The probabilities of all the branches from a single point must sum to 1 – in other words, be mutually exclusive *and* exhaustive.

Let's draw a probability tree to illustrate this scene from 'conditional probability'. Two prisoners are to be chosen from a group, which consists of seven robbers and five fraudsters.

Examiner's secrets

If the probabilities of any of the sets of branches from one joint do not add up to 1 then you've either written them on wrong, or left one out!

Making a start

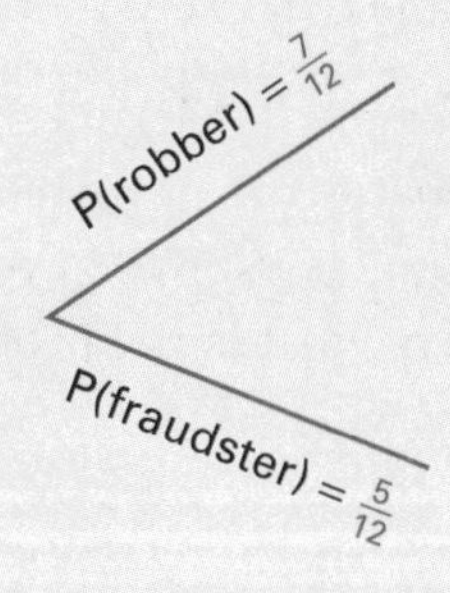

The two outcomes from the first choice have their own branch. Write the probabilities on them.

Checkpoint 1

Draw up the probability tree to show all outcomes when a coin is flipped, and a four-sided spinner is spun. What is the probability that a head and a 3 are obtained?

The next event

From each branch, two more must be drawn, to represent each possibility at stage two (the second choice). Keep to columns!

Checkpoint 2

A bag contains 8 red and 4 blue marbles. Three marbles are taken from the bag one at a time and without replacement. Draw a tree diagram to show all the possibilities.

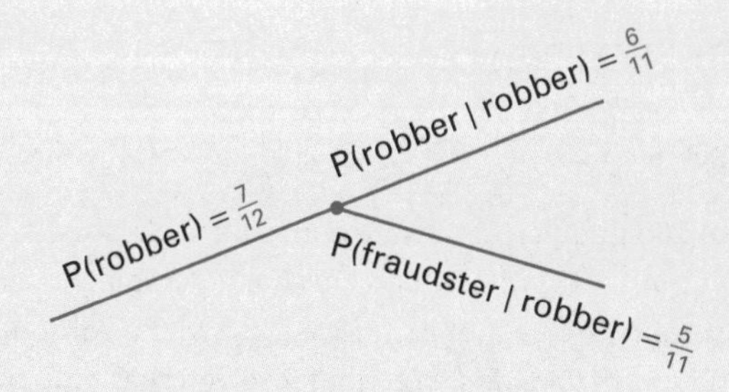

Notice that the probabilities from each pair of branches sum to 1 – so they must be both mutually exclusive *and* exhaustive.

What you can find easily using probability trees

Here is the information that can be obtained from trees more easily than most other methods. It all refers to the complete tree diagram, for our prisoner data, drawn below.

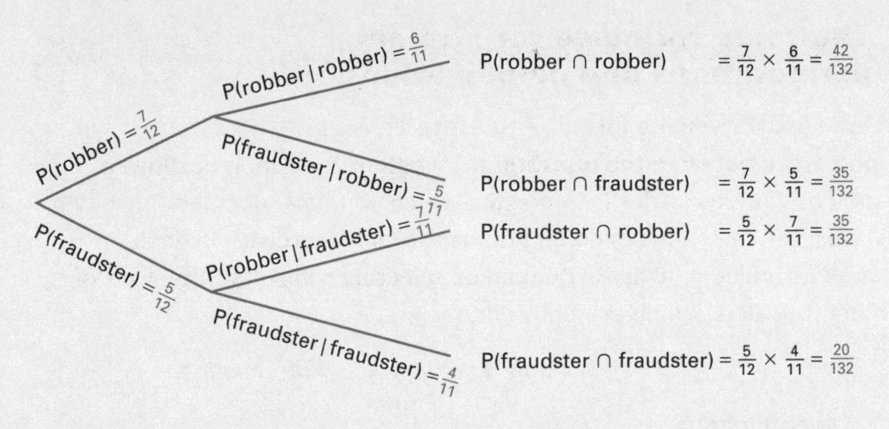

- ➜ The probabilities on each branch after the second event point are all *conditional* probabilities; so on the top branch, we have P(R), followed by P(R | R) – illustrating the point that conditional probabilities relate to choices *as* they are being made.
- ➜ In contrast, $P(R \cap R)$ refers to the probability of *ending up with* two robbers, *after* both choices have been made.
- ➜ The probabilities at the end of each final branch have been multiplied, showing that $P(R)P(R \mid R) = P(R \cap R)$.
- ➜ Supposing we knew that our second choice had been a robber – what is the probability that our first choice was a fraudster? Here, we want

  P(F1st | R2nd), and we use $\dfrac{P(F1st \cap R2nd)}{P(R2nd)}$

  where $P(F1st \cap R2nd)$ can be read off the tree, and P(R2nd) is found by calculating $P(F1st \cap R2nd) + P(R1st \cap R2nd)$. This is extremely difficult to imagine without using a tree diagram!

Exam question answer: pages 158–9

In the town of Spenville on any given day, the newsagent serves three times as many men as women. Only two papers are available: the *Spenville Times* and the *Late Press*. Customers either have the right amount of money or they need change. 65% of the men and 35% of the women buy the *Late Press*.

Of the men buying the *Spenville Times*, 85% do not have the correct change, and of the men buying the *Late Press*, 65% do have the correct change.

Of the women buying the *Spenville Times*, 50% do not have the correct change, and of the women buying the *Late Press*, 40% do have the correct change.

By drawing a tree diagram to represent this information, find the probability that (a) a customer buys the *Spenville Times*, (b) a customer has the right change, (c) a woman customer has the right change, and (d) a customer who has the right change for the *Late Press* is a man. (20 min)

Examiner's secrets

When you draw your tree, a common mistake is to make it too small. Don't! Use as much space as you like. Also, be sure to write on each branch exactly what the probability is. If it's on the second level, the probability will always be conditional on what happened before. Writing it all on as you go gets the problem clear in your mind.

Checkpoint 3

I take the bus or walk to school. The probability I walk is 0.2. If I walk, I am late with probability 0.4. If I catch the bus I am late with probability 0.1. Draw up a tree diagram to represent my journey.

Examiner's secrets

It's always a good idea with a probability question to read it all the way through once. This will give you a better idea of what you've got to do and how all the elements of information in the question fit into your formulae. If you want to, as you go, write next to each part of the question the particular formula you think might apply!

Permutations and combinations

Watch out!

This area has now been absorbed into probability, but is hard enough to need special attention! A typical question will expect you to use this information to calculate probabilities that will then fit into your formulae so practise to get confident in both.

The jargon

The exclamation mark means 'factorial', and $n!$ means $n(n-1) \times (n-2) \times (n-3) \times \ldots \times 3 \times 2 \times 1$ Try to remember up to about 6 factorial, to save calculator time.

Examiner's secrets

Your calculator will have ${}^nP_r$ and ${}^nC_r$ functions on it already – save yourself time now by learning how to use them!

Action point

Try to think up, or research, three different examples matching each of the three different situations.

Checkpoint 1

Six people are going to the cinema. How many ways can they queue for tickets? Later, in a restaurant, how many ways can they sit at a circular table?

This spread presents formulae for counting the number of ways things can be arranged or selected. Arrangements are often called permutations, and selections are generally combinations. The toughest thing about these questions is deciding on the right formula to use!

The three formulae for straight permutations and combinations

There are three main formulae to learn. To decide which to use, you must work out from the question (a) whether repetition is allowed, and (b) whether order is important. Once you have decided these two issues, the questions become number-crunch exercises. In each of these formulae, you are arranging or selecting r objects from a set of n in total, all of which are different.

	Without repetition	*With repetition*
Order important	${}^nP_r = \dfrac{n!}{(n-r)!}$	n^r
Order unimportant	${}^nC_r = \dfrac{n!}{r!(n-r)!}$	

Examples:

- Making up three-digit numbers, picking digits from random number tables; 'order important', 'with repetition', use the n^r formula.
- Picking the first, second and third runners in a race to receive gold, silver and bronze medals respectively; 'order important', 'without repetition', use the ${}^nP_r$ formula.
- Picking the first three runners in a race to qualify for the next heats; 'order unimportant', 'without repetition', use the ${}^nC_r$ formula.

The three special cases of permutations

There are three main special cases, of which you must be aware.

When the *n* objects are not all the same

If we had a set of n objects, and of those objects, s are alike, t are alike, u are alike, and so on, then the number of arrangements of all n objects is: $\dfrac{n!}{s!t!u!\ldots}$.

Examples:

- To find the number of arrangements of the word UNIQUE, work out $\frac{6!}{2!}$ (6 letters, 2 U's) to get 360 in total.
- To find the number of arrangements of the word STATISTICS, work out $\frac{10!}{(3!3!2!)}$ (10 letters, 3 S's, 3 T's, 2 I's) to get 50 400 in total.

When the *n* objects are in a circle

The classic case here is finding the number of ways n people can be seated at a circular table. Answer: seat the first one anywhere, and *then* the others – to get $(n-1)!$ ways.

When the n objects are in a ring

Here, the n objects are in a circular pattern, but the difference is that they can now be viewed from both sides, making exactly twice as many *repeated* arrangements. The formula becomes $\frac{(n-1)!}{2}$.

Selecting from two or more sets of objects

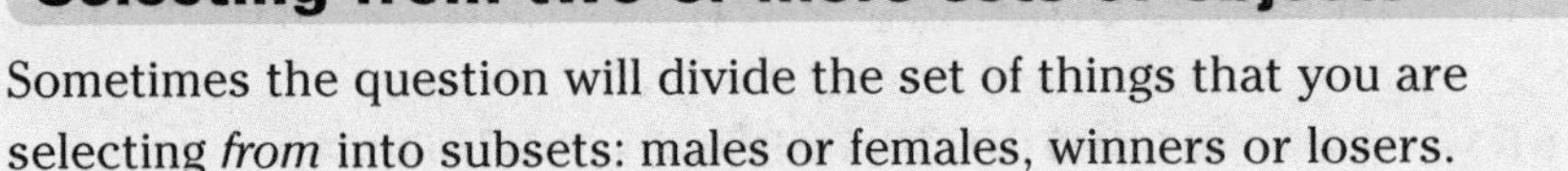

Sometimes the question will divide the set of things that you are selecting *from* into subsets: males or females, winners or losers. Here, you simply select objects from each set until you are done.

Example: A football manager makes up a team of 11 players, consisting of 1 goalkeeper and 10 others. Her pool of players has 3 goalkeepers, and 13 others.

(a) How many teams can she make up if all members are available?

(b) For the next match, she *must* pick the 2 best strikers, and 1 goalkeeper is injured. How many teams can now be made up?

Answers:

(a) To pick the goalkeeper: ${}^{3}C_{1} = \frac{3!}{1!(3-1)!} = 3$

To pick the others: ${}^{13}C_{10} = \frac{13!}{10!(13-10)!} = 286$

$\Rightarrow$ total number of teams $= 3 \times 286 = 858$.

(b) To pick the goalkeeper: ${}^{2}C_{1} = \frac{2!}{1!(2-1)!} = 2$

To pick the others: ${}^{11}C_{8} = \frac{11!}{8!(11-8)!} = 165$

$\Rightarrow$ total number of teams $= 2 \times 165 = 330$.

Dealing with restrictions

Examples using arrangements of the words STATISTICS:

- Arrangements where the last two letters are both T's – fill these two spaces first, then use the rest to fill the remaining eight spaces.
- Arrangements where all three S's must be together – put them together as *one object*. Then arrange the *eight* objects we now have (SSS, and the seven other letters).
- Arrangements where the I's *cannot* be next to each other – work out the total number *without* any restrictions, and then subtract the number of arrangements which are 'illegal' – i.e. all those where the I's *are* together.

Exam questions answers: page 159

1 In how many different ways can the letters of the word MILLENNIUM be arranged? Find the probability that an arrangement chosen at random has the two I's next to each other. (5 min)

2 A charitable organisation decides to hold a lottery to raise funds. The lottery works as follows: each of the letters A to Z will be written on 26 counters, and five will be drawn at random without being replaced. The order of the letters drawn is not important. Tickets are £1 each. (a) If all possible tickets are sold, how much will they expect to take? (b) You can win £5 if you match any three letters. What is the probability that you win £5 with one ticket? (c) The prize for four matching numbers is £10, and five matching wins £100. Assuming all tickets are sold, how much will they raise? (15 min)

Examiner's secrets

To remember the difference between circles and rings, just ask yourself, would it be rude to look from underneath? It certainly would be if you think about a *circle* of friends sat at a table!

Checkpoint 2

A group of friends decide to play a team game. There are ten boys and six girls. How many different teams of eight people can be made if there are to be three girls on each?

Examiner's secrets

It's essential to show *at least* this much working, just in case you hit the wrong buttons on your calculator. You may be asked to leave your answers in factorial form – in which case this can be left as it is.

Checkpoint 3

I have 12 books on a shelf. How many ways can I arrange them so that my five maths books are together? How many ways can I arrange them if my two science books are split, one at each end?

Action point

Try to find an older statistics text book for some really great questions on this topic, as they will treat it separately as we have here. This will really give your skills a boost.

Discrete random variables

Random variables are simply the outcomes of experiments or situations. Discrete random variables (DRVs) are outcomes which can only take integer (whole number) values. Common examples are the score on a die, the number of accidents on a stretch of road, or the number of heads obtained when some coins are flipped.

Conditions for discrete random variables

There are essential conditions which a variable, X, has to satisfy before it can be called a DRV. They are:

- ➜ it can only take integer values;
- ➜ each outcome has a probability attached to it;
- ➜ the sum of all the associated probabilities is 1.

If you need to prove that a discrete variable is a random variable, then it is enough to show that the sum of probabilities is 1, or, more formally, that $\sum P(X = x_i) = 1$.

Example: Let X be the outcome when a die is rolled. Then X is a DRV since the probabilities of all outcomes add to 1.

The jargon

The Greek symbol Σ means 'sum of' – so the statement $\Sigma P(X = x_i) = 1$ reads 'the sum of all of the probabilities (associated with each value X can take) equals 1'.

PDFs and CDFs

PDF stands for **probability density function**. It normally gives a simple formula in x for working out each probability.

Example: Let X be the discrete variable 'the score on a five-sided spinner'. Its probability density function is given by

$$P(X = x) = \frac{6 - x}{15}$$

Form a table of probabilities for X, and confirm that it is a discrete random variable.

Answer: The first two look like this:

$$P(X = 1) = \frac{6 - 1}{15} = \frac{5}{15} \quad \text{and} \quad P(X = 2) = \frac{6 - 2}{15} = \frac{4}{15}$$

Here's the completed table.

X	1	2	3	4	5
$P(X = x)$	$\frac{5}{15}$	$\frac{4}{15}$	$\frac{3}{15}$	$\frac{2}{15}$	$\frac{1}{15}$

To confirm X is a discrete random variable, add up the probabilities, to get:

$$\frac{5 + 4 + 3 + 2 + 1}{15} = 1$$

OK!

You will normally be given the probabilities in the form of a table, like the one above. This is called the **probability distribution of X**.

CDF stands for **cumulative distribution function**. It works just like an accumulating total; the probabilities are added up as they go along. Formally, it's written as $F(t) = P(X \leq t)$, i.e. the sum of all probabilities up to $X = t$.

Example: Let X be 'the score on a ordinary die'. Then the cumulative distribution function is given by $F(x) = \frac{x}{6}$. So, $P(X \leq 3) = \frac{3}{6}$. Does that make sense? The probability of each outcome on a die is $\frac{1}{6}$, so

$$P(X = 1) + P(X = 2) + P(X = 3) = \tfrac{1}{6} + \tfrac{1}{6} + \tfrac{1}{6} = \tfrac{3}{6}$$

Examiner's secrets

With any probability question, laying out your solution with variables clearly defined will impress the examiner. It doesn't take long, and looks very professional indeed.

Checkpoint 1

A discrete random variable has PDF $P(X = x) = kx^2$ for $0 \leq x \leq 5$. Using the fact that the probabilities sum to 1, find the value of k, and write out the probability distribution of x.

Action point

Your calculator is an invaluable tool – find out the best way to perform these longer calculations. Sometimes you may need to use the memory and work in stages, if your machine doesn't like long calculations.

Expectation and variance of a DRV

In the example above, what score would you *expect* to get when rolling a die? We work this out by weighting each outcome with its associated probability, and adding up the results. Formally, the **expected value of *X***, $E(X) = \sum P_i x_i$.

If we are thinking about dice, the probability of each outcome will be $\frac{1}{6}$, and so the expected score, $E(X)$, will be:

$$(1 \times \tfrac{1}{6}) + (2 \times \tfrac{1}{6}) + (3 \times \tfrac{1}{6}) + (4 \times \tfrac{1}{6}) + (5 \times \tfrac{1}{6}) + (6 \times \tfrac{1}{6}) = \tfrac{21}{6} = 3\tfrac{1}{2}$$

Note that we would expect to get around three and a half anyway – this is the middle of the symmetrical data set. Note also that the expected value is often a number which cannot appear in the original data set.

The **variance of *X***, written $\mathrm{Var}(X)$, can be found using a simple formula: $\mathrm{Var}(X) = E(X^2) - E^2(X)$. Note the difference between the two terms: for the first term, find the expected value of X^2, for the second, find the expected value of X and square *that*. The best way to tackle this is to copy out the table of X and $P(X = x)$, and add an X^2 row.

The first term for the variance of our die score would be:

$$E(X^2) = (1 \times \tfrac{1}{6}) + (4 \times \tfrac{1}{6}) + (9 \times \tfrac{1}{6}) + (16 \times \tfrac{1}{6}) + (25 \times \tfrac{1}{6}) + (36 \times \tfrac{1}{6}) = \tfrac{91}{6}$$

making our variance $\frac{91}{6} - (\frac{21}{6})^2 = \frac{546}{36} - \frac{441}{36} = \frac{105}{36}$.

Transformations of *X*, of the type *ax* + *b*

If you take a DRV X, and apply a transformation to it to get a new variable, Y, then there are two basic formulae which we use. If $y_i = ax_i + b$ then $E(Y) = aE(X) + b$ and $\mathrm{Var}(Y) = a^2\mathrm{Var}(X)$.

Example: X is a discrete random variable where $E(X) = 10$ and $\mathrm{Var}(X) = 2$. Y is a transformation of X of the form $y_i = 3x_i + 4$. Then $E(Y) = (3)(10) + 4 = 34$ and $\mathrm{Var}(Y) = 3^2\,\mathrm{Var}(X) = (9)(2) = 18$.

Exam questions

answers: pages 159–60

1 A gambler plays a game in which two dice are rolled and the scores added. If the total is 2, she loses £10. If the total is 6, she loses £7, but if either die show a 6, the gambler wins £6.
 (a) Draw up the probability distribution table of the gambler's losses.
 (b) Find the expected gain per game for the gambler.
 (c) How much would the gambler gain if she played the game 200 times? Would you say the game is fair? (15 min)

2 The probability of there being X faulty light bulbs in a pack is given by the following: $P(X = 0) = 8k$; $P(X = 1) = 5k$; $P(X = 2) = P(X = 3) = k$.
 (a) Calculate the value of the constant, k, and find the expected number and variance of the number of faulty light bulbs.
 (b) Find the probability that in two packs of bulbs there are more than four faulty light bulbs altogether, and *without further calculation*, state the expected number and variance of the number of faulty light bulbs in the two packs. (15 min)

The jargon

The expected value of a variable is outlined in the section opener. It's literally like saying to yourself, 'What would I expect?' The variance is very similar, in that we are asking ourselves, 'How far away from the expected value should we expect our value to be?'

Examiner's secrets

Remember this formula in words as 'expected X squared minus expected squared X'.

Checkpoint 2

Draw up the table of probabilities associated with X where X is the total score when two dice are rolled. Find the expected score and variance.

Checkpoint 3

You are expected to be able to quote the results of these types of transformations! A DRV is such that $E(X) = 3$, and $\mathrm{Var}(X) = 2$. Write down the mean and variance if (a) 4 is added to every X, (b) every X is multiplied by 5, and (c) every X is transformed into Y by the equation $y_i = 3x_i - 5$.

Action point

Get together with a friend and make up questions like checkpoint 3 using your own numbers – just change the ones given and go!

The binomial distribution

This distribution models situations having only two outcomes, hence the name *bi*nomial. Common examples are passing or failing a test, or rolling a die and getting a six or not.

Examiner's secrets

You will be expected to be able to quote and use these conditions – make a list of all the distributions you need to know, including their conditions, formulae, means and variances!

Link

Check out the Permutations and combinations section for a recap of 'n choose r' and the associated calculation.

Checkpoint 1

Work out $P(x \leq 3)$ if $X \sim \text{Bin}(6,0.2)$. Start by working out the individual probabilities (0, 1, 2 and 3), and then add them. Once you've done that, what is $P(X > 3)$?

Examiner's secrets

Unless the numbers are really awkward, leaving any fractions in their unsimplified form makes it easier to remember whether they represent p or q (i.e. success or failure) – it also makes any additions that bit simpler.

Checkpoint 2

A bag contains five red and six blue balls. A ball is drawn and replaced. This is repeated eight times. Write down the probability distribution of X, the number of blue balls.

Examiner's secrets

Where decimals are very long *don't* round them during the calculations, as you may lose accuracy marks. When showing them in working, write '. . .' after the first few digits, just to indicate what you are doing.

Examiner's secrets

Practise using your calculator for these probabilities – each type of machine does it differently! In particular, make sure you are fully familiar with your ^nC_r button!

Conditions for the binomial distribution

There are three conditions which have to be satisfied for the binomial distribution to be an appropriate model:

- → there must only be two **mutually exclusive** outcomes – heads or tails, pass or fail, win or lose;
- → there must be a **finite number** of trials – usually denoted by n;
- → each trial must be **independent** of all others – the probability of success must remain constant.

If all conditions are satisfied we write $X \sim \text{Bin}(n,p)$.

Calculations involving the binomial formula 

The formula for calculating probabilities for any events satisfying the three conditions is:

$$P(X = x) = {}^nC_r\, p^x q^{(n-x)}$$

where:

n is the total number of trials
r is the number of successes, ranging from 0 to n
p is the probability of success and
q is the probability of failure: note that $q = 1 - p$.

Example: A card is to be drawn from a well-shuffled pack, and then replaced. The experiment is repeated ten times.

(a) What is the probability of drawing an ace on six of these trials?
(b) What is the probability that an ace will be drawn at least twice?
(c) Explain why the binomial distribution would not be a suitable model for this situation if the cards were not replaced.

Answer:

(a) The probability of drawing an ace on one trial is $\frac{4}{52}$. Let p be the probability of success: then $p = \frac{4}{52}$, and so $q = 1 - \frac{4}{52} = \frac{48}{52}$. There are ten trials, so $n = 10$, and we want six successes, so $r = 6$. We now substitute this information into the formula:

$$P(x = 6) = ({}^{10}C_6)(\tfrac{4}{52})^6(\tfrac{48}{52})^4$$
$$= (210)(0.000\ldots)(0.726\ldots) = 0.000\,031\,6 \text{ (3 s.f.)}$$

(b) $$P(X \geq 2) = 1 - P(X < 2) = 1 - [P(X = 0) + P(X = 1)]$$
$$= 1 - [({}^{10}C_0)(\tfrac{4}{52})^0(\tfrac{48}{52})^{10} + ({}^{10}C_1)(\tfrac{4}{52})^1(\tfrac{48}{52})^9]$$
$$= 1 - [0.449\ldots + 0.374\ldots] = 1 - 0.823\ldots = 0.177 \text{ (3 s.f.)}$$

(c) If the cards are not replaced, then the probabilities at each trial are not constant. For example, the denominator in the probability goes down one each time. There is no *independence* between trials.

Expectation and variance 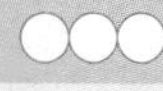

In the example above, how many aces would you *expect* in the ten trials? On average, you effectively draw '$\frac{4}{52}$' of an ace each time you pick a card – which you do ten times. So you would expect $\frac{4}{52} \times 10 = \frac{40}{52}$ aces!

- The expected value of X, $E(X)$, is given by: $E(X) = np$
- The variance of X, $Var(X)$, is given by: $Var(X) = npq$

Note that, since $q = 1 - p$, that $Var(X) = np(1 - p)$.

You may only be given n and the values of $E(X)$ and $Var(X)$, and be asked to find p, and then use it to calculate probabilities.

Using tables of cumulative binomial probabilities

In many cases, it is adequate to use tables of cumulative probabilities rather than make lengthy calculations. These will give $P(X \leq x)$, or the sum $P(X = 0) + P(X = 1) + P(X = 2) + \ldots + P(X = x)$. You will need to know the values of n and p, which are the parameters on the top row and left-hand column.

- *To find an individual probability using tables*, all you need to do is a quick calculation. For example, to find $P(X = 3)$, do $P(X \leq 3) - P(X \leq 2)$.
- *To find* $P(X \geq x)$, you do the calculation $1 - P(X \leq x - 1)$ for your distribution – for example, to find $P(X \geq 3)$, do $1 - P(X \leq 2)$.
- *When your value of* p *is bigger than 0.5* you need think about the *symmetrical* 'partner' distribution to the one in question. The two are linked as follows: if $X \sim Bin(n,p)$ then $Y \sim Bin(n,q)$. Also, the following equation holds and is very useful!

  $P(X \leq x) = P(Y \geq n - x)$

This means that if you wanted $P(X \leq 4)$ in a binomial distribution where $n = 9$, $p = 0.9$, you work out $P(Y \geq 5)$ in a binomial distribution where $n = 9$, $p = 0.1$, which happens to be $1 - P(Y \leq 4)$.

The jargon

When we say we expect a certain value of X to occur, we are really saying that we expect it to fall near our estimate. The amount by which we could be wrong is the variance.

Examiner's secrets

To practise using tables for probabilities, use them to check your calculated answers – remembering of course that they have been rounded to 4 or 5 d.p.

Checkpoint 3

Draw a bar chart to represent the probability distribution of X if $X \sim Bin(5,0.1)$.

Action point

If you have any doubts about how symmetrical distributions work, draw out a bar chart of the probabilities of the two distributions, for example where $n = 9$ and $p = 0.3$.

Exam questions

answers: page 160

1 A test has two sections. Section one is a multiple-choice section consisting of eight questions, each having four possible answer options, of which only one is correct. Assume I guess randomly at the answers.
  (a) Write down a formula for calculating the probability of getting r questions right out of the eight, fully defining all terms.
  (b) What is the probability that I get less than two questions correct?
  (c) What is the probability that I get two or more questions correct?
  Section two has four questions. Each question has five possible answers, of which two are correct.
  (d) What is the probability that I get question 1 completely correct?
  (e) What is the probability that I get more than half the questions in this section completely correct? (25 min)

2 On average, 20% of the washers produced by a machine are defective. A sample of ten washers is to be taken from the day's output, each washer being measured and returned before the next washer chosen. The machine is reset if more than two are faulty.
  (a) What is the probability that the machine requires resetting?
  (b) What is the expected value and the variance of the number of acceptable bolts in any sample? (10 min)

The Poisson distribution

This distribution models situations where events are spread out at random over an interval. Common examples are the number of flaws in a length of metal bar, the number of phone calls in a minute, or the number of cars passing a particular point on a road.

Action point

Make sure you can quote these conditions. Make a list of as many different examples you can imagine, of things that follow a Poisson distribution.

Conditions for the Poisson distribution

We have already noted that events have to be *randomly* scattered in time or space – we must have independence between each occurrence of the event. We call the mean number of these events for a particular interval λ. Occurrences of that event will follow a Poisson distribution, which is written $X \sim \text{Po}(\lambda)$.

Checkpoint

Find $P(X \leq 3)$ if $X \sim \text{Po}(5)$. Start by working out each probability for 0, 1, 2 and 3 and then adding them. Without further calculation, what is $P(X > 3)$?

Calculations involving the Poisson formula

The probability distribution function for the Poisson distribution having mean λ is:

$$P(X = x) = e^{-\lambda}\frac{\lambda^x}{x!} \quad \text{for } x = 0, 1, 2, 3 \ldots \text{ to infinity.}$$

Examiner's secrets

Some of the numbers in these calculations get very small. *Don't* round them before you have finished, and avoid having to write them out in full by using '. . .' in your written workings.

Example: The mean number of accidents per week at a factory is 2.6. Find the probability that in a given week there will be: (a) exactly one accident; (b) no accidents at all; (c) at least two accidents.

Answers: Assume accidents are independent, let X be the event 'an accident occurs', then $X \sim \text{Po}(2.6)$, i.e. $\lambda = 2.6$ in the formula.

(a) $P(X = 1) = e^{-2.6}\frac{2.6^1}{1!} = 0.193\ldots = 0.193$ (3 d.p.)

(b) $P(X = 0) = e^{-2.6}\frac{2.6^0}{0!} = 0.074\ldots = 0.074$ (3 d.p.)

(c) $P(X \geq 2) = 1 - P(X \leq 1) = 1 - [P(X = 0) + P(X = 1)]$
$= 1 - [0.193\ldots + 0.074\ldots] = 0.733$ (3 d.p.)

Examiner's secrets

Practise these calculations yourself, to really get to know your own calculator. You don't want to drop accuracy marks!

Expectation and variance

The expected value of X, $E(X) = \lambda$.
The variance of X, $\text{Var}(X) = \lambda$ as well!

Action point

Make up a table, with the name of the distribution, its PDF (probability density function) and the mean and variance, for all the standard models you need to know – this information is best kept at your fingertips!

Combinations of Poisson distributions

The unit interval. Suppose there were, on average, 10 flaws in one square metre in a particular piece of carpet. If you had 3 square metres of carpet you would expect around 30 flaws, since $3 \times 10 = 30$. Therefore, if $X \sim \text{Po}(\lambda)$ then $2X \sim \text{Po}(2\lambda)$. In other words, if we transform the interval, we transform the value of λ for the distribution.

When combining two independent Poisson distributions we can simply add the two values of λ. In other words, if $X \sim \text{Po}(m)$ and $Y \sim \text{Po}(n)$ then $X + Y \sim \text{Po}(m + n)$.

Example: Two sports cars develop faults in a race. Car A develops five faults per race, and car B develops four faults. The total number of faults developed by the two cars follows a Poisson distribution, with mean (and variance) $5 + 4 = 9$, i.e. $A + B \sim \text{Po}(9)$.

Using Poisson cumulative probability tables

To use Poisson cumulative probability tables you need a value of λ and a value of X. Looking at the table, you simply find your λ row (or column), and follow it along (or across) until you reach the desired X value.

Approximating the binomial distribution

The calculations for binomial probabilities are fine when n is small. When n becomes large (around 50 or more) and p is quite small (less than 0.1) we can use the Poisson distribution to *approximate* the probabilities which we cannot calculate. The mean of the binomial distribution would have been np, and so we use this value for λ.

So if $X \sim \text{Bin}(n,p)$, $n > 50$ and $p < 0.1$, then $X \sim \text{Po}(np)$.

Example: A machine produces components, of which 98% pass the inspector's tests. A batch of 100 are produced. To find the probability of x faulty components in the batch, we would start by stating that X, the number of faulty components, follows a binomial distribution. Let p be the probability that a component is faulty, then $p = 0.02$; also, $n = 100$, so that $X \sim \text{Bin}(100,0.02)$ and the probability that exactly x components are faulty would be:

$$P(X = x) = {}^{100}C_x(0.02)^x(0.98)^{100-x}$$

For any value of x, this is one horrible calculation! The numbers are so small that rounding errors produced by the length of your calculator's display could make the final answer very wrong indeed. We choose instead to *approximate* this distribution with $X \sim \text{Po}(\lambda)$ where $\lambda = 100 \times 0.02 = 2$. The calculation now becomes:

$$P(X = x) = e^{-2}\frac{2^x}{x!}$$

Exam questions answers: page 160

1 An observer counts the number of cars per minute passing her on a quiet stretch of road during the period 7 am to 9 am. The results were:

Cars per minute	0	1	2	3	4	5	6	7	8
Frequency	7	16	28	30	23	10	4	2	1

Show, by calculating the mean and variance of this distribution, that a Poisson distribution would be a good model.

Find the probability that:

(a) there are five cars passing during one minute, and

(b) that there are three or more cars passing in any minute. (25 min)

2 A newspaper has two sections, one for news and one for sport. Pages in the news section have, on average, 2.8 misprints each, while pages in the sport section have only 2.1. Assuming they both follow a Poisson distribution, calculate the probability that:

(a) two news pages have no misprints;

(b) a news and a sports page combined have exactly three misprints;

(c) a news and a sports page combined have more than five misprints. (10 min)

Examiner's secrets

Get your hands on a copy of the type of tables you will be using in the exam and practise to be perfect!

Action point

You can also calculate a string of probabilities using a simple formula, if you have no tables around.
The formula is:
$P(X = x + 1) = \lambda/(x + 1)\, P(X = x)$.
Try it!

Links

See 'The binomial distribution', page 130.

Action point

The conditions under which this approximation is valid are quite strict. Find out exactly what your exam board says and learn them – you are expected to be able to quote and use them!

Examiner's secrets

Remember to write down for the examiner exactly what you are doing – if you use an approximation, unless you have been directly told to do so, you should prove that the situation demands it, even if it's just by saying that n is very large and p is very small.

Continuous random variables

Random variables are simply the outcomes of experiments or situations. Continuous random variables (CRVs) are outcomes which can take any whole number, fractional or decimal values. Common examples are heights, weights, times – in fact, most of the continuous random variables you meet will be measurements of some kind.

Conditions for continuous random variables

There are essential conditions which a variable, X, has to satisfy before it can be called a CRV. They are:

- it has only been measured to a certain degree of accuracy;
- its probabilities are defined by a function of X;
- the integral of the probability function over all X is 1.

If you need to prove that a continuous variable is a random variable, then it is enough to show that the sum of all associated probabilities is 1, or, more formally, that $\int_{\text{all } x} f(x)\,dx = 1$, where f($x$) is the probability density function of X.

Examiner's secrets

Your differentiation and integration skills need to be fairly sharp to deal with these questions quickly – so practise lots of them!

PDFs and CDFs

PDF stands for **probability density function**. It normally gives a simple formula in x for working out probability. In most cases this will be a low-order polynomial, and x will be defined by inequalities.

Example: A continuous random variable has PDF $f(x) = cx^2$, $0 \le x \le 2$. To find the value of the constant, c, we integrate cx^2 between 0 and 2, and set the result equal to 1:

$$\int_0^2 cx^2\,dx = \left[\frac{cx^3}{3}\right]_0^2 = c\frac{8}{3} = 1, \text{ so } c = \frac{3}{8}$$

The area under the curve $y = f(x)$ represents probability. The integrated function, F(t) (in this example $\frac{1}{8}t^3$), becomes the **cumulative distribution function** (CDF) for $x \le t$. Formally, it is $F(t) = P(X \le t) = \int f(x)\,dx$, this integral going from the lowest value of x, up to $x = t$. To use the CDF to find the probability of x falling in a particular interval, between x_1 and x_2, we use the equation $P(x_1 \le X \le x_2) = F(x_2) - F(x_1)$.

Checkpoint 1

A CRV has PDF $f(x) = kx^2$ for $0 \le x \le 3$. Find the value of k, and sketch the PDF of x.

Example: A continuous random variable X has PDF as follows:

$$f(x) = \begin{cases} 0 & \text{for } x < 0 \\ \dfrac{x}{2} & \text{for } 0 \le x < 2 \\ 0 & \text{for } x > 2 \end{cases}$$

The jargon

This might look confusing, but all it says is that f(x) doesn't exist outside if 0 and 2; if you think about it, that's exactly what we want!

To find the cumulative distribution function, F(t), we integrate between 0 and t to get:

$$F(t) = \int_0^t \frac{x}{2}\,dx = \left[\frac{x^2}{4}\right]_0^t = \frac{t^2}{4}$$

To find the probability that x lies between 0 and $\sqrt{2}$, we do this:

$$P(0 \le x \le \sqrt{2}) = F(\sqrt{2}) - F(0) = \tfrac{2}{4} - \tfrac{0}{4} = \tfrac{2}{4} = \tfrac{1}{2}$$

In these examples we have seen an important relationship at work: that $f(x) = \frac{d}{dx}F(x) = F'(x)$. So, to go from F($x$) to f($x$) we differentiate, and we integrate to go from f(x) to F(x).

Examiner's secrets

Take extra special care when reading the question – a common error is to mistake F(x) for f(x) and integrate when you should differentiate.

Finding the median and quartiles using the DF

In the previous example, the probability that x was between 0 and $\sqrt{2}$ was exactly 0.5, making $x = 2$ the middle value – or *median* – of the distribution (the total area under $y = f(x)$ is 1). Formally, the median of any distribution is given by $X = m$, where $F(m) = 0.5$, and the nth percentile by $X = p$, where $F(p) = \frac{n}{100}$.

Sketching density functions

Here's a quick guide, using the example above, where $f(x) = \frac{3}{8}x^2$, and $0 \le x \le 2$. First find the end points: $f(0) = (\frac{3}{8})(0^2) = 0$; $f(2) = (\frac{3}{8})(2^2) = \frac{12}{8}$. Then we look for any turning points: $\frac{d}{dx}\frac{3}{8}x^2 = \frac{6}{8}x$ $= 0$ at $x = 0$. This function is quadratic, so it's easy to draw. When drawing other types of functions, just look for the term with the highest index and sketch according to that.

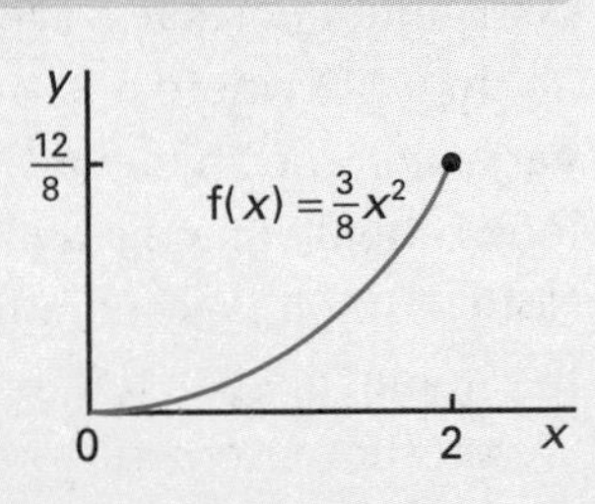

Expectation and variance of a CRV

How tall would you *expect* someone to be? The expected value of any random variable is given by $\sum x_i P_i$. This leads us to the formula for the expected value of a continuous random variable: $E(X) = \int_{\text{all } x} x f(x)\, dx$. (*Note*: $E(X)$ is sometimes denoted by μ, or the mean of X.)

The variance of X gets the same treatment – as long as you remember 'expected X squared minus expected squared X' you'll be OK! Formally, it looks like this:

$$\text{Var}(X) = \int_{\text{all } x} x^2 f(x)\, dx - \mu^2$$

Example: A continuous random variable X has PDF $f(x) = \frac{x^2}{9}$, $0 \le x \le 3$.

The expected value of X, $E(X)$, $= \int_0^3 x\frac{x^2}{9}dx = \left[\frac{x^4}{36}\right]_0^3 = \frac{3^4}{36} = \frac{81}{36} = 2\frac{1}{4}$

The variance of X would look like this:

$$\text{Var}(X) = \int_0^3 x^2\frac{x^2}{9}dx - \left(\frac{81}{36}\right)^2 = \left[\frac{x^5}{45}\right]_0^3 - \left(\frac{81}{36}\right)^2 = \frac{243}{45} - \left(\frac{81}{36}\right)^2 = \frac{27}{80}$$

Exam questions

answers: page 161

1 A random variable X has the following probability density function: $f(x) = ax(5 - x)$ for $0 \le x \le 5$ and $f(x) = 0$ elsewhere.
(a) Find the value of the constant a. (b) State and sketch the density function, $F(x)$. (10 min)

2 A continuous random variable has probability density function $f(x)$, where $f(x) = 0$ if $x < 0$ and $x > 3$, and between these two values it takes the form shown in the graph. (a) Find the value of C. (b) Express $f(x)$ algebraically and obtain the mean and variance of X. (c) Find the median of X. (20 min)

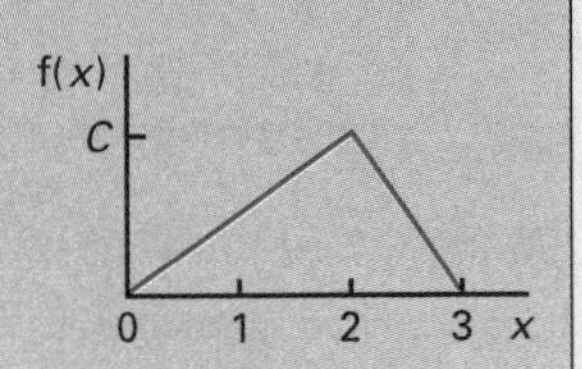

Action point

Many questions will ask you to sketch the graph of the probability distribution function. Look up graph sketching (page 20) for more details!

Checkpoint 2

A CRV X has PDF $f(x) = x$ for $0 \le x \le 1$, and $f(x) = 2 - x$ for $1 \le x \le 2$. Sketch the graph of the PDF and find the expected value of X.

The jargon

Remember this? The expected value of a variable is like saying to yourself 'What would I expect?' The variance is very similar, in that we are asking ourselves, 'How far away from the expected value should we expect our value to be?'

Examiner's secrets

The expected value and variance of a CRV are quotable formulae – and as mentioned, they still follow the same rules as with DRVs and the good old 'expected X squared minus expected squared X^2 still holds!'

The normal distribution

The normal distribution is probably the most wide-ranging and important model in statistics. It will appear, as you go on through your course, that just about everything follows that familiar bell-shaped curve. For now, good examples are heights, weights, lengths of rods, the amounts of milk in pint bottles and IQs.

Characteristics of the normal distribution

Checkpoint 1

If $X \sim N(25,64)$, then how much of the distribution will lie within $X = 17$ and $X = 33$? How much will lie between $X = 9$ and $X = 41$? Illustrate each answer with a sketch.

Let X be a random variable which is distributed normally. We say that $X \sim N(\mu,\sigma^2)$ where μ is the mean and σ^2 the variance. The shape of the distribution will look roughly like the figure on the right. Notice that it is symmetrical. This means that the mode, median and mean all have the *same value*.

Another interesting fact is that around 68% of all values lie within ± 1 standard deviation of the mean, and around 95% within 2 standard deviations. This means that if the mean were 10 and the standard deviation 2, then 68% of the data are between 8 and 12, and 95% are between 6 and 14.

Finding probability under the normal distribution 

Examiner's secrets

The hardest part of most questions is the final step from Z value to probability.

The actual normal probability density function is way too complicated to manipulate, and so we use tables to help. The tables are of *standardised* values, or Z values. We use a simple expression to convert X values into Z values:

$$Z = \frac{X - \mu}{\sigma}$$

Fact: If $X \sim N(\mu,\sigma^2)$ then $Z \sim N(0,1)$.

Note also that the tables only give the positive values of the cumulative distribution function of Z, known as $\Phi(z)$. So $\Phi(z) = P(Z < z)$, as long as this z is to the right of 0. Sketches are *essential* tools to make sure we're on the right lines!

Examiner's secrets

Drawing the diagrams will really help to clear your thoughts on exactly what it is you are trying to work out. Remember that any value to the left of the mean will have a negative Z value!

Using the standard normal distribution tables

There are four basic cases, depending on the probabilities required.

Case 1: Wanted: $P(X < a)$, where $Z = \frac{a - \mu}{\sigma} > 0$. Work out your Z value, and read $\Phi(z)$ off the table.

Case 2: Wanted: $P(X > a)$, where $Z = \frac{a - \mu}{\sigma} > 0$. Work out your Z value, then do $1 - \Phi(z)$.

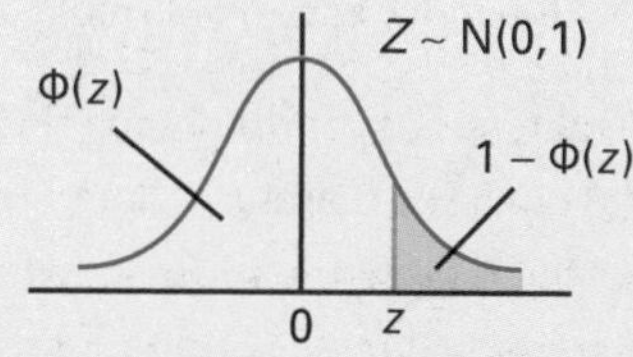

Checkpoint 2

If $X \sim N(7,100)$, find (a) $P(X < 9)$, (b) $P(X < 6)$, and (c) $P(X > 5)$. Don't forget the sketches!

The areas of interest have been shaded in each case. Case 3 uses the fact that if z is negative, we must use the positive value $\Phi(-z)$.

Case 3: Wanted: $P(X < a)$, where $Z = \frac{a - \mu}{\sigma} < 0$. We see that we need to find $1 - \Phi(-z)$.

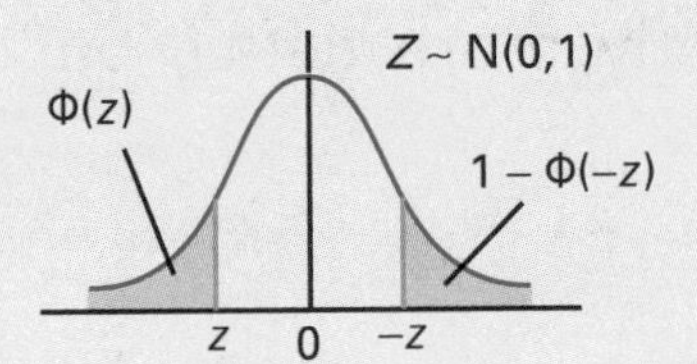

Case 4: Wanted: $P(X > a)$, where $Z = \frac{a - \mu}{\sigma} < 0$. Work out your Z value, then do $\Phi(-z)$.

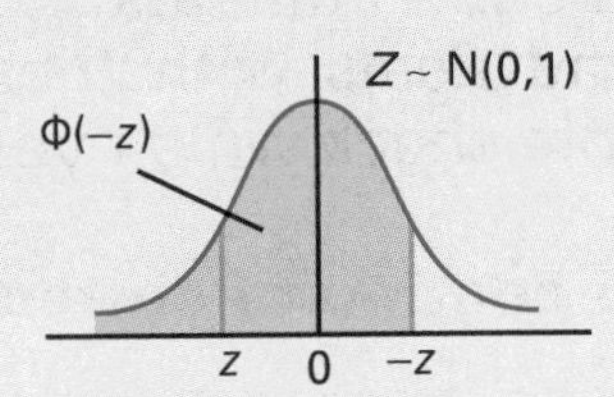

These diagrams are *essential* if you want to minimise any mistakes. You will also be required to be able to de-standardise – where you are given a probability and need to find (via Z) one or more of x, μ or σ.

Problems involving simultaneous equations

Sometimes we need to set up a pair of equations involving μ and σ.

Example: A machine fills large bags of sand, the weights of which are distributed normally with mean μ and variance σ^2. It is known that 5% of the bags are heavier than 111.515 kg, while 33% are lighter than 96.92 kg. Find the values of μ and σ.

Answer: 5% heavier than 111.515 kg means that 95% are lighter, and the Z value corresponding to a probability of 0.95 is 1.645. This means that $\frac{111.515 - \mu}{\sigma} = 1.645$; in other words

$$111.515 - 1.645\sigma = \mu \qquad ①$$

Also, 33% lighter than 96.92 kg puts our Z value into the negative region. The symmetrical alternative is the negative of the Z value for 67%, which turns out to be −0.44. Therefore $\frac{96.92 - \mu}{\sigma} = -0.44$, which put more usefully becomes

$$96.92 + 0.44\sigma = \mu \qquad ②$$

Equations ① and ② give $111.515 - 1.645\sigma = 96.92 + 0.44\sigma$; so $\sigma = 7$. Substituting $\sigma = 7$ into ① gives $\mu = 100$.

Exam questions answers: pages 161–2

1 A number of chimpanzees were given IQ tests. Their IQ was discovered to be normally distributed, with mean 80 and variance 25.
(a) Find the probability that a chimpanzee picked at random has an IQ score higher than 88.
(b) Find the probability that a chimpanzee chosen at random has an IQ score less than 73.
(c) Find the probability that a chimpanzee picked at random has an IQ score between 75 and 84.
(d) What are the approximate values within which the central two-thirds of the IQ scores lie. (20 min)

2 A group of 60 people are asked to close their eyes and estimate the length of one minute. Three people overestimated by at least 5 seconds, while nine people underestimated by at least 3 seconds. Estimate the number of people who estimated within plus or minus 5 seconds of exactly one minute, by finding the mean and variance of this distribution. (15 min)

Examiner's secrets

Once you have found your answer, take a second to write a full sentence which puts the final figure in terms of the question – something like 'so the probability that a value will be greater than 64 is 0.98 (2 d.p.)'.

Examiner's secrets

To de-standardise a variable, look up its Z value, and set up the equation you would have to find Z. You will then have a simple equation to solve!

Action point

Turn over the page, and try to draw the four standard diagrams. Think of the type of probability first – start with $x - \mu > 0$.

Examiner's secrets

This is more about algebra than statistics! A good tip when attacking *any* A-level question is to ask yourself 'What *can* I do now?', rather than 'What *should* I do now?' Here, all you *can* do is get to the point where you have two equations. By then it's all over!

Checkpoint 3

If X has a mean of 6 and variance 25, find the value of k such that
(a) $P(X < k) = 0.87$ and
(b) $P(X > k) = 0.66$.

Normal approximations

Many examination boards expect you to be familiar with the notion that the normal distribution, under the right circumstances, is a very accurate approximation for the binomial and Poisson probability models. This spread deals with these in slightly more detail, but double check which you need.

Action point

Practise these, by just making up your own examples using $X \le a$, $X \ge b$ and $X = c$, where a, b and c are any numbers.

Examiner's secrets

Avoid silly mistakes by getting into the habit of drawing these little diagrams when working out a continuity correction. It's time well spent!

Action point

Make a table up of the distributions you need to know – including mean, variance, PDF, *and* the conditions under which they can be approximated. You need to know this stuff!

Checkpoint 1

50 people each flip a coin. Find the probability that there will be (a) exactly 25 heads and (b) less than 30 heads.

Examiner's secrets

When approximating a binomial distribution, always *justify* your approximation by showing that both np and $nq > 5$, and then tell the examiner that this is enough evidence to proceed.

The continuity correction

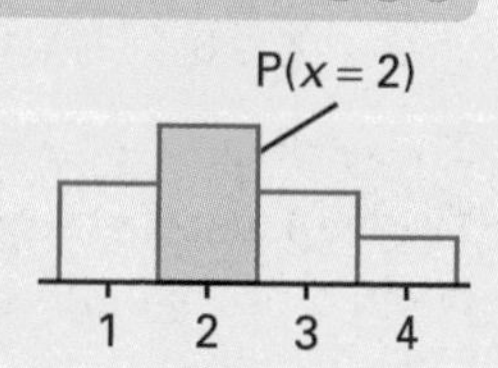

The normal distribution generally involves the *continuous* variable X. Both the binomial and Poisson distributions deal with *discrete* variables. As the histogram shows, to find the probability $X = 2$, we must include everything from $X = 1.5$ to 2.5, since all these values would have rounded to 2 anyway.

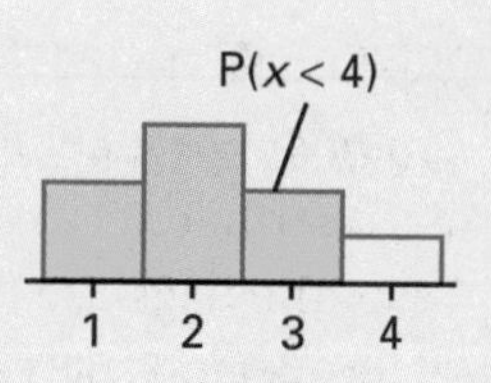

With cumulative probabilities, we can see that to find $P(X \ge 3)$ we would have to start at $X = 2.5$, so that all values associated with $X = 3$ are included. $P(X < 4)$ would start at 3.5.

Approximating the binomial distribution

As a rough guide, if $X \sim \text{Bin}(n,p)$ and both np and nq are bigger than 5, then we can approximate very closely using the normal distribution. This new distribution of X will have mean np and variance npq, and we now write $X \sim N(np,npq)$. In general terms, if a question asks you to do something that would involve some incredibly long or complicated calculation, then 'approximation' type alarm bells should be ringing!

Example: I am going to roll 30 dice, and count the number of sixes obtained (please don't ask *why*!). The probability I get exactly 10 sixes would look like this: $P(X = 10) = ({}^{30}C_{10})(\frac{1}{6})^{10}(\frac{5}{6})^{20}$. Not friendly! Instead, use the normal approximation, where $X \sim N(np,npq)$, i.e. $X \sim N(5, \frac{25}{6})$. Now, I need to find $P(X \le 10.5)$ minus $P(X \le 9.5)$, which would give the probability of any X value that might round up or down to 10. The two corresponding Z values are

$$\frac{10.5 - 5}{\sqrt{\frac{25}{6}}} = 2.69 \text{ (2 d.p.)}$$

and

$$\frac{9.5 - 5}{\sqrt{\frac{25}{6}}} = 2.20 \text{ (2 d.p.)}$$

The next step is to calculate $P(\text{'}X = 10\text{'}) = \Phi(2.69) - \Phi(2.20)$ $= 0.996\,4 - 0.986\,1 = 0.010\,3$.

Imagine we had wanted $P(X \le 10)$ – we would have to have done 11 separate calculations, one for each of the X values 0 to 10, and then added them up. We can see directly from the above that $P(X \le 10)$, which corresponds to $P(X \le 10.5)$, is $\Phi(2.69) = 0.996\,4$. The ease with which this type of calculation is carried out is the true strength of the approximation.

Approximating the Poisson distribution

When faced with a Poisson distribution which has a parameter $\lambda > 10$ we look to an approximation using the normal distribution in which the mean and variance are both λ. Formally, we write that $X \sim N(\lambda, \lambda)$.

Example: On a particularly dangerous stretch of road, the average number of accidents per year is 215. Find the probability that, in any one month, there will be less than 25 accidents. Also, find the probability that, in any two months, there will be exactly 30 accidents.

Answer: If there are 215 accidents per year, then the mean number per month will be $215 \div 12 = 17.92$ (2 d.p.). As this value is greater than 10 we are justified in approximating this distribution normally. This means that $X \sim N(17.92, 17.92)$ is a suitable model for X.

The probability that $X < 25$ corresponds to a corrected probability that $X \leq 24.5$. We get that $Z = \dfrac{24.5 - 17.92}{\sqrt{17.92}} = 6.58 \div 4.233 = 1.55$. From tables, $\Phi(1.55) = 0.939\,4$, and so *the probability there are no more than 25 accidents in one month is 0.939 4.*

In two months, we expect there to be about $215 \div 6 = 35.83$ (2 d.p.) accidents. We approximate the original distribution of X using $X \sim N(35.83, 35.83)$, and find the probability that X is between 29.5 and 30.5 (using the continuity correction). The two corresponding Z values we need are $\dfrac{30.5 - 35.83}{\sqrt{35.83}} = -0.89$ (2 d.p.) and $\dfrac{29.5 - 35.83}{\sqrt{35.83}} = -1.06$ (2 d.p.).

A sketch (you do it!) of what we're after shows that instead of using these negative values, we use the positives; what's more, we only need to do $\Phi(1.06)$ minus $\Phi(0.89)$, which is $0.855\,4 - 0.813\,3 = 0.042\,1$. So *the probability that, in two months, there will be exactly 30 accidents is 0.042 1.*

Examiner's secrets

Making the right decision about when to use an approximation depends on learning the criteria. Different books say different things – so learn what your exam board wants. A safe rule is that when numbers become unfriendly to use on your calculator – approximate!

Checkpoint 2

A piece of material has about 15 faults per metre. By using a normal approximation, find the probability that there will less than 12 faults in a particular 1 m^2 piece.

Examiner's secrets

Forgetting the continuity correction is probably the most common mistake made with this kind of work.

Exam questions

answers: page 162

1 When a certain species of flatfish have developed, their eyes shift from their conventional position on each side of their heads, to both being on one side. This allows them to lie flat on the bottom of the sea. By six months of age, 45% of flatfish have developed. A diver discovers a shoal of 200 flatfish all around six months of age. Find the probability that in this sample the number of developed flatfish is: (a) exactly 90; (b) less than 85; (c) more than 90; (d) between 82 and 91 inclusive. (20 min)

2 The number of seeds produced by a particular plant follows a Poisson distribution where $\lambda = 200$. Find the probability that:

(a) more than 160 seeds are produced;

(b) more than 240 seeds are produced;

(c) between 180 and 230 (inclusive) seeds are produced.

It is known that one seed in ten will germinate and grow.

(d) Show that the number of germinated seeds produced by the plant follows a Poisson distribution with parameter 20.

(e) Find the probability that more than 30 germinated seeds are produced.

(25 min)

Hypothesis testing

We have already seen tests that involve finding whether or not a single value could have reasonably come from a particular distribution. The emphasis now is not on the single observation; a typical test looks at the mean of a sample and says, 'Hey, this value looks a bit unusual for this distribution – maybe our idea of the distribution parameters are wrong.' Our focus is primarily on normally distributed variables.

Examiner's secrets

This recipe for hypothesis testing must be followed if you want to guarantee as many marks as possible – and it looks very professional to the examiner!

Action point

Turn the page over and try to list the ingredients for a successful hypothesis test.

The jargon

If you were to take lots of independent samples from a distribution, then the means of all these samples follow the 'sampling distribution of the means'. See 'The central limit theorem' (page 142).

Examiner's secrets

n will not be less than 'large' as it's not on the A-level syllabus! If *n* is less than 30, don't panic. This is only a rough rule, and the result still holds as long as you aren't going too small!

Examiner's secrets

As with all normal distribution and significance test questions, these sketches are essential if you want to minimise possible errors. Here, they show whether or not your *Z* value is going to be negative when you look it up, *and* they show clearly where the critical region starts!

Checkpoint

Imagine you are carrying out a hypothesis test. Draw the diagrams corresponding to testing for (a) $\mu > 10$, (b) $\mu < 10$ and (c) $\mu \neq 10$.

Testing a mean

These tests all follow much the same method, that is:

- State the null and alternate hypotheses.
- Consider the distribution under H_0.
- Decide on the level of the test.
- State the rejection criteria.
- Calculate the value of the test statistic.
- Make your conclusion.

By the central limit theorem, the **sampling distribution of the means** follows a normal distribution with mean $\bar{X}$ and variance $\frac{\sigma^2}{n}$. Our test statistic is the standardised variable $Z = \frac{\bar{X} - \mu}{\frac{\sigma}{\sqrt{n}}}$, where $Z \sim N(0,1)$. This corresponds to a **null hypothesis** H_0 that the true population mean is μ.

If we don't know the variance, $\hat{\sigma}^2$, of the population, then we estimate it using $\hat{\sigma}^2 = \frac{n}{n-1}S^2$. However, if n is larger than, say 30, we see that $\frac{n}{n-1} \approx 1$, and so we can simply use $\hat{\sigma}^2 = S^2$. Our test statistic is then $Z = \frac{\bar{X} - \mu}{\frac{S}{\sqrt{n}}}$ where $Z \sim N(0,1)$. Again, this corresponds to a null hypothesis H_0 that the true population mean is μ.

The **alternate hypothesis** depends on the question. If it says that the true mean is thought to be say, 15, but you are testing to see if it's higher, then you must write your alternate hypothesis as H_1: $\mu > 15$. If you are testing to see if it's lower, then you must write H_1: $\mu < 15$. If you are simply arguing about the population mean, without stating what you believe it to be, then it's a two-tailed test. This corresponds to an alternate hypothesis that the true population mean does not equal μ, that is H_1: $\mu \neq$ value given.

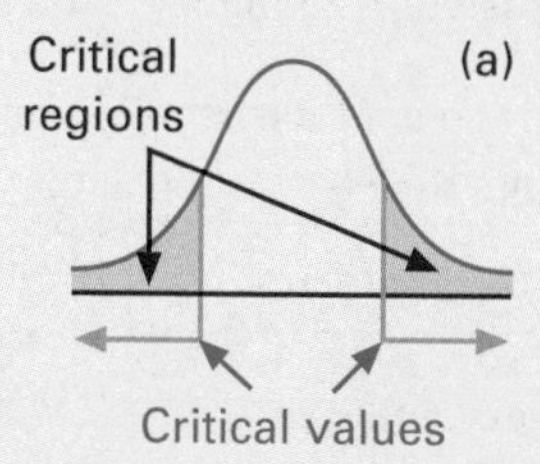

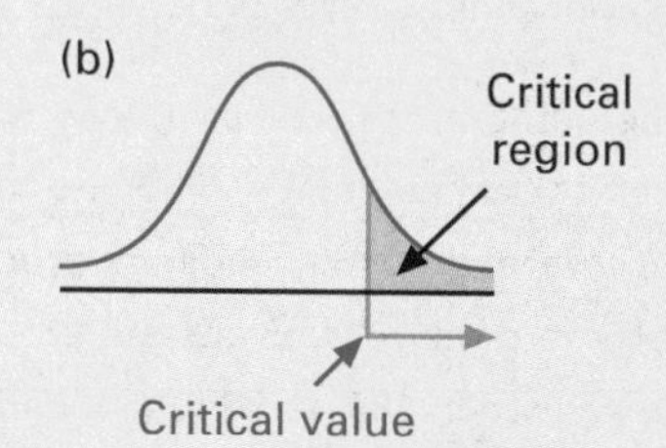

The level of significance will be given in the question, and the critical values obtained from tables of the variable Z. The two diagrams show (a) a two-tailed test and (b) a one-tailed test, where the alternate hypothesis was H_1: $\mu > 15$. The case for a one-tailed test where $\mu < 15$ would be the opposite end of the tail. Let's do an example.

Example: A normal distribution is thought to have $\mu = 15$. A random sample of size 100 was taken having a mean of 17.6 and standard error 14.5. Is there evidence, at the 5% level of significance, that the population mean has increased?

Answer: Let the population mean be μ, and the population variance be σ^2. Then we have:

H_0: $\mu = 15$ (the population mean has not increased)
H_1: $\mu > 15$ (the population mean has increased)

Under H_0, $\bar{X} \sim N(\mu, \frac{\sigma^2}{n})$, but σ^2 is unknown, and as n is large we use S^2 instead, so that $\bar{X} \sim N(\mu, \frac{S^2}{n})$. It's a one-tailed test, at the 5% level, so we reject H_0 if $Z > 1.645$ (see diagram). To work out Z, we use

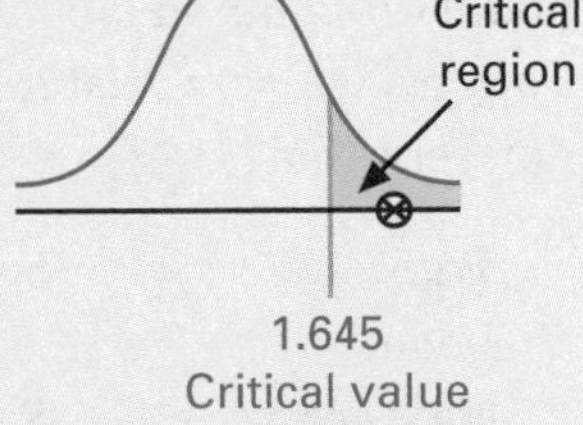

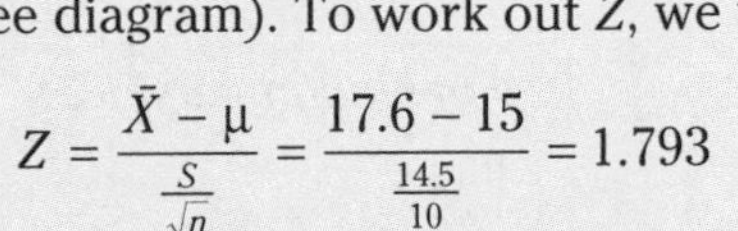

$$Z = \frac{\bar{X} - \mu}{\frac{S}{\sqrt{n}}} = \frac{17.6 - 15}{\frac{14.5}{10}} = 1.793$$

The calculated value is marked with a ⊗ and it is clearly within the critical region. We conclude that, as $Z > 1.645$, there is evidence to reject H_0 and accept the alternate hypothesis that the population mean has increased.

Testing the difference between two means

In actual fact the test follows exactly the same recipe, apart from the bit where you calculate the Z value. You should now use the formula for the *sampling distribution of the difference between means*.

If

$$X_1 \sim N(\mu_1, \sigma_1^2)$$

and

$$X_2 \sim N(\mu_2, \sigma_2^2)$$

then

$$Z = \frac{\bar{X}_1 - \bar{X}_2 - (\mu_1 - \mu_2)}{\sqrt{\frac{\sigma_1^2}{n_1} + \frac{\sigma_2^2}{n_2}}}$$

Examiner's secrets

See how the recipe is followed, with all terms and distributions identified and clearly stated. This gives you a path to stick to which will help keep your confidence to its maximum and mistakes at a minimum!

Examiner's secrets

What do you do if the distribution you are taking samples from is not normal? You approximate using a normal approximation, stating clearly why you think the approximation is valid. So, for the case where the original distribution follows a binomial pattern, you would say that n is quite large, p is quite small, and the approximation is that $X \sim N(np, npq)$.

Action point

Check out for yourself whether this formula is in your exam board's formula book – if not, you'll need to learn it!

Exam questions

answers: page 162

1. The manufacturer of a particular brand of video tape claims that their 180-minute tape actually lasts, on average, 186 minutes with standard deviation 5.4 minutes. A magazine decides to investigate the manufacturer's claim and using a sample of size 49 finds that, at the 5% level, the mean playing time of the tapes is less than 186 minutes. What can be said about the value of the sample mean for the investigator to have taken this decision? (15 min)
2. The time taken for 65 pigeons to fly from Trafalgar Square to their home had a mean value of 98.4 minutes and standard error 3.6 minutes. One particular pigeon owner said that they were faster than usual; for many years the average time had been 99.2 minutes. Test, at the 5% level, if the pigeon owner's boasts are well founded. (15 min)

Sampling, estimates and confidence

When we need to find something out about a distribution, usually the most cost-effective way to do so is through taking a sample. This is not normally the whole of the population, but some subset of it. The behaviour of the sample statistics, as we take more and more samples, is quite interesting indeed.

Sample statistics versus population parameters

Usually, what we need to know are the mean and variance of the population, but in fact all we can find are these values for the sample. They are closely related, but not quite the same, and so we call them different things.

Statistic:	*Sample version:*	*Population version:*
Mean	%	μ
Variance	S^2	σ^2

Examiner's secrets

Don't mix up your names – it is very important that you understand the difference between $\bar{x}$ and μ and how they are related, or easy marks will be lost!

Action point

Turn the page over, and try to remember the symbols for population and sample means and variances!

Best estimators

We can only make guesses, based on samples, of what the true μ and σ actually are. We call these calculated guesses the **best estimates** (or sometimes **unbiased estimates**). Formally, we say that for a population with unknown mean, μ, and unknown variance, σ^2, the most efficient estimator for μ is $\bar{x}$, and the most efficient estimator for σ^2 is $\frac{n}{n-1}S^2$, where n is the sample size. Your calculator should do both the sample standard deviation, σ_n, and the estimated population standard deviation, σ_{n-1}.

Here is where *keywords in the question* are essential clues: if it says 'find the variance of the sample' then you find σ_n. If it says 'find an estimate for' – or 'use the sample to estimate' the population standard deviation – then use σ_{n-1}. Making the right decision is the whole point of the question!

Examiner's secrets

It's all about decisions – make sure you look for the magic word 'estimate' before applying this theory! Just ask yourself – 'Am I going from the sample to the population? Or just finding L?'

Example: In a sample of 12 men's heights, it was found that the mean height was 1.8 m and the standard deviation 2. The best estimate for the population's height would be $\hat{\mu} = 1.8$ m, and the best estimator for the variance, $\hat{\sigma}^2 = \frac{12}{11}2^2 = 4.36$ (2 d.p.).

The jargon

A symbol with a 'hat' means it's a best estimator for that parameter – so $\hat{\mu}$ means the best estimator for μ.

The central limit theorem

The central limit theorem says that when taking samples of size n from a population which has mean μ and variance σ^2, then the mean of the samples themselves follow a normal distribution having mean μ and variance $\frac{\sigma^2}{n}$. Formally, if $X \sim N(\mu,\sigma^2)$, then $\bar{X} \sim N(\mu,\frac{\sigma^2}{n})$. Dividing by n tells us that as the sample size gets bigger, our estimate, $\bar{X}$, gets closer to the true mean. This is in fact true for a population following *any* distribution, as long as the sample size is greater than about 30.

Checkpoint 1

Estimate the mean and variance of the weight of parakeets if a sample of 16 had mean 55 g and standard error 4 g.

Example: A particular gorilla population is known to have normally distributed IQ, having mean score 73 and standard deviation 23. The probability that in a sample of 36 gorillas the mean IQ score is less than 83 is calculated as follows: $\bar{X} \sim N(73, \frac{23^2}{36})$. We want $P(\bar{X} < 83)$. We find the Z value:

$$Z = \frac{83 - 73}{\frac{23}{6}} = \frac{10}{3.8} = 2.61 \text{ (2 d.p.)}$$

$\Phi(2.61) = 0.995\,5$, so the *probability that the mean of the sample is less than 83 is 0.995 5.*

Confidence limits

To get round the fact that sample means vary, we define a region around the mean which we are confident the population mean will fall into. The wider the region, the more confident we become. The most common **confidence interval** is the 95% version, where we are 95% sure that the mean is within the two numbers.

When we know σ, the actual interval is given by $\bar{x} \pm 1.96\frac{\sigma}{\sqrt{n}}$. If we don't know σ, the interval becomes $\bar{x} \pm 1.96\frac{S}{\sqrt{n}}$. The 1.96 comes from the Z value for 97.5% – so that 2.5% is taken from the top and bottom of the distribution. The diagram shows why it's plus or minus – because one side is positive, the other negative.

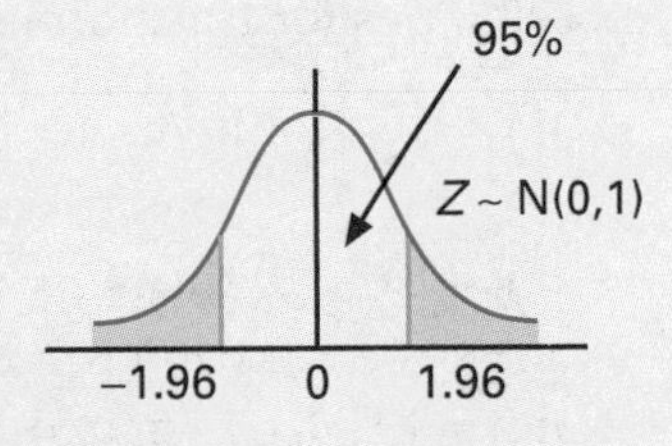

Other values of Z which come up a lot are 2.575 for 99% confidence intervals, and 2.326 for 98% confidence intervals.

Example: A sample of 25 people measured how long their arms were. The mean length of their arms was 60 cm, with standard deviation 5 cm. We can be 99% confident that the mean of the population's arm lengths lies in the interval $60 \pm 2.575(\frac{5}{5})$ cm, that is, between 57.425 cm and 62.575 cm.

Checkpoint 2

The lengths of phone calls for a particular town is known to follow a normal distribution with mean 5 min and variance 16 min. Find the probability that a sample of 25 calls has mean less than 5.5 min.

Examiner's secrets

There are two formulae here – you need to know them! The numbers (e.g. 1.96) in the formulae refer to Z values; the three Z values given are the most popular, but be aware that you may have to find your own!

Checkpoint 3

A sample of 64 bottles of milk were found to have mean 0.98 pints and variance 0.4 pints. Find a 90% confidence interval for the mean of all bottles of milk, based on this sample.

Examiner's secrets

Technical point – when you don't know s then you can replace it by S in the formula when n is large, say greater than 30. Strictly speaking, when n is less than 30 it gets less accurate.

Action point

Write down the Z values corresponding to probabilities of 95%, 98% and 99%.

Exam questions

answers: pages 162–3

1 A lift is designed to carry a maximum of 12 people. A sample of 12 people who used the lift had mean weight 86 kg with standard deviation 3 kg. Assuming a normal distribution, use this sample value to estimate the mean and standard deviation of all of the people who work in the building. Find 95% confidence limits for the upper and lower boundaries of the weight that the lift can carry. (15 min)

2 An investigation into the sizes of portions of chips sold by various chip shops was carried out. It was found that for one sample the variance of the sizes was 45 g, the mean was 124.5 g and $\Sigma x = 6\,225$. Find the probability that a particular shop from this sample served a portion of chips weighing at most 130 g. Write down estimates for the mean and standard deviation of all portions of chips served. (15 min)

Independent normal distributions

This section looks at what happens to the mean and standard deviation of data sets under certain changes. We may simply be transforming one set of data. We may be combining different sets of data. Conversely, we may even be taking more than one sample value from a single distribution. To find probabilities under these conditions we need to know how the final transformed, combined or repeated observations are themselves distributed.

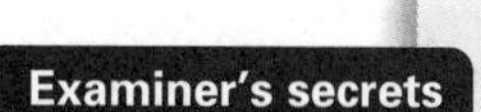

Examiner's secrets

These transformations follow the same basic ideas as transforming means of samples, so they should be familiar from there!

Transformations of the type $ax + b$

You will be very happy indeed to learn that when the data in a normally distributed set are transformed, they behave exactly as we would want them to. In other words, if $y_i = ax_i + b$ then $\mu_y = a\mu_x + b$ and $\sigma_y^2 = a^2\sigma_x^2$. What happens is that the mean goes through exactly the same transformation as each x_i value, but the variance only gets multiplied by the square of the a term in the equation. Here are most of the possibilities for transforming $X \sim N(40,5^2)$:

Action point

To really get to grips with these, try writing out the possibilities with your own values of a and b – *remember that negatives become positive when it comes to transforming the variance!* If you're stuck, try $a = 5$, $b = 4$ and $a = -3$, $b = -7$.

- → if $y_i = x_i + 10$ then $\mu_y = \mu_x + 10 = 40 + 10 = 50$,
  and $\sigma_y^2 = \sigma_x^2 = 5^2 = 25$;
- → if $y_i = 8x_i + 10$ then $\mu_y = 8\mu_x + 10 = 8 \times 40 + 10 = 330$,
  and $\sigma_y^2 = 8^2\sigma_x^2 = 8^2 \times 5^2 = 1\,600$;
- → if $y_i = 10 - 8x_i$ then $\mu_y = 10 - 8\mu_x = 10 - 8 \times 40 = -310$;
  and $\sigma_y^2 = (-8)^2\sigma_x^2 = (-8)^2 \times 5^2 = 1\,600$.

Combinations of the type $ax + by$

Suppose X is a random variable, and $X \sim N(\mu_x,\sigma_x^2)$. Also, Y is a random variable, and $Y \sim N(\mu_y,\sigma_y^2)$. What can we expect when we take a sample from each and add them? Luckily enough, we once again get a very friendly result. If X and Y are both independent normally distributed variables, then $X + Y \sim N(\mu_x + \mu_y,\sigma_x^2 + \sigma_y^2)$. What's even better is that *any* combination of X and Y follows the same rules.

So, $aX + bY \sim N(a\mu_x + b\mu_y,a^2\sigma_x^2 + b^2\sigma_y^2)$. Here are all the possibilities when $X \sim N(10,2^2)$ and $Y \sim N(20,4^2)$:

Action point

Again, the best way to really get to grips with these is to try your own numbers for a and b. Try $a = 2$, $b = 5$; also try $a = -2$, $b = 5$, and lastly try $a = 2$, $b = -5$.

- → $5X + Y$ has mean $5\mu_x + \mu_y = 5 \times 10 + 20 = 70$,
  and variance $5^2\sigma_x^2 + \sigma_y^2 = 5^2 \times 2^2 + 4^2 = 116$;
- → $X + 7Y$ has mean $\mu_x + 7\mu_y = 10 + 7 \times 20 = 150$,
  and variance $\sigma_x^2 + 7^2\sigma_y^2 = 2^2 + 7^2 \times 4^2 = 788$;
- → $5X + 7Y$ has mean $5\mu_x + 7\mu_y = 5 \times 10 + 7 \times 20 = 190$,
  and variance $5^2\sigma_x^2 + 7^2\sigma_y^2 = 5^2 \times 2^2 + 7^2 \times 4^2 = 884$;
- → $5X - 7Y$ has mean $5\mu_x - 7\mu_y = 5 \times 10 - 7 \times 20 = -90$,
  and variance $5^2\sigma_x^2 + -7^2\sigma_y^2 = 5^2 \times 2^2 + 7^2 \times 4^2 = 884$.

A common mistake is to forget that $\text{Var}(X + Y) = \text{Var}(X) + \text{Var}(Y)$ but $\text{Var}(X - Y) = \text{Var}(X) + \text{Var}(Y)$ as well!

Combinations of x_1 and x_2

This is a very tricky area. Imagine we had a random variable, X which follows a normal distribution having mean μ and variance σ^2. One particular observation taken from this distribution, say x_i, would have mean μ and standard deviation σ. What about if we take another observation, say x_j, and add it to our first one? The mean of $x_i + x_j$ must be $\mu + \mu = 2\mu$. But what about the variance? What we have here is *not* two lots of one observation – we have not taken a value and doubled it. We have taken two *separate* and *independent* values, and so we must add the two variances, to get $\sigma^2 + \sigma^2 = 2\sigma^2$. Make sure, before proceeding in a question, whether you are taking a single observation and transforming it, or taking many observations and combining them.

Examples:

- → Lifts normally have weight limits defined by the number of people they can carry. Weighing one person and multiplying by 20 is not the same as weighing 20 people. The first case gives $20x_1$, the second gives $x_1 + x_2 + x_3 + \ldots + x_{20}$.
- → Weighing the contents of a bag of sugar and multiplying by 12 is not the same as taking 12 bags and weighing them. The first case gives $12x_1$, the second gives $x_1 + x_2 + x_3 + \ldots + x_{12}$.

Formally, if $x_1, x_2, x_3 \ldots x_n$ are independent observations from the same normal distribution having mean μ and variance σ^2, then we can say that $x_1 + x_2 + x_3 + \ldots + x_n \sim N(n\mu, n\sigma^2)$. So the 20 different people in the lift example follow a normal distribution having mean 20μ, and variance $20\sigma^2$; as opposed to 20 lots of one person's weight having mean 20μ and variance $20^2\sigma^2$.

We can see that when we multiply the variances they quickly get huge, but when we add them they stay manageable. This tells us that any predictions based on only one observed value are not as reliable as those made using many observed values, since the (scaled) variance for the first prediction will be much larger than the variance of the second.

Examiner's secrets

As a rule, *you should always read the whole question before attempting any part of it.* You will normally find it much easier to decide whether it's a multiple observation or an observation multiplied once you can see all the parts in context!

Action point

Try to think up at least three examples of your own, and write them up in exactly the same way as they have been here. You could start with 'measuring how tall one person is and multiplying their height by ten is not the same as . . .'.

Examiner's secrets

The results for two normal distributions will *only* work if they are independent of each other. The same goes for repeated observations. You will usually be told that they are independent in the question. If not, once you are sure, remember to state that you are assuming independence before you continue with this part of the question.

Exam question answer: page 163

In a school canteen, the portions of beans are normally distributed with mean 100 g and variance 16 g, while chips follow a normal distribution having mean 150 g and variance 25 g.

(a) State the distribution you would use and calculate the probability that one portion of beans and one portion of chips together weigh more than 265 g.

(b) State the distribution you would use and calculate the probability that a double portion of chips plus one portion of beans weigh less than 380 g.

(c) State the distribution you would use and calculate the probability that the portions of chips that I and five of my friends buy have a combined weight of more than 1 kg. (20 min)

Goodness of fit tests

Imagine you had some data, and you thought you could model it very well with, say, a binomial distribution. The role of the goodness of fit test is to tell you just how good your model is. As usual, the accuracy of the conclusion is in some part decided by the level of significance. The test is sometimes named after the variable which makes it all possible: the χ^2 goodness of fit test.

Action point

Turn the page over and try to write out the formula for working out X^2_{calc}.

Examiner's secrets

Drawing up the table is, as usual, essential for making this work as error free as possible.

The jargon

The *degrees of freedom* is simply the number of things that are 'free' to vary. Here, the first four class frequencies were free (we could have tried any numbers really) but the last one is restricted in that it must make the total up to exactly 100. See below for further cases.

Examiner's secrets

Suppose the conclusion had been reached that the spinner was not fair. The test does not actually tell us what the actual distribution of scores is – only what it isn't, and so we are really quite limited as to how detailed our conclusion can be.

The test: a practice run

An experiment into the fairness of a five-sided spinner yielded the following results:

Score	1	2	3	4	5
Frequency	27	21	17	22	13

The numbers in the frequency row are called the observed values, and the ith observed value is written as O_i. Were the spinner to be fair, then we would expect a frequency of 20 in each, as the total is 100. The ith expected value is called E_i. It just so happens that when we find the value of $\sum \frac{(O_i - E_i)^2}{E_i}$ we get a variable which follows the χ^2 distribution.

Score	1	2	3	4	5
Frequency	27	21	17	22	13
Expected	20	20	20	20	20
$(O_i - E_i)^2$	$7^2 = 49$	$1^2 = 1$	$3^2 = 9$	$2^2 = 4$	$7^2 = 49$

So $\sum \frac{(O_i - E_i)^2}{E_i} = \frac{49}{20} + \frac{1}{20} + \frac{9}{20} + \frac{4}{20} + \frac{49}{20} = \frac{112}{20} = 5.6.$

We call this final figure χ^2_{calc}. The only restriction on our calculation was that the total of the frequencies has to be the same. We had 5 classes, so our **degrees of freedom** is $5 - 1 = 4$. We see from the tables that $\chi^2_{5\%}(4) = 9.49$, and since $\chi^2_{calc} < \chi^2_{5\%}(4)$ we conclude that the scores follow a uniform distribution, i.e. that the spinner is fair.

The only thing we did not state was our null hypothesis. We should have written H_0: the spinner is fair. We don't really need an alternate hypothesis with this test as our distribution either fits or not!

Testing different distributions

The example above illustrates the essential layout and basic procedure that every goodness of fit test should follow. The next example looks at one involving the binomial distribution. The only difference here is that the expected values require a bit more work.

Goodness of fit test for a binomial distribution

In order to check the postal service of Flunstone, 100 samples of four letters were sent and the number arriving the next day were recorded in the following table. We will test at the 5% level to see if the observed data follows a binomial pattern.

Number arriving next day	0	1	2	3	4
Frequency	2	7	22	40	29

H_0: the distribution follows a binomial pattern. To be able to calculate our expected frequencies, we first need the value of p for this distribution. We do that by finding the mean, remembering that the mean of a binomial distribution is np. Our mean turns out to be 2.87, and $n = 4$, which gives us $p = 0.72$. The expected value in each class is then given by the probability × total frequency, so for the first one this would be $P(X = 0) \times 100 = 0.6$ (1 d.p.). The table can now be extended to include all relevant rows.

Number arriving next day	0	1	2	3	4
Frequency	2	7	22	40	29
Expected (to 1 d.p.)	0.6	6.3	24.4	41.8	26.9
$(O_i - E_i)^2$	1.96	0.49	5.76	3.24	4.41

So $\chi^2_{\text{calc}} = 3.826$. There are five classes, and two restrictions (the totals agree, and the means agree), so our degrees of freedom = 5 – 2 = 3. From tables: $\chi^2_{5\%}(3) = 7.82$. Since $\chi^2_{\text{calc}} < \chi^2_{5\%}(3)$ we conclude that the distribution follows a binomial pattern.

The test for any other distribution will work the same. If you are given probabilities or means and variances then use them. If not, they will need to be calculated and the degrees of freedom adjusted accordingly.

Action point

When fitting your own expected distribution, you need the facts: the mean of a binomial distribution is np; the mean of a Poisson distribution is λ; and you will need to calculate the mean *and* variance if you are trying to fit a normal distribution. Make a list of distributions and their associated means.

Checkpoint

A section of a test has three questions. Out of 20 people no one got them all right, only 7 people got one right, and 13 people got two out of three. Carry out a Chi-squared (χ^2) test to see if these scores follow a binomial distribution.

Examiner's secrets

Degrees of freedom can be tricky. If you are told something, then you don't need to worry about it and it does not affect the degrees of freedom. The basics are this: any time the totals agree then minus one; any time you work out the mean then minus one; any time you work out the variance then minus one.

Exam questions answers: page 163

1 A group of 120 people were asked by a marketing company to give a particular fragrance on test a mark out of four. Test their responses, at the 5% level, to see if they had an overall favourite. (15 min)

Mark	1	2	3	4
Frequency	35	32	25	28

2 At a particular shop, polo neck jumpers are sold in four colours. The manager thinks that sales of the four colours red, green, black and white are split in the proportion 4:4:5:12. Last week, she sold 800 jumpers: 110 red, 130 green, 150 black and 410 white. Do these figures differ significantly from expectations at the 5% level? (15 min)

Correlation

When two variables seem to share a relationship, we say that they are correlated. This means that the variation in one can be explained, to some degree, by variation in the other. We restrict our interest to straight line or linear relationships. Common examples are an increase in rainfall being associated with bigger harvests, or a decrease in calorie intake being associated with losing weight.

The product moment correlation coefficient

When we think two variables are related in some way, it seems a good idea to measure the *strength* of that relationship. One way to do this is to calculate the **product moment correlation coefficient**, ***r***. It is based on three numbers: the variances of X, and of Y; and the *covariance* of X and Y – which is the combined variance between the two, from both their means. These are called S_x, S_y and S_{xy} respectively. Now,

$$S_x^2 = \frac{\Sigma x^2}{n} - (\bar{x})^2,\ S_y^2 = \frac{\Sigma y^2}{n} - (\bar{y})^2 \text{ and } S_{xy} = \frac{\Sigma xy}{n} - \bar{x}\bar{y}.$$

Formula: $r = \dfrac{S_{xy}}{S_x S_y}$. It must always lie between −1 and 1.

Example: For a set of bivariate data X and Y, $S_x = 4$, $S_y = 6$, and $S_{xy} = 12$. The product moment correlation coefficient will be given by $S_{xy} \div (S_x \times S_y) = 12 \div 24 = 0.5$.

The jargon

The 'product moment correlation coefficient' might sometimes be called the 'Pearson's product moment correlation coefficient', after the mathematician who invented it while investigating the relationship between the number and size of potatoes in his vegetable patch!

Examiner's secrets

You need to learn what the terms S_x, S_y and S_{xy} stand for – their meanings will not be given in the formula book! Like all of A-level maths, the best way to learn them is to use them – so get practising!

Your calculator 

Most calculators available have statistical functions built in. You will be expected to use these wherever possible.

On my calculator, the correlation coefficient is given by hitting COR; I can enter data in pairs, and then the calculator does the rest. You should be fully familiar with how your machine does things. Here's a quick example to check.

Example: Find the correlation coefficient for this data.

x	2	4	7	9	12	16	23	30
y	43	39	33	28	20	15	9	2

From your calculator, you should get the following information: $\Sigma x = 103$, $\Sigma y = 189$, $\Sigma x^2 = 1\,979$, $\Sigma y^2 = 5\,953$, $\Sigma xy = 1\,472$. You should also get that $r = -0.975\,426\,571$, or -0.975 (3 d.p.). It would be a good idea to write:

$$r = \frac{\dfrac{1\,472}{8} - 12.875 \times 23.625}{\sqrt{\left(\dfrac{1\,979}{8} - 12.875^2\right)}\sqrt{\left(\dfrac{5\,953}{8} - 23.625^2\right)}}$$

as well, to show working (just in case you've entered a wrong value).

Checkpoint 1

For a set of bivariate data, $S_x = 10$, $S_y = 2$ and $S_{xy} = -18$. Calculate the product moment correlation coefficient and comment on the result.

Examiner's secrets

Even if you've done most of the work on your calculator, it's still a good idea to show the final form of any formulae with the numbers in, to guarantee method marks, even if you've made a mistake.

Interpreting *r* and r^2 values 

What does it mean when $r = -0.975$? As mentioned above, r can go from −1 to 1. In addition, calculating r^2 will give a figure that tells us how much of the variation in y is due to x *in percentage terms*. So here, $r^2 = -0.975^2 = 0.951$, which means that over 95% of the variation in the y values can be 'explained' by variation in the x values. Here's a rough guide:

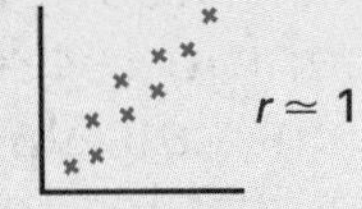

- r close to 1: there exists *very strong* positive correlation, so as one goes up, the other is very likely to go up with it.

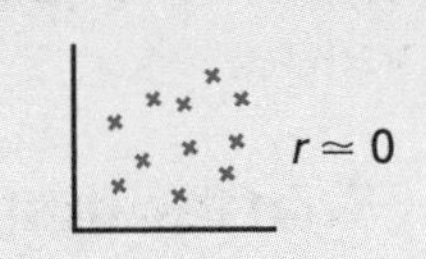

- r close to 0: there exists *very weak* correlation if any, so we might conclude that the two variables are not linked.

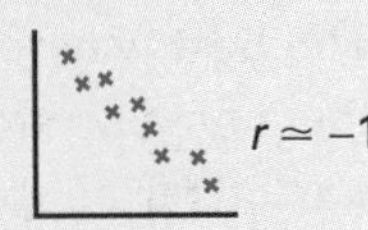

- r close to −1: there exists *very strong* negative correlation, so as one goes up, the other is very likely to go down.

Notice that r values near plus or minus 1 still give a very high value of r^2, and that the sign of r tells us the direction of any relationship.

Limitations of the theory

There are two main points that you need to know.

Firstly, the fact that two variables are correlated does not imply that changes in one has *definitely* caused the other to change. *Coincidental* correlation is a problem you need to bear in mind.

Secondly, r and r^2 are measures of *linear* correlation. A very low correlation coefficient can only indicate that these data do not form a straight line – they might, however, form a perfect quadratic curve. A scatter diagram would show this immediately; if you haven't drawn one, then you can only point out that it could be a possibility.

Action point

These little sketches are very useful as they sum up the strength and direction of the correlation present. Draw your own diagrams for the following cases of correlation: strong positive; weak positive; zero; weak negative; and strong negative.

Watch out!

On his birthday, Jude drinks pints of beer. Is the correlation between alcohol units consumed and reaction time positively or negatively correlated? If you don't see why it's positively correlated, sketch a graph with units consumed on the x-axis and reaction time on the y-axis.

Checkpoint 2

A set of bivariate data had a product moment correlation coefficient value of 0.5. What conclusions can you draw, and what are the limitations of those conclusions?

Watch out!

Correlation *does not* imply cause.

Exam question

answer: pages 163–4

The marketing manager of a large company is investigating the effects of advertising on his sales. Over the last seven weeks various numbers of pages in a particular magazine have been purchased. The corresponding sales figures, in thousands of units, are shown in the table below.

x (pages)	15	8	22	11	25	18	20
y (000s)	16	11	20	15	26	32	20

Plot these data on a scatter diagram.

(a) Calculate, to three decimal places, the product moment correlation coefficient.

(b) Comment on your results, using your scatter diagram and the analysis carried out so far.

(c) Choose six points from the given data which you believe would possess a much higher level of correlation. Explain your choice of points.

(d) Calculate the product moment correlation coefficient for the six chosen points, and comment on your results and their implications for the marketing manager. (30 min)

Examiner's secrets

For their A-level Statistics project, someone once compared the number of ducks on their local pond with the number of people passing through a particular door at Harrod's – and got almost perfect correlation! This is a typical case of *coincidental* or *spurious correlation* and was not taken up by Harrod's as a predictor for their future profits!

Regression

Drawing a line of best fit by eye is not as precise as it could be. The line of best fit models a distribution of bivariate data, or data that come in pairs. The idea of the 'most efficient' line of best fit is the driving force behind regression. It describes a process whereby we can calculate the equation for this line. Common examples are height versus weight or age versus value.

Scatter diagrams

The best way to view a relationship between two variables is to plot them on a **scatter diagram**. The data will be in pairs: while it may be obvious from the question, choosing the right variable to use as X is very important. X is normally know as the **independent** or **explanatory variable**. It is usually the one item out of each pair which we *choose*. The variable selected for Y is called the **dependent** or **response variable**. We are trying to see if changes in the *response* variable are *dependent* on changes in the *explanatory* or *independent variable*. Y is usually the one we measure, once we know X. If in doubt, try to think about which would cause the other.

Examples:

- Height versus weight: does height (X) explain weight (Y)?
- Age versus value: does the age of an object (X) affect its value (Y)?
- Blood alcohol level versus reaction time: does blood alcohol level (X) affect a person's reaction time (Y)?

The least squares regression line

There are always going to be gaps between the line of best fit and plotted points. The **least squares regression line** minimises the squares of the sizes of those gaps. It's a bit like finding the standard deviation of the points from the line, and then shifting the line until that deviation is as small as possible. The ideal line has equation $y = ax + b$, where a and b are constants. Fortunately you will only need to calculate three numbers: the variances of X, and of Y; and the *covariance* of X and Y – which is the combined variance between the two, from both their means. These are called S_x, S_y and S_{xy} respectively.

$$S_x^2 = \frac{\Sigma x^2}{n} - (\bar{x})^2, \quad S_y^2 = \frac{\Sigma y^2}{n} - (\bar{y})^2 \quad \text{and} \quad S_{xy} = \frac{\Sigma xy}{n} - \bar{x}\bar{y}$$

In our equation, $a = S_{xy} \div S_x^2$. To find b, note that the best line of best fit passes through $\bar{x}$ and $\bar{y}$, since we wouldn't want to look at the average X and find anything with it but the average Y. We find b using the equation $\bar{y} = a\bar{x} + b$ with known values of everything but b.

Example: In a set of data, $S_x = 5$, $S_y = 3.2$ and $S_{xy} = 15$. Also, the mean values of X and Y are 32 and 48 respectively. The equation for the line of best fit, $y = ax + b$, is calculated as follows. Find a using $a = S_{xy} \div S_x^2 = 15 \div 25 = 0.6$. Then, to find b, we solve $\bar{y} = a\bar{x} + b$, which gives us $48 = (0.6)(32) + b$, i.e. $b = 48 - 19.2 = 28.8$. So the line of best fit is $y = 0.6x + 28.8$.

Examiner's secrets

You should have had plenty of practice at drawing scatter diagrams from GCSE. At A-level you just need to be more careful about the scales for the axes. Choose them so that you can fit everything in – but make sure that enough detail and accuracy can be shown.

Examiner's secrets

Remember – marks are allocated for correct scales and labels as well as accurate plotting.

Action point

Try to think up five different pairs of variables which you believe to be positively correlated. Then do the same for variables which could be negatively correlated.

Examiner's secrets

You will be expected to use the statistical functions on your calculator wherever possible – *but* make sure that the final stages are written out, so you are showing the examiner that you know how to use the formulae!

Checkpoint

A set of bivariate data x has a mean of 45, while the mean of y is 2. If $S_x = 6.7$, $S_y = 8$ and $S_{xy} = 73$, find the equation of the least squares regression line of y on x.

To draw the least squares regression line of y on x, start at the intercept, $(0,b)$, and then plot the means (as our line must go through both of them). Join them up to get the least squares regression line!

Making predictions

The point of the least squares regression line is that it makes our vague idea of seeing a 'trend' in the data more real. We don't even need a *drawing* of the line; just its equation will do. *When making predictions*, unless you are actually *asked* to use a line, stick to substitution of the variables into the equation. In our last example, the equation was $y = 0.6x + 28.8$. If we knew that x was 100, then we can *predict* that y was $(0.6)(100) + 28.8 = 88.8$. If we knew that y was 57.6, then we *predict* that x would have to be $(57.6 - 28.8) \div 0.6 = 48$.

Examiner's secrets

The functions on your calculator might also include finding the intercept and slope of the line of best fit – if so, then use them as well as the method above, as a double check that there are no mistakes!

Action point

Make up a table, showing the formulae for the coefficient of x in the least squares regression line on one side, and the product moment correlation coefficient on the other. This will help remember them and how they are applied!

The absurd region

We have to be very careful about how far we believe the predictions made using the least squares regression line. For example, the diagram below shows a sample of heights of children of various ages. We see that the regression line slopes upwards. This means that, on average, a person gets taller as they get older. But what happens when they hit 45? Extending the line up to age 45 would give a height in excess of 3 m! It is clear from this example that there are severe limitations to the predictive power of the least squares regression line. That part of the line beyond the main body of the data is called the **absurd region**, and any predictions that involve going this far up (or down – negative height?) need the point clearly written: *the least squares regression line becomes unreliable outside of the range of the data you are given.*

Examiner's secrets

When making a prediction, stop and think: 'Am I within the range of the data, or not?' You can tell quite easily from the table, and *don't forget to write it down if you think it matters.*

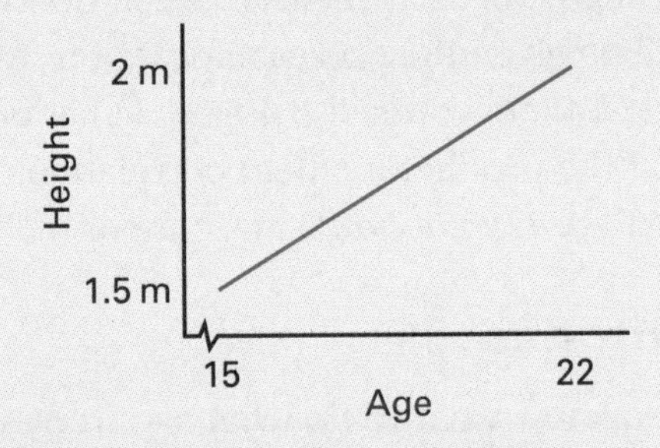

Action point

Draw a sketch to represent how you think a person's weight changes as they go from 1 to 20 years old. What weight does your sketch predict for a 90-year-old? Is this reasonable? Comment on your result.

Exam question

answer: page 164

The ages, in years, and weights, in kg, of a random sample of dogs are shown in the table below.

Age	11	12	12	13	13	13	14	14	15
Weight	14.4	15.5	15.5	16.4	14.7	17.2	17.6	17.9	18.4

(a) Calculate, to three decimal places, the coefficients a and b of the least squares regression line, $y = ax + b$, of weight on age for this sample. Interpret the meaning of a and b.

(b) Use this equation to estimate the mean weight of a dog aged 17 years.

(c) State clearly any reservations you may have about your estimate, with reasons. (30 min)

Rank correlation

Names for children are just as fashionable as fabrics or colours; each new year sees in the latest favourites. Question: are the top ten most popular names this year the same as last year's? To answer this, we look for correlation between the sets of ranks – i.e. how much last year's positions match up with this year's.

Spearman's rank correlation coefficient

The mathematician Spearman discovered that we can plot ranks on a scatter diagram, in pairs, to see if there is correlation. He went on to develop a correlation coefficient which gives us a measure of how closely one set of rankings matches up with another.

Spearman's rank correlation coefficient, $r_S = 1 - \dfrac{6\Sigma d^2}{n(n^2 - 1)}$

where d is the difference between the ranks.

Example: The table below shows the ranks given by two judges in a contest to judge children for a television advert:

Child	*Rank (1)*	*Rank (2)*	$\lvert d_i \rvert$	d_i^2
Abigail	2	1	$\lvert 2 - 1 \rvert = 1$	1
Belinda	4	5	$\lvert 4 - 5 \rvert = 1$	1
Caroline	3	2	$\lvert 3 - 2 \rvert = 1$	1
Deidre	1	3	$\lvert 1 - 3 \rvert = 2$	4
Edwina	5	4	$\lvert 5 - 4 \rvert = 1$	1

$\Sigma d^2 = 1 + 1 + 1 + 4 + 1 = 8$, so $r_S = 1 - \dfrac{(6)(8)}{5(25 - 1)} = 0.60$. This shows that there is a fair degree of agreement between the two judges. Sometimes you will be given the scores allocated to each item. You will then need to rank them yourself before calculations can begin. It is worth noting that the product moment correlation coefficient of the raw scores is actually a *more accurate* measure of their correlation.

Examiner's secrets

It might seem like a waste of time, but writing out the table with the extra lines is essential – not just for getting the answers right (any mistakes as easy to trace) but also for showing working.

The jargon

A very common mistake when working out Σd^2 is to find Σd and then *square that. This cannot be true.* Let's look at our example: $\Sigma d = 6$, so $(\Sigma d)^2 = 36$, which is *not the same* as $\Sigma d^2 = 8$, as worked here.

Checkpoint 1

The sum of squared differences of a set of 10 ranks, Σd^2, was 87.2. Calculate Spearman's rank correlation coefficient and comment on your result.

Problems with ties

You are expected to know what to do when confronted with a tied situation. A **tie** occurs when two items are ranked equally. Suppose the two items are joint second – the next item then ranked fourth. Now, we need to assign 'second' *and* 'third' – not just two seconds, and so we say that they each have the average rank of 2.5.

Example: Two judges give the following results in a writing contest:

Writer	*Rank (1)*	*Actual rank (1)*	*Rank* (2)	$\lvert d \rvert$	d^2
Alan	5	5	3	2	4
Brian	3 } tied ranks	3.5	4	0.5	0.25
Christine	3 } tied ranks	3.5	2	1.5	2.25
David	1	1	1	0	0
Euan	2	2	5	3	9

Examiner's secrets

If you get a question involving ties, but get into trouble, then try this method of finding what to do: rank them from first downwards, up to the tied ranks; then start at the last (if there are n numbers the last is always going to be nth) and work backwards. The missing rank numbers are then spread evenly over the remaining items.

Judge 1 gave two third ranks, instead of third and fourth – so they get 3.5 each. $\sum d^2 = 4 + 0.25 + 2.25 + 0 + 9 = 15.5$, so $r_S = 1 - \frac{(6)(15.5)}{5(25-1)} = 0.225$, so there is very little agreement between the judges.

Limitations of the theory

As with any measure in statistics, there are limitations on what we can expect from Spearman's rank correlation coefficient. One particular problem with this measure of correlation is that it only uses consecutive numbers attached to each pair. If the original data is available, then this might give us a more accurate measure of the strength of any relationships. It may be that the original data are quite well correlated, but the rankings are not. Alternately, the original data may not be linearly correlated at all, but is still quite highly rank correlated (see exam question 2 below).

We must also note that the sample sizes are often small; in general terms, as the sample gets smaller, we need the correlation measure to be much higher to indicate a good level of correlation. So 5 people showing 75% agreement is not as good as 100 people showing 50% agreement.

Interpreting r_S values

Spearman's rank correlation coefficient has been designed to always fall between −1 and 1. An r_S value of 1 means that the judges are in perfect agreement. If you get an r_S value of around zero, don't make the common mistake of thinking that this has told you nothing; in fact, it has told you the very important fact that the judges do not agree *at all* on what they think. Lastly, an r_S value of around −1 tells us that the judges rank the items in *perfectly opposite agreement* – so the best for one judge is exactly the worst for the other.

Exam questions answers: page 164

1 Two judges in a photography contest ranked the entries. Their results are shown in the table below.

Judge 1	6	4	5	3	1	2
Judge 2	5	6	3	4	2	1

Calculate Spearman's rank correlation coefficient. (5 min)

2 In a study to investigate a possible link between spelling ability and arithmetical ability, six people were selected at random and given two tests. The first test asked them to spell 100 words; the second test asked them to perform 100 mental calculations. The results are shown in the table.

Spelling score (%)	9	16	59	79	95	100
Arithmetic score (%)	49	52	17	52	66	96

By first numbering each candidate with their appropriate rank, calculate the Spearman's rank correlation coefficient for this data. Comment on your results. Then plot the original data on a scatter diagram and calculate their product moment correlation coefficient. State, with reasons, why the rank correlation coefficient may be misleading. (25 min)

Action point

Make a list of the two types of correlation coefficient (rank and product moment) and their limitations. You will need an awareness of when these limitations apply.

Action point

Try some examples for yourself – imagine two judges are ranking eight photographs. Calculate the r_S values for them if they are in total agreement (first to first, etc.), totally opposite agreement (first to last, etc.), or in random order. Make sure you get $r_S = 1$, $r_S = -1$ and $r_S = 0$.

Checkpoint 2

Two judges ranked 15 girls' names and their r_S value was 0.99. If judge A ranked the name 'Ameeta' as 1st, what did judge B rank it as? How sure can you be? Would you be as sure if their r_S value was −1?

Examiner's secrets

To avoid unnecessary errors, quickly check that you have copied the table out correctly before doing any calculations. It's also a common mistake to work out the 'six sigma *d* squared' stuff, but then forget to take your answer away from one. A dead giveaway for this kind of mistake is coming up with an r_S value bigger than 1!

Testing correlation coefficients

When you have worked out your r or r_S value you might well say to yourself, 'OK, so I've got some correlation – but am I satisfied that there is enough correlation?' To answer this question, we look to testing the values of r and r_S to see how significant they are in a firm statistical sense.

The jargon

The greek letter ρ stands for the population correlation coefficient. So what we actually test is whether or not there is significant correlation present in the *population*, based on the evidence produced by our sample.

Checkpoint 1

What are the roles of the null and alternate hypotheses in a significance test?

Examiner's secrets

As with all tables, make sure you get plenty of practice at using them well before your exam – you don't want to waste thinking time with unnecessary processing time.

Examiner's secrets

The conclusions drawn from any significance tests have to be quite carefully worded. The thing to remember is that you have not produced solid evidence since this particular test statistic may have been a fluke! So don't commit yourself, OK?

Setting up a good test: hypotheses

When running a hypothesis test, it's essential to follow the recipe. First of all, we want to know exactly what we are testing *for*. We set up our **hypotheses** – there are two of them. We either settle for one or the other. The **null hypothesis** (called H_0) says that, if the test fails, then correlation is zero. We interpret this more realistically as saying that there is not significant correlation. The **alternate hypothesis** (called H_1) states what we believe might be true; that there is evidence for positive correlation, negative correlation, or simply any correlation.

You must write: H_0: $\rho = 0$

Then follow with *one* of these:
H_1: $\rho > 0$ (positive correlation)
H_1: $\rho < 0$ (negative correlation)
H_1: $\rho \neq 0$ (any correlation)

The significance level

The next step is the **significance level**. This is a measure of how sure we want to be about the outcome of the test. The higher the significance level, the less chance of our being wrong about our conclusion. The most common levels are 95%, 98% or 99%. You must write down the significance level of your test!

The critical region and conclusion

You now need to use the tables. Be sure to look up on the correct page of your formula book, since different numbers are used for the Spearman's (r_S) and product moment correlation coefficients (r). The number you find will define the **critical region**. It's a good idea to sketch a number line showing the start of the critical region, and where your value is in relation to that (remember to take the modulus of your value of r).

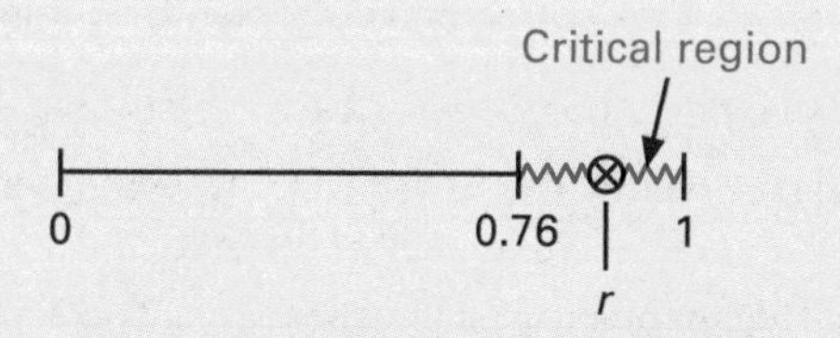

Suppose the critical region starts at 0.76 (the number obtained from the tables). If our value lies to the right of the critical value, within the critical region, we have enough correlation. Therefore, we would write something very carefully worded, like this:

> 'The calculated value of r lies within the critical region. This indicates that there is sufficient evidence, at the 95% significance level, to reject H_0 and conclude that positive correlation exists.'

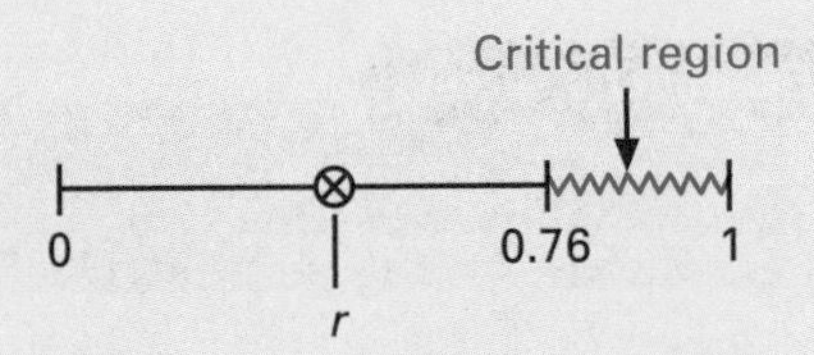

This diagram shows the calculated value of r to the left of the critical value, well outside the critical region. In this case, there is not enough correlation, and so we again write a carefully worded sentence:

> 'The calculated value of r lies outside the critical region. This suggests that, at the 95% significance level, there is not sufficient evidence to reject H_0 and we conclude that significant correlation is not present.'

The conclusion is a very important part of the test, but follow the recipe, don't over commit yourself, and count up the exam marks!

Example: A sample of the heights and weights of 50 women is taken. The product moment correlation coefficient is calculated at 0.33. Here is the significance test for this value, using the 95% significance level.

Null hypothesis:	H_0: $\rho = 0$
Alternate hypothesis:	H_1: $\rho > 0$ (positive correlation)
Significance level:	95%
Critical value:	0.235 3

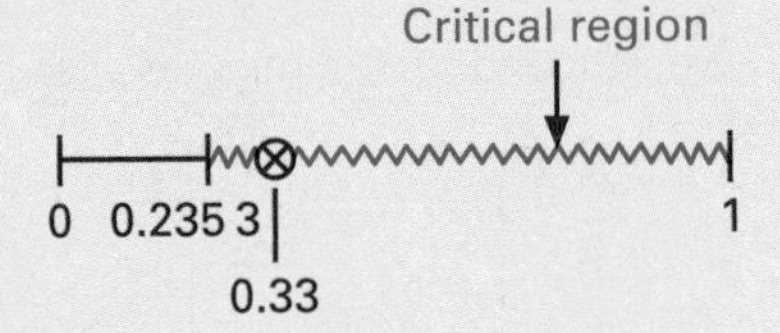

We see from the diagram that the calculated value of r lies to the right of the critical value, well inside the critical region, and so there is sufficient evidence, at the 95% significance level, to reject H_0 and conclude that positive correlation exists.

Exam questions answers: pages 164–5

1 Two people were selected at random and asked to rank six different chocolate bars in order of preference. Their choices are shown in the table. (20 min)

Person 1	1	2	3	4	5	6
Person 2	2	1	3	5	6	4

Test, at the 1% level, how closely their selections agree.

A further 58 people were asked to perform the same task. Their results, not including the two people above, were such that $\Sigma d^2 = 2\ 140$. Test, at the same significance level, how closely their selections agree.

2 The reaction times, y, of 50 people were measured after they had consumed x units of alcohol, to investigate a possible link. The results can be summarised as follows:

$$\Sigma x = 25.6;\ \Sigma y = 458;\ \Sigma x^2 = 16.64;\ \Sigma xy = 248;\ \Sigma y^2 = 4\ 516$$

Test, at the 5% level, the strength of the relationship between reaction time and alcohol consumed. Comment on your result, in terms of the original investigation. (15 min)

Examiner's secrets

These diagrams may be a bit basic but *remember, you will be working under quite stressful conditions and anything that makes it easier helps build up those marks!*

Examiner's secrets

The recipe for the test is essential practice – you should always follow it to maximise your marks. Also, you may be asked to explain the purpose of the null and alternate hypotheses as well as interpret the result.

Checkpoint 2

A sample of ages and lengths of trunks of 50 elephants showed an r value of 0.22. Test to see if significant correlation exists at the 95% level, using 0.235 3 as the critical value.

Answers
Statistics

The mode

Checkpoints

1

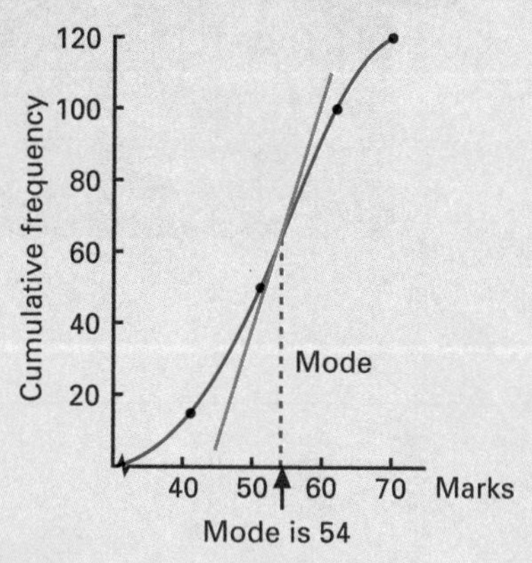

2

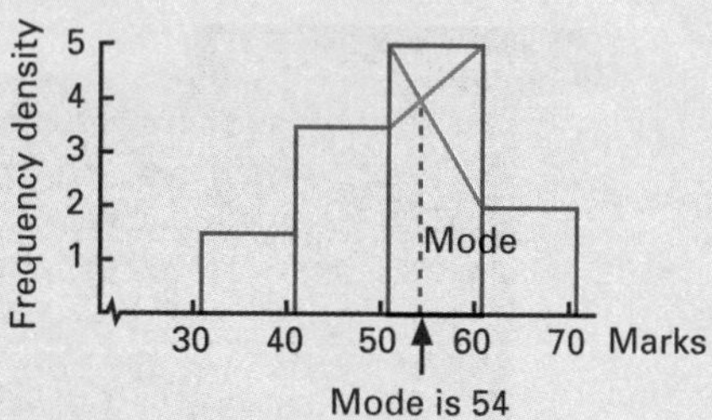

Exam question

(a) The cumulative frequency curve, with line drawn at the steepest point, looks like this:

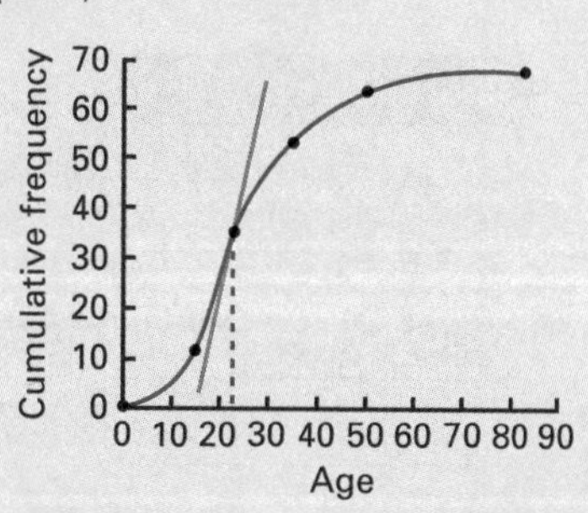

The mode is 23 years.

(b) The first class in the table actually goes from 9 to 16, or rather $9 \leq X < 16$, as your age is 15 right up to the exact moment you turn 16, as opposed to being rounded up to 16 from 15.5 onwards.

Ages (years)	9–15	16–24	25–35	36–49	50–80
Frequency	12	23	22	8	2
Class width	7	9	11	14	31
Frequency density	$\frac{12}{7} = 1.7$	$\frac{23}{9} = 2.6$	$\frac{22}{11} = 2$	$\frac{8}{14} = 0.6$	$\frac{2}{31} = 0.1$

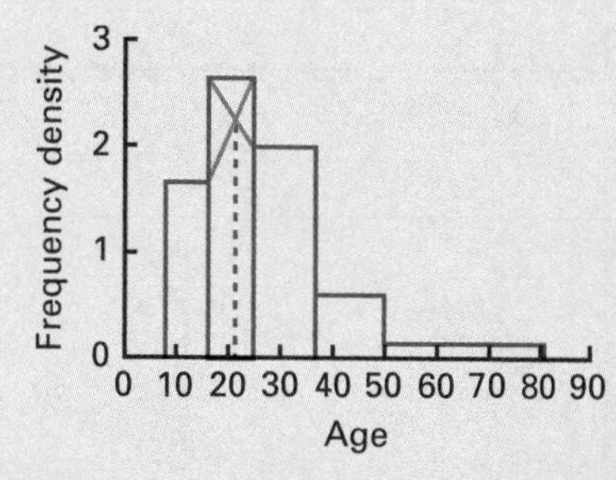

The mode is 21 years.

Examiner's secrets

Leave all lines drawn as they are. They are the only way to show how you worked it out!

The median

Checkpoints

1 Arranged in order the numbers are: 1, 2, 4, 5, 7. The median is 4. Doubling all numbers gives a median of 8; adding 6 to them gives a median of 10 – so whatever you do to the numbers, you do to the median as well!

2 An estimate for the median would be: 31.5 kg.

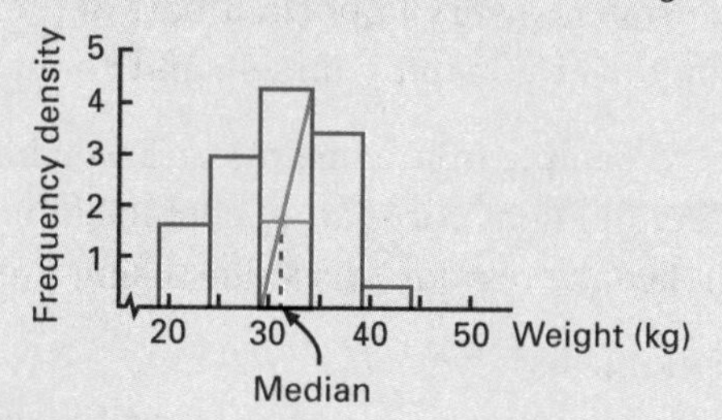

Exam question

(a) The first class of this table goes from 2.5 minutes to 4.5 minutes, as these are the times within which you would round to 3–4 minutes. The first class width is 2.

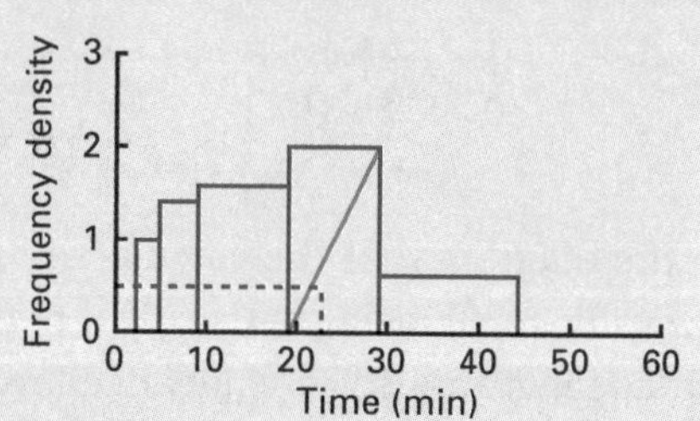

The median is read off at 21 minutes.

(b) To calculate the median, we draw the diagram showing the lower and upper class boundaries, how many items of data in the class, and how far the median is across the interval.

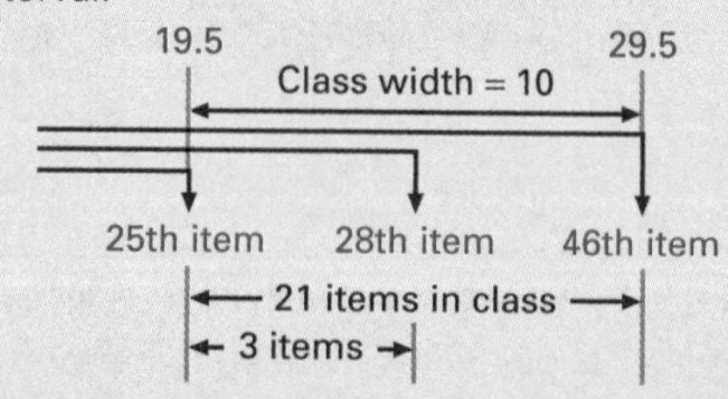

The median will be $19.5 + (\frac{3}{21})10 = 20.9$. The calculation is the more accurate. Errors can easily be introduced with drawings.

Cumulative frequency

Checkpoints

1

Heights (cm)	1–4	5–8	9–12	13–16	17–20
Frequency	2	8	12	9	3
Upper class boundary	4.5	8.5	12.5	16.5	20.5
Cumulative frequency	2	10	22	31	34

2

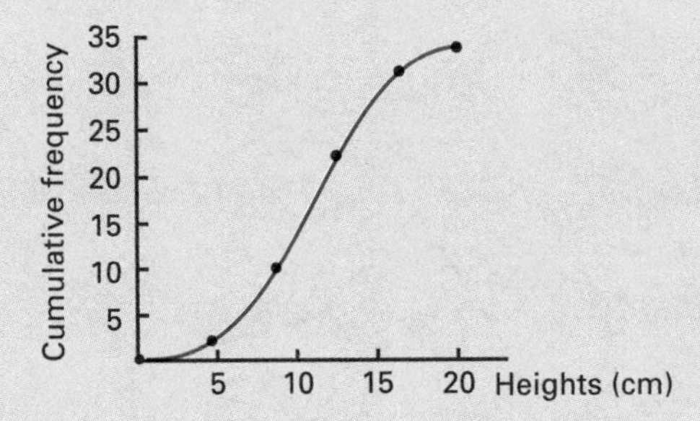

3 An estimate for the median would be 2.3. An estimate for the interquartile range would be 2.8 – 1.5 = 1.3.

4 The median and interquartile range are both unaffected by extreme values, and so if the distribution is quite skewed or has outliers, they would be better to use than the mean and standard deviation.

Exam question

The cumulative frequency for the recruits, with median and upper and lower quartiles shown, would look like this:

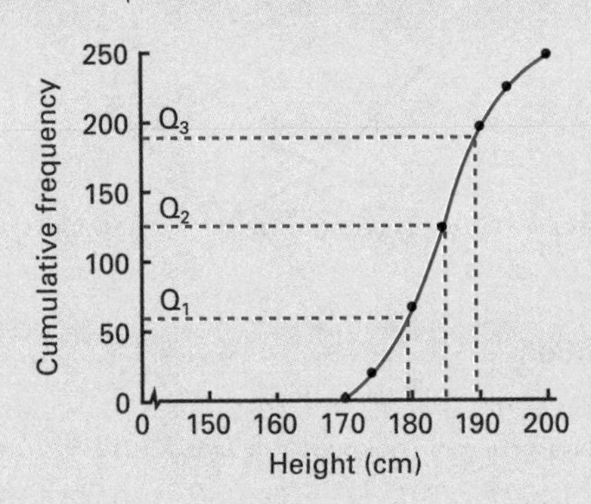

The estimated median is 185 cm.
The estimated interquartile range is 189 – 178 = 11 cm.
For the special squad, the estimated median is 178 cm and the estimated interquartile range is 181 – 175 = 6 cm.

The mean and standard deviation

Checkpoints

1 The mean was 22, and there were 10 numbers, so the total of the numbers was 10 × 22 = 220. We add 30 to get a new total of 250, and the mean becomes 250 ÷ 11 = 22.72 (2 d.p.).

2

m_i	2.5	6.5	10.5	14.5
$f_i m_i$	12.5	26	84	43.5
m_i^2	6.25	42.25	110.25	210.25
$m_i^2 f_i$	31.25	169	882	630.75

Mean = $\frac{166}{20}$ = 8.3

Standard deviation = $\sqrt{\frac{1713}{20} - 8.3^2}$ = 4.09 (to 2 d.p.)

Exam question

We estimate the value of X_i by using the midpoint of each class. The midpoint of the first class is going to be 4.4, the second will be 5.2, and so on. We assume equal class widths, making the last midpoint 8.4. The estimate for the mean is 6.656, and the standard deviation is 0.745 16.

From the first sample, $\Sigma m = 332.8$, $\Sigma m^2 = 2\,242.88$ and $n = 50$. In the new sample, $\Sigma x = 6.24 \times 40 = 249.6$ and $\Sigma x^2 = (0.25^2 + 6.24^2)(40) = 1\,560.004$. So, for the combined sample, $\Sigma x = 332.8 + 249.6 = 582.4$ and $\Sigma x^2 = 2\,242.88 + 1\,560.004 = 3\,802.884$. The combined mean is $582.4 \div 90 = 6.471 = 6.5$ (to 1 d.p.) and variance is $3\,802.884 \div 90 - (6.471)^2 = 0.380\,4\ldots = 0.4$ (to 1 d.p.).

Statistical diagrams

Checkpoint

The box and whisker diagram would look like this:

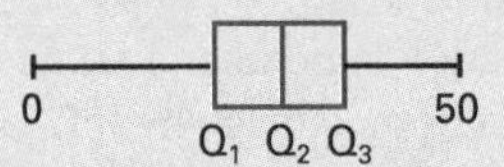

Exam question

The assumption we make is that the final class was of the same width as the majority of the others. Again, with ages something strange happens with class boundaries, because you are 19 right up to the point that you turn 20, rather than being rounded to 20 years old at 19.5 years. Here's the final histogram:

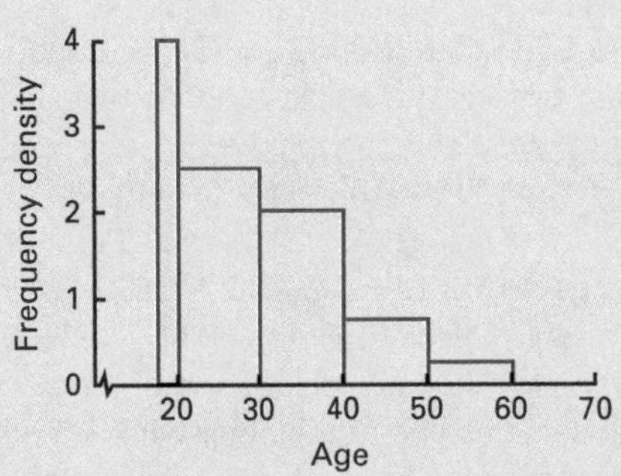

Probability

Checkpoints

1 We will use P(A ∩ B) = P(A) + P(B) – P(A ∪ B), where P(A) = 0.8, P(B) = 0.3 and P(A ∪ B) = 0.9. We get P(A ∩ B) = 0.8 + 0.3 – 0.9 = 0.2, i.e. the probability I watch *Eastenders* and do the ironing is 0.2.

2 You could have:
A = Drawing an ace from a pack,
B = Drawing a queen from a pack; or
A = getting 3 on a nine-sided spinner,
B = getting a 4 on a nine-sided spinner; or
A = watching BBC 1 at 9:00 pm,
B = watching Channel 4 at 9:00 pm.

3 You could have:
A = drawing an ace from a pack,
B = drawing a card worth at least 2; or
A = getting a 6 on a die,
B = getting any number from 1 to 5 on a die; or
A = getting a number bigger than 0 on an eight-sided spinner,
B = getting a number less than 10 on an eight-sided spinner.

4 With examples like this we can state directly that if we've taken a green, then there are 15 sweets left, of which 12 are red, so P(red | green) = $\frac{12}{15}$. The probability of getting two greens is $\frac{4}{16} \times \frac{3}{15} = \frac{12}{240}$.

Exam questions

1 Let C be the event a person owns a computer. Let M be the event a person owns a mobile phone.

So $P(C) = \frac{131}{200}$, $P(M) = \frac{157}{200}$ and $P(C \cap M) = \frac{104}{200}$

(a) We need $P(C \cup M) = P(C) + P(M) - P(C \cap M)$
$= \frac{131}{200} + \frac{157}{200} - \frac{104}{200}$
So the probability someone owns a computer or a mobile (or both) is $\frac{184}{200}$.

(b) The probability that someone owns a mobile or a computer but not both is given by $P(C \cup M) - P(C \cap M)$ which is $\frac{184}{200} - \frac{104}{200}$ since $C \cup M$ is computer or mobile or both, and $C \cap M$ is only both, and not one or the other. The final answer is $\frac{80}{200}$.

(c) $P(C \mid M) = P(C \cap M) \div P(M)$
$= \frac{104}{200} \div \frac{157}{200} = \frac{104}{157} = 0.66$ (2 d.p.)

(d) 27 have computer but no mobile phone.
$\therefore \frac{27}{200} \div \frac{131}{200} = \frac{27}{131}$
or $P(M \mid C) = P(M \cap C)/P(C) = \frac{104}{200} \div \frac{131}{200} = \frac{104}{131}$,
$P(M \mid C) = 1 - \frac{104}{131} = \frac{27}{131}$

2 We are given that $P(A) = 0.7$, $P(B) = 0.2$ and $P(A \mid B) = 0.1$. Reading through the whole question reveals that we are better off answering part (c) first:
$P(A \cap B) = P(B)P(A \mid B) = 0.2 \times 0.1 = 0.02$.

(b) is easiest next: $P(A \cup B) = P(A) + P(B) - P(A \cap B)$ so we get $P(A \cup B) = 0.7 + 0.2 - 0.02 = 0.88$.

(d) $P(B \mid A) = P(B \cap A) \div P(A) = 0.02 \div 0.7 = 0.029$ (2 d.p.).
Lastly, we'll do (a): exactly one of the events is the same as one or the other without both. This corresponds to $P(A \cup B) - P(A \cap B) = 0.88 - 0.02 = 0.86$.

Examiner's secrets

Don't assume that the question must be answered in the order it's given; as long as you leave some space so that it's written in the right order, you can do whichever part you think best!

Probability trees

Checkpoints

1

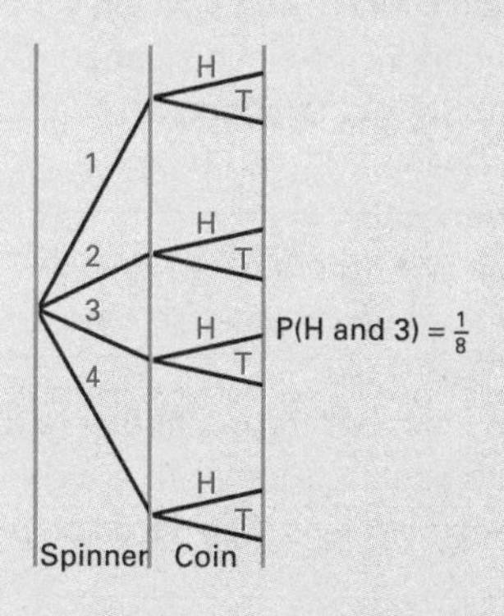

2

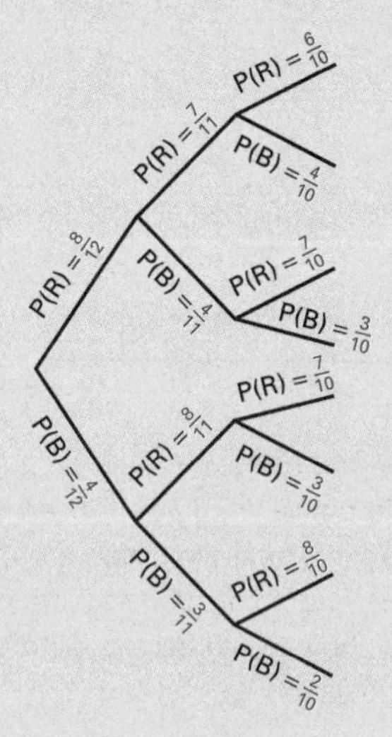

3

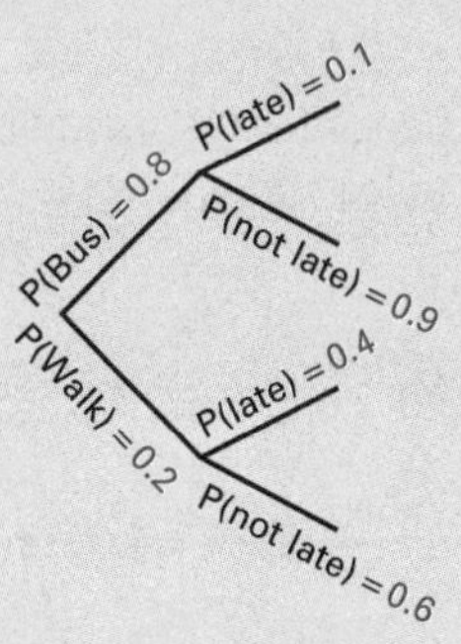

Exam question

Sort out the events first, having read the whole question. Customers can only be male (M) or female (F); buy the *Spenville Times* (T) or the *Late Press* (P); can have the right (R) or wrong (W) change. The best way forward is to draw the tree – it has two branches at each level in this order: M/F; T/P; R/W. It should initially look like the following tree.

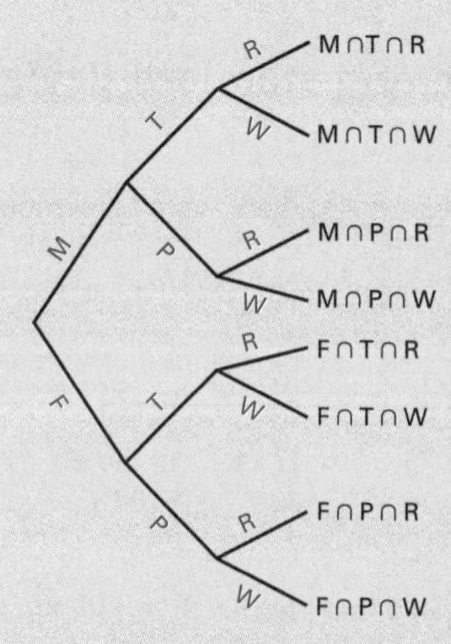

Now, by following the story, we add on the probabilities. A man buys the *Late Press* with probability 0.65, so he buys the *Spenville Times* with probability 0.35. A woman will buy the *Late Press* with probaility 0.35, and so buys the *Spenville Times* with probability 0.65. The last level has been included.

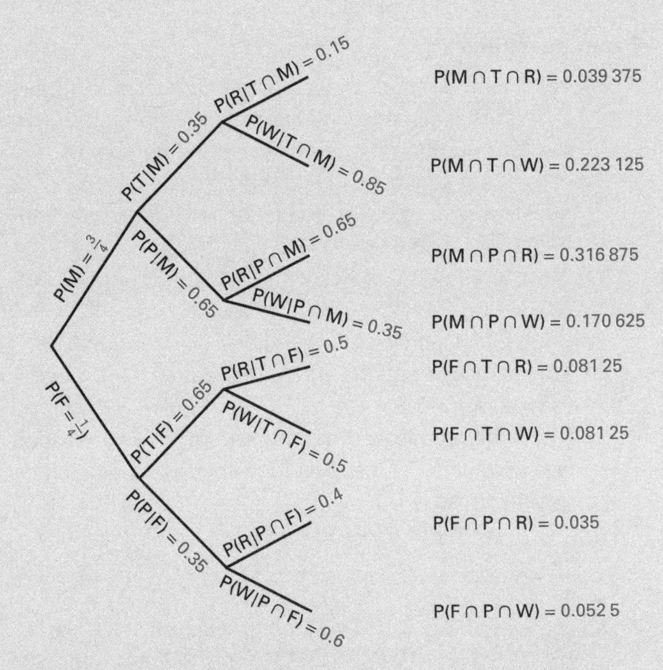

So the answers are now read off.

(a) We find the probability that a customer buys the *Spenville Times* by adding up every outcome where it's been mentioned, to get 0.425.

(b) To find P(R) we add up all outcomes where R has been listed, to get 0.472 5.

(c) $P(F \cap R) = P(F \cap T \cap R) + P(F \cap P \cap R) = 0.116\ 25$.

(d) We need $P(M \mid R \cap P) = P(M \cap R \cap P) \div P(R \cap P)$
$= 0.316\ 875 \div 0.351\ 875$
$= 0.886$ (3 d.p.).

We got $P(R \cap P)$ by $P(M \cap R \cap P) + P(F \cap R \cap P)$.

Permutations and combinations

Checkpoints

1 Six people can queue up in $6! = 720$ ways; they can sit at a circular table in $(6 - 1)! = 120$ ways.

2 $n = 16$; 10 boys, 6 girls. Teams are of eight people, with 3 girls on each, so there are ${}^6C_3 \times {}^{10}C_5 = 5\ 040$ ways. We only need to count the ways of choosing one team, as the other team is made up of who's left!

3 Arrangements are to be ordered without repetition. The maths books are together, so if we treat them as one lump there are actually eight objects, the maths lump and seven other books. The number of ways is which the eight things can be ordered is 8!. But then the five maths books can be ordered in 5! ways, so the total becomes $8! \times 5! = 4\ 838\ 400$ ways.

If the two science books are at each end, we have two ways they can be ordered. The other ten books are divided into five maths and five others. The five maths are treated as one lump, which can be ordered in 5! ways. The five others, with the lump of maths books, can be ordered in 6! ways, so the total is $2 \times 5! \times 6! = 172\ 800$.

Exam questions

1 There are 10 letters, with 2 M's, 2 I's, 2 L's, 2 N's and one each of E and U. Order is important, and there is no repetition. We'll use total $= 10! \div (2 \times 2 \times 2 \times 2 \times 1 \times 1)$ $= 10! \div 16 = 3\ 628\ 800 \div 16 = 226\ 800$.

If the two I's are together, then we count them as one item, and the total becomes $9! \div (2 \times 2 \times 2 \times 1 \times 1 \times 1)$ $= 9! \div 8 = 362\ 880 \div 8 = 45\ 360$.

So the probability that an arrangement chosen at random has the two I's together is $\dfrac{45\ 360}{226\ 800}$.

2 (a) $n = 26$, five being chosen. 'Order unimportant', 'without repetition'. If all combinations are sold, then they will get ${}^{26}C_5 = £65\ 780$.

(b) We split the numbers being selected from into two sets: the five winning numbers, and the rest. The number of ways to match three numbers out of the winners is ${}^5C_3$. Then we have to choose two losing numbers, which gives us ${}^{21}C_2$. The total number of ways we can match three is therefore ${}^5C_3 \times {}^{21}C_2 = 2\ 100$. So the probability of matching three numbers is $\dfrac{2\ 100}{65\ 780} = 0.032$ (3 d.p.)

(c) Number of ways of matching four numbers is ${}^5C_4 \times {}^{21}C_1 = 105$. So they pay out $105 \times £10 = £1\ 050$. Number of ways of matching five numbers is ${}^5C_5 = 1$. So they pay out $1 \times £100 = £100$. For three matches, they pay out $2\ 100 \times £5 = 10\ 500$. Altogether they pay out $1\ 050 + 100 + 10\ 500 = 11\ 650$, and they take 65 780, and so they make $65\ 780 - 11\ 650 =$ £54 130 profit.

Examiner's secrets

It's OK to make assumptions, but you must tell examiners about them otherwise they won't know why you've taken the line you have. If in doubt with any question, write a few examples out to get the problem into your own terms. In the last question, that would mean writing out a few winning tickets, to imagine what they look like.

Discrete random variables

Checkpoints

1 We know that $P(0 \le x \le 5) = 1$, and so we get $k(0^2) + k(1^2) + k(2^2) + k(3^2) + k(4^2) + k(5^2) = 1$
i.e. $k + 4k + 9k + 16k + 25k = 1 \Rightarrow 55k = 1$, so $k = \frac{1}{55}$.
The distribution of x is now:

$P(x = 0) = 0$; $P(x = 1) = \frac{1}{55}$; $P(x = 2) = \frac{4}{55}$;
$P(x = 3) = \frac{9}{55}$; $P(x = 4) = \frac{16}{55}$; $P(x = 5) = \frac{25}{55}$.

2 You should have made a six-by-six table, with the scores from 2 to 12 entered. Then, write out another table with the score and its probability, e.g. $x = 2$

$P(2) = \frac{1}{36}$.

Put another line on your table to show $xP(X = x)$, which for this column is $xP(X = x) = \frac{2}{36}$.
The expected value is given by $\Sigma xP(X = x) = 7$.

If you add another line, $x^2P(X = x)$, you can then calculate the variance.

You should get $\Sigma x^2P(X = x) - E^2(x) = 54\frac{5}{6} - 49 = 5\frac{5}{6}$.

3 $E(x) = 3$, $Var(x) = 2$. If $y = x + 4$, then $E(y) = 7$, $Var(y) = 2$.
If $y = 5x$, then $E(y) = 15$, $Var(y) = 50$.
If $y = 3x - 5$, then $E(y) = 4$, $Var(y) = 18$.

Exam questions

1 (a) The table looks like this:

X	-10	-7	-6
$P(X = x)$	$-\frac{1}{36}$	$-\frac{5}{36}$	$\frac{11}{36}$

(b) We add this row onto the table:

$xP(X = x)$	$-\frac{10}{36}$	$-\frac{35}{36}$	$\frac{66}{36}$

The expected gain, $\Sigma xP(X = x) = \frac{21}{36}$.

(c) If she played 200 times, she would win $200 \times \frac{21}{36}$. This game is good for her, but not fair, as the expected gain is greater than zero.

2

x	0	1	2	3
$P(X = x)$	$8k$	$5k$	k	k

(a) As X is a random variable $\sum_{\text{all } x} P(X = x) = 1$.

So $8k + 5k + k + k = 1$. $\therefore k = \frac{1}{15}$.

(b) $E(X) = 0 \times \frac{8}{15} + 1 \times \frac{5}{15} + 2 \times \frac{1}{15} + 3 \times \frac{1}{15} = \frac{2}{3}$.

$Var(X) = [0^2(\frac{8}{15}) + 1^2(\frac{5}{15}) + 2^2(\frac{1}{15}) + 3^2(\frac{1}{15})] - (\frac{2}{3})^2 = \frac{34}{45}$.

$P(\text{more than 4 faulty}) = (\frac{1}{15} \times \frac{1}{15}) + (\frac{1}{15} \times \frac{1}{15}) + (\frac{1}{15} \times \frac{1}{15}) = \frac{1}{75}$.

$E(X) = \frac{4}{3}$. $Var(X) = \frac{68}{45}$.

Examiner's secrets

Drawing up the table takes time, but will make any mistakes very easy to spot, as well as making what you've got to do more straightforward!

The binomial distribution

Checkpoints

1 $P(X = 0) = 0.262\,144$; $P(X = 1) = 0.393\,216$; $P(X = 2) = 0.245\,76$; and $P(X = 3) = 0.081\,92$.
So $P(X \le 3)$ is all these added together, which comes to $0.983\,04$. $P(X > 3) = 1 - P(X \le 3) = 0.016\,96$.

2 $P(X = 0) = 0.007\,836$; $P(X = 1) = 0.052\,237$; $P(X = 2) = 0.152\,358$; $P(X = 3) = 0.253\,929$; $P(X = 4) = 0.264\,51$; $P(X = 5) = 0.176\,34$; $P(X = 6) = 0.073\,475$; $P(X = 7) = 0.017\,494$; $P(X = 8) = 0.001\,822$.

3 The bar chart would look roughly like this:

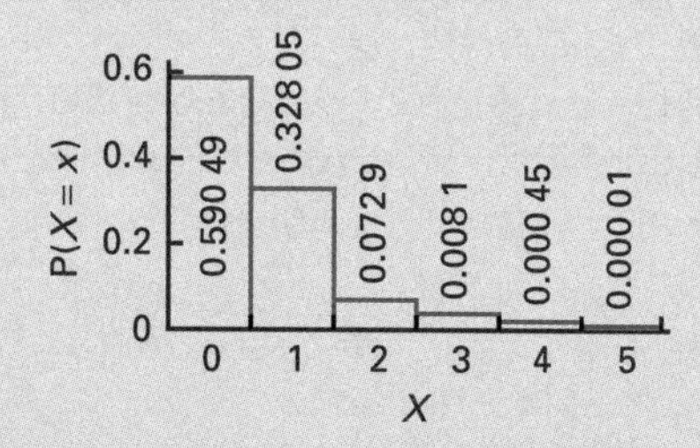

Exam questions

1 (a) Let X be the random variable 'the number of correctly answered questions in section one of the test'. $n = 8$, the total number of questions; r is the number of correct questions. $p = \frac{1}{4}$ is the probability I get a question right; $q = \frac{3}{4}$ is the probability I get a question wrong. We use $P(X = r) = {}^8C_r(\frac{1}{4})^r(\frac{3}{4})^{8-r}$.

(b) $P(X < 2) = P(X = 0) + P(X = 1) = 0.100\,113 + 0.266\,968 = 0.367\,081$.

(c) $P(X \ge 2) = 1 - P(X < 2) = 1 - 0.367\,081 = 0.632\,919$.

(d) Two particular options are correct out of the five so $P(X = 2) = \frac{2}{5} = 0.2$.

(e) By (d) the probability I get any one question in section two correct is 0.2, i.e. $p = 0.2$. Section two has four questions, so $n = 4$.

We want $P(X > 2) = P(X = 3) + P(X = 4)$
$= 0.025\,6 + 0.001\,6$
$= 0.027\,2$.

2 (a) We want $P(X > 2) = 1 - P(X \le 2)$ where $n = 10$, and $p = 0.20$. Now, $P(X \le 1) = P(X = 0) + P(X = 1) + P(x = 2)$ $= 0.107\,374 + 0.268\,435 + 0.301\,989 = 0.677\,798$.
So $1 - P(X \le 1) = 1 - 0.677\,798 = 0.322\,202$.

(b) Now X has become the number of acceptable washers, making $p = 0.80$ since 80% are OK. So that gives us $E(X) = np = 10 \times 0.80 = 8$ and $Var(X) = npq$, i.e. $10 \times 0.20 \times 0.80 = 1.6$.

The Poisson distribution

Checkpoint

Under $X \sim Po(5)$, $P(X = x) = e^{-5}5^x \div x!$

$P(X \le 3) = P(X = 0) + P(X = 1) + P(X = 2) + P(X = 3)$
$= 0.006\,738 + 0.033\,69 + 0.084\,224 + 0.140\,374$
$= 0.265\,026$.

$P(X > 3) = 1 - P(X \le 3) = 1 - 0.265\,026 = 0.734\,974$.

Exam questions

1 The mean of this distribution turns out to be 2.89, while the variance is 2.56. The closeness of these values would suggest that a Poisson model is valid.

(a) $P(X = 5)$ under $X \sim P(2.89) = 0.093$.

(b) $P(X \ge 3) = 1 - P(X \le 2)$
$= 1 - P(X = 0) - P(X = 1) - P(X = 2)$
$= 1 - 0.055\,576 - 0.160\,615 - 0.232\,089$
$= 1 - 0.448\,28 = 0.551\,72$.

2 When we add two independent Poisson distributions, we get another Poisson distribution whose mean is the sum of the two individual means. Here, the errors on one page from each section combined follow a Poisson distribution having mean $2.8 + 2.1 = 4.9$.

(a) $P(X = 0) = e^{-4.9}4.9^0 \div 0! = 0.007\,447$.

(b) $P(X = 3) = e^{-4.9}4.9^3 \div 3! = 0.146\,014$.

(c) $P(X > 5) = 1 - P(X \le 5)$
$= 1 - 0.633\,501$
$= 0.366\,499\ldots$

Continuous random variables

Checkpoints

1 We firstly need to integrate kx^2 between 0 and 3 and set the answer equal to 1, which gives $\frac{27k}{3} = 1$, i.e. $k = \frac{3}{27}$ or $\frac{1}{9}$. The PDF of x looks roughly like:

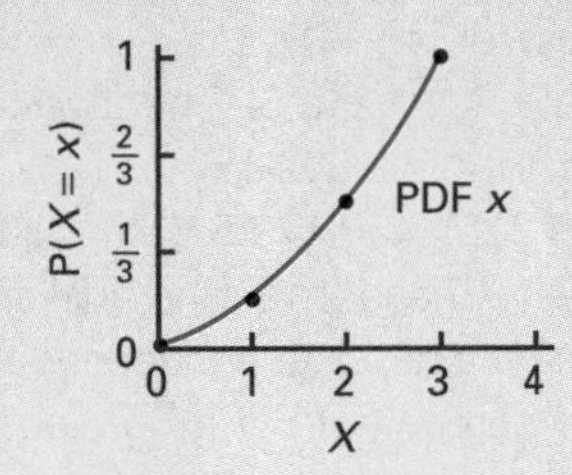

2 The graph of f(x) will look like this:

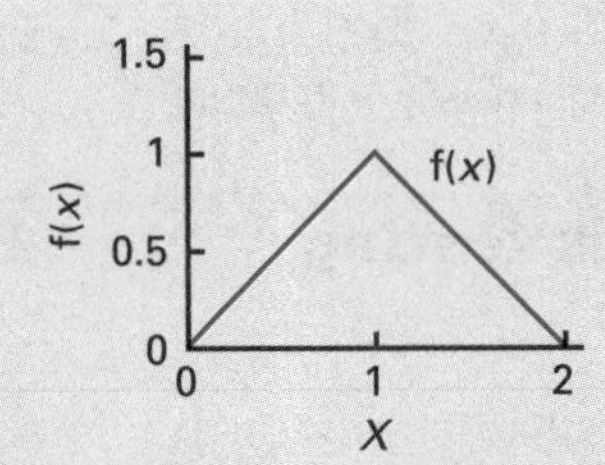

To find the expected value of x, we do two separate integrations, one for $0 \le x \le 1$ and one for $1 \le x \le 2$. Both parts turn out to be $\frac{1}{2}$ each, so when added give the expected value of 1 – which makes sense as it's the central value of a symmetrical distribution.

Exam questions

1 (a) To find a, integrate f(x) between 0 and 5 and set the result equal to 1. We get $\frac{5}{2}a5^2 - \frac{1}{3}a5^3 = 1$. This gives us $\frac{125}{6}a = 1$, or that $a = \frac{6}{125}$.

(b) The density function is the integrated version of the PDF, which, from (a) is $F(x) = \frac{5}{2}ax^2 - \frac{1}{3}ax^3$, which simplifies to $F(x) = \frac{3}{25}x^2 - \frac{2}{125}x^3$.
Here's the sketch:

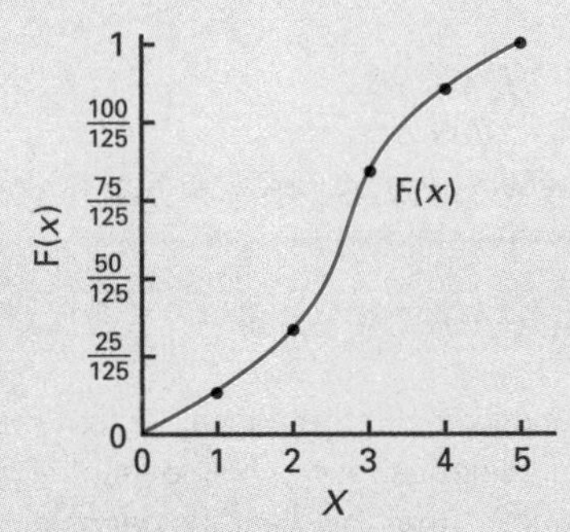

2 (a) The area under the curve is the probability; therefore, the area must equal 1. It's a triangle, of base 3, so $\frac{1}{2}$ base × height $= \frac{1}{2} \times 3C = 1$, i.e. $C = \frac{2}{3}$.

(b) $f(x) = \frac{1}{3}x$ for $0 \le x \le 2$, and $2 - \frac{2}{3}x$ for $2 \le x \le 3$. To find the mean, integrate $x\mathrm{P}(X = x)$ for each part separately and add the results. So integrate $\frac{1}{3}x^2$ between 0 and 2, and add the integral of $2x - \frac{2}{3}x^2$ between 2 and 3, which gives the sum:

$$(\tfrac{8}{9}) - 0 + [(9 - 6) - (4 - \tfrac{16}{9})] = \tfrac{5}{3}$$

To find the variance, we need to integrate $x^2\mathrm{P}(X = x)$, i.e. $\frac{1}{3}x^2$ between 0 and 2, and $2x^2 - \frac{2}{3}x^2$ between 2 and 3. Once we've done this, we subtract the squared mean (*remember*: expected X-squared minus expected-squared X). The first integration gives us the sum

$$\tfrac{16}{12} + \tfrac{54}{3} - \tfrac{162}{12} - \tfrac{16}{6} + \tfrac{32}{12} = \tfrac{19}{6}$$

The variance is therefore $\frac{19}{6} - (\frac{5}{3})^2 = \frac{7}{18}$.

(c) The median of X occurs where the area under the curve is exactly 0.5. We could integrate but it's easier to note that it's going to fall in the left-hand triangle, making the area of that triangle $\frac{1}{2}$ base × height or $\frac{1}{2}B \times \frac{1}{3}B = \frac{1}{2}$, which gives $B = \sqrt{3}$, i.e. the median is at 1.732.

The normal distribution

Checkpoints

1

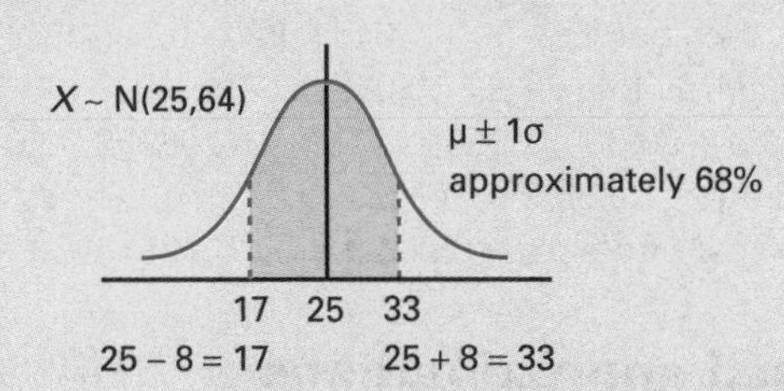

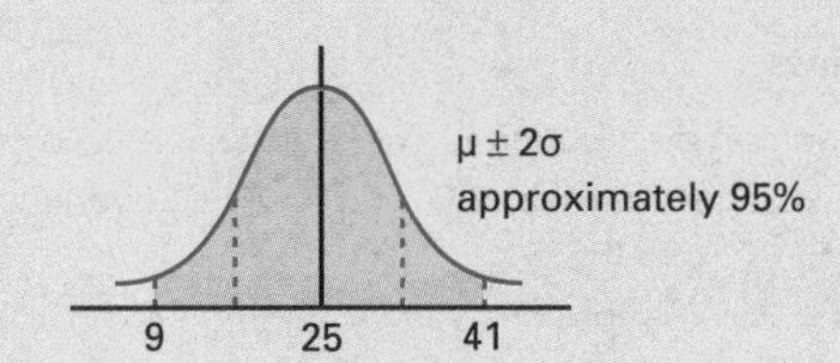

2 $X \sim N(7,100)$.

(a) The Z value for $P(X < 9)$ is $\frac{9-7}{10} = 0.2$. It's positive, i.e. to the right of 0, and so we simply look up the corresponding probability, which is 0.579 3.
$P(x < 9) = 57.93\%$.

(b) The Z value for $P(X < 6)$ is $\frac{6-7}{10} = -0.1$. It is to the left of 0, so we look up the positive value and take it away from 1. $P(x < 6) = 1 - 0.539\,8 = 0.460\,2 = 46\%$ (to 2 s.f.).

(c) $P(x > 5) = 57.93\%$.

3 $X \sim N(6,25)$.

(a) $P(X < k) = 0.87$. Look up 0.87 in the tables to find Z value of 1.127. $\therefore 1.127 = \frac{x-6}{5}$.
$\therefore P(X < 11.635)$, i.e. $k = 11.635$.

(b) $P(X > k) = 0.66$. Look up 0.66 in the tables to find Z value of 0.412. As $X > k$ the value is to the left of 0 and is negative. $\therefore -0.412 = \frac{x-6}{5}$.
$\therefore P(X > 3.94)$, i.e. $k = 3.94$.

Exam questions

1 Let X be the IQ of a chimpanzee; then we are told that $X \sim N(80,5^2)$.

(a) We need $P(X > 88)$; our Z value is given by $(88 - 80) \div 5 = 1.6$.
    We need $1 - \Phi(1.6) = 1 - 0.945\,2 = 0.054\,8$.
(b) We need $P(X < 73)$ with Z value $(73 - 80) \div 5 = -1.4$. If you sketch the distribution you will see that we actually do $1 - \Phi(1.4) = 1 - 0.919\,2 = 0.080\,8$.
(c) We need $P(75 < X < 84)$. The lower Z value is -1 and the upper is 0.8. We'll work out $\Phi(0.8)$, and subtract from that $1 - \Phi(1)$. We end up with $0.788\,1 - (1 - 0.841\,3) = 0.629\,4$.
(d) Approximately two-thirds of the population lie within ± 1 standard deviation, which is 80 ± 5, i.e. in the range 75 to 85.

2 Three out of 60 overestimated by 5 seconds, so $P(\text{Guess} < 65) = \frac{51}{60}$. The Z value for this is 1.645, so $(65 - \mu) \div \sigma = 1.645$. Also, $P(\text{Guess} > 57) = \frac{51}{60}$, so that $(57 - \mu) \div \sigma = -1.035$ (negative since it's to the left of the mean, 0). We have two simultaneous equations, which we can solve by eliminating σ. The first gives us $(65 - \mu) \div 1.645 = \sigma$ and second $\sigma = (57 - \mu) \div (-1.035)$.
  We then get $(65 - \mu) \div 1.645 = (57 - \mu) \div (-1.035)$ which solves to give $\mu = 60.089\ldots$ Therefore $\sigma = 2.985\ldots$ $A[(65 - 60.089)/2.985] - A[(55 - 60.089)/2.985] = 0.95 - (1 - 0.955\,9) = 0.905\,9$.
  Number of people $= 0.905\,9 \times 60 = 54.354$.

Normal approximations

Checkpoints

1 This binomial distribution is well approximated by a normal distribution having mean $50 \times 0.5 = 25$ and variance $50 \times 0.5 \times 0.5 = 12.5$.
(a) $P(X = 25)$ is approximated by finding $P(24.5 < X < 25.5)$, i.e. double the area up to 0.5 each side of the mean. The Z value is $(25 - 25.5) \div \sqrt{12.5} = 0.141$. $\Phi(0.141) = 0.556\,1$ and we just need the slice between 0.5 and 0.556 1 which is 0.056 1. Double this to get 0.112 2.
(b) $P(X < 30) \Rightarrow P(X < 29.5)$. The corresponding $Z = (29.5 - 25) \div \sqrt{12.5} = 1.27$ and $\Phi(1.27) = 0.898$.

2 The approximation used is $X \sim N(15, 15^2)$. We require $P(X < 12) \Rightarrow P(X < 11.5)$. The corresponding Z value is $(11.5 - 15) \div 15 = -0.233$. We find $1 - \Phi(0.233) = 0.408\,8$.

Exam questions

1 We have a binomial distribution where $p = 0.45$ and $n = 200$. We approximate using $X \sim N(90, 49.5)$.
(a) $P(X = 90) \Rightarrow P(89.5 < X < 90.5)$. As in checkpoint 1(a) above, we find the slice of area between the mean and mean +0.5 and double it.
    Our Z value is $0.5 \div \sqrt{49.5} = 0.071$.
    $\Phi(0.071) = 0.528\,3$ so the slice is about 0.028 3.
    We double it to get 0.056 6.
(b) $P(X < 85) \Rightarrow P(X < 84.5)$.
    The associated Z value is $(84.5 - 90) \div \sqrt{49.5} = -0.782$.
    We have to find $1 - \Phi(0.782) = 0.217\,1$.
(c) $P(X > 90) \Rightarrow P(X > 90.5)$.
    $Z = 0.5 \div \sqrt{49.5} = 0.071$, $\Phi(0.071) = 0.528\,3$.
    $\therefore P(X > 90) = 1 - 0.528\,3 = 0.471\,7$.
(d) $P(82 \le X \le 91) \Rightarrow P(81.5 \le X < 91.5)$.
    Lower Z value is -1.208 and upper Z value is 0.213.
    The final answer is 0.470 5.

2 The approximation we use is $X \sim N(200, 200)$.
(a) $P(X > 160) \Rightarrow P(X > 160.5)$. The Z value is -2.79, and the final answer is 0.997 4.
(b) $P(X > 240) \Rightarrow P(X > 240.5)$. The Z value is 2.86, and the final answer is 0.002 1.
(c) $P(180 \le X \le 230) \Rightarrow P(179.5 \le X < 230.5)$. The two Z values are -1.45 and 2.16. The final answer is 0.911 1.
(d) The germinating seeds are independent; germination is randomly spread out per m^2 of seed – one in ten of 200 is 20.
(e) We use the approximation $X \sim N(20, 20)$. We want $P(X > 30) \Rightarrow P(X > 30.5)$. The Z value is 2.347 . . ., and the final answer is 0.009 4.

Hypothesis testing

Checkpoint

μ > 10
(a)
μ < 10
(b)
μ ≠ 10
(c)

Exam questions

1 The critical region for a 5% test starts at -1.645 for this one-tailed test. The mean of the sample was so low that it caused the test statistic to fall below -1.645.

$$\frac{\bar{x} - 186}{5.4/7} < -1.645\ \text{. So}\ \bar{x} < 184.7$$

2 H_0: $\mu = 99.2$. H_1: $\mu < 99.2$.
Under H_0, $\bar{X} \sim N(99.2, \frac{3.6^2}{65})$.
One-tailed test, at the 5% level: critical region starts at -1.645. The test statistic is

$$\frac{98.4 - 99.2}{3.6/\sqrt{65}} = -1.792 < -1.645$$

The test statistic lies in the critical region. We conclude that there is evidence, at the 5% level of significance, to support the claim that this year's pigeons are faster.

Sampling, estimates and confidence

Checkpoints

1 Estimate $\mu = 55$, $\sigma^2 = \frac{4^2}{16} = \frac{16}{16} = 1$.
2 $\bar{x} \sim N(5, \frac{16}{25})$. We need $P(\bar{x} < 5.5)$. Our Z value is 0.625, and the final answer is 0.734.
3 We use $0.98 \pm 1.645(\frac{0.63}{8})$, to get the interval 0.850 5 to 1.109 5.

Exam questions

1 Estimate $\mu = 86$; $\sigma = 3x\sqrt{\frac{12}{11}}$. 95% confidence interval has boundaries 12 multiplied by $86 \pm 1.96(\sqrt{12} \times 3 \times \sqrt{\frac{12}{11}})$, and the interval goes from 84.30 kg to 87.70 kg.

2 Chips from shops in this sample have mean 124.5 and standard deviation $\sqrt{45}$. We need $P(X < 130)$; the Z value is 0.820 and so the probability is 0.793 9.
We estimate the mean of the population as 124.5 g. We need the number in the sample to estimate the variance, and we get that from $\Sigma x = 6\,225$; because $6\,225 \div n = 124.5$, which makes $n = 50$. Our estimate for the variance is $\frac{45}{50} = 0.9$.

Examiner's secrets

Look for key words in the questions like 'estimate using the sample values' to tell you what to do. Think carefully about the numbers you are given: do you have population mean or sample mean, population variance or sample standard error?

Independent normal distributions

Exam question

(a) One portion of beans and one of chips would follow a normal distribution having mean 100 + 150 = 250 g and variance 16 + 25 = 41 g. We need $P(X + Y > 265)$, our Z value is 2.34, and the final answer is 0.009 6.

(b) Double portions of chips follow a normal distribution with mean 300 g and variance $2^2(25) = 100$ g. We add to that the beans, to get a total mean of 400 g and variance 116 g. We then need $P(2X + Y < 380)$. The Z value for this is −1.857, and the final answer is 0.031 7.

(c) This time we have five independent samples, the total weight of which has mean 6(150) = 900 g, and variance 6(25) = 150 g. The Z value for this is 8.16; tables quote that anything bigger than 3 or 4 has associated probability of 1, so the probability of the chips weighing more than 1 kg is zero!

Examiner's secrets

Be very careful about when to think of things as separate samples or repeats of a single sample. Do plenty of past paper questions to really nail it down!

Goodness of fit tests

Checkpoint

H_0: The distribution follows a binomial pattern.
mean of binomial = np
mean = $[(0 \times 0) + (1 \times 7) + (2 \times 13) + (3 \times 0)]/20 = 1.65$.
So, $1.65 = 3p$. $\therefore p = 0.55$.
$\chi^2_{calc} = 8.024$ (to 3 d.p.). $\chi^2_{5\%}(2) = 5.99$.
$\chi^2_{calc} > \chi^2_{5\%}(2)$. $\therefore$ Reject H_0. It does not follow a binomial distribution.

Exam questions

1 We are testing to see if the responses were evenly distributed or not. If so, then the expected frequency for each would be 30. H_0 = the data are evenly spread. The first $(O_i - E_i)^2 = (35 - 30)^2 = 25$.
$\chi^2_{calc} = \frac{58}{30} = 1.93$.
There are four classes, and one restriction, so the degrees of freedom = 4 − 1 = 3. From tables, $\chi^2_{5\%} = 7.815$, and since 1.93 < 7.815 we conclude that, at the 5% level, there is evidence to suggest there is no clear favourite.

2 She expected to sell 128 red, 128 green, 160 black and 384 white jumpers.
The first $(O_i - E_i)^2 = (110 - 128)^2 = 324$.
$\chi^2_{calc} = 4.947\,9$. There are four classes, and one restriction, so the degrees of freedom = 4 − 1 = 3. From tables, $\chi^2_{5\%} = 7.815$, and since 4.947 9 < 7.815 we conclude that, at the 5% level, there is no evidence to suggest that the actual sales differed significantly from expected levels.

Examiner's secrets

Always draw up the table of expected and observed values and their differences squared – it makes life so much easier!

Correlation

Checkpoints

1 The product moment correlation coefficient will be $(-18) \div 20 = -0.9$. There exists strong negative correlation between the two variables; so as one goes up, the other is very likely to go down.

2 There exists some positive correlation, but the strength of that correlation depends on the size of the sample. The larger the sample, the stronger this result will be. Also, it may be that the data are not linearly correlated, and we would have to visually inspect a scatter plot to see if this were the case.

Exam question

The scatter diagram looks like this:

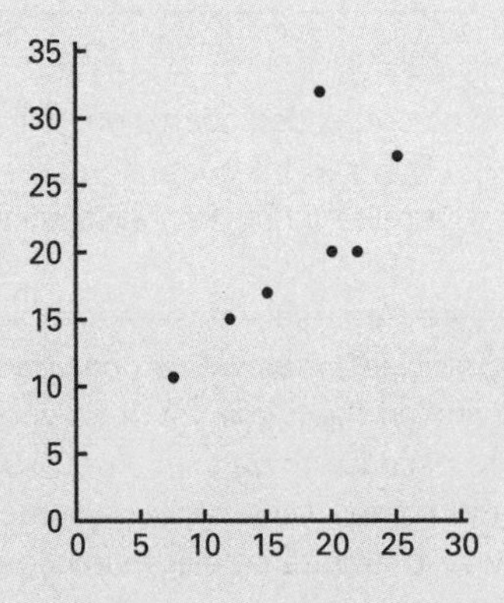

(a) Using my calculator, I got these figures: $\Sigma xy = 2\,559$, $\Sigma x^2 = 2\,243$ and $\Sigma y^2 = 3\,102$. It also gave me the correlation coefficient as 0.69 (2 d.p.).

(b) For such a small sample, this is not really a very high correlation coefficient. Looking at the scatter plot, it could be said that there is one item of data, (18,32) which is way off the general trend.

(c) I would choose the six points other than (18,32) because if you cover it with your finger you see that the others form a nice line.

(d) The product moment correlation coefficient for these six is 0.96 (2 d.p.) which is very much higher. We can say that the week where 18 pages gave a good response was something of a fluke; however, even without this result there is still a very high positive correlation between advertising space and sales returns.

Regression

Checkpoint

In the least squares regression line, $y = ax + b$, a is found by $73 \div 6.7^2 = 1.626$. To find b, note that the line passes through both means, so that $2 = 1.626(45) + b$, making $b = -71.17$. The equation is $y = 1.626x - 71.17$.

Exam question

(a) My calculator gives me the following: $\Sigma xy = 1\,931.3$, $\Sigma x^2 = 1\,533$. It then gave me the values of a as 1.042, and b as 2.858. The slope of the line, 1.042, gives us an indication of the amount of weight put on by the average dog in one year of its life. The intercept, 2.858, tells us that a newborn pup would weigh around 2.9 kg.

(b) The mean weight of a dog aged 17 years would be, based on this sample, $1.042(17) + 2.858 = 20.572$ kg.

(c) Although there are no problems with the methods used to get this estimate, the fact that it is outside the range of the data means that it should be treated with suspicion. This is because it is quite unlikely that a dog's weight will increase linearly for its entire life. In fact its weight will probably even out after a while – otherwise, when the dog was 30, its weight would be literally double that when it was 15!

Rank correlation

Checkpoints

1 Spearman's rank correlation coefficient will be $1 - [6(87.2) \div 990] = 0.472$. There exists some agreement between the rankings, but quite a few inconsistencies must have occurred.

2 If judge A ranked 'Ameeta' 1st, with such a high correlation coefficient, then we can be very sure that judge B also ranked the name 1st or maybe 2nd. We can't be 100% sure because there was 'only' 0.99 agreement; if however, there was 100% negative agreement, then judge A ranking it 1st must mean that judge B ranked the name 15th.

Exam questions

1 The squared differences give the sum: $1 + 4 + 4 + 1 + 1 + 1 = 12$.
Spearman's rank correlation coefficient is $1 - \frac{72}{200} = 0.657$ (3 d.p.).

2 The table of ranks looks like this:

Rank (1)	6	5	4	3	2	1
Rank (2)	5	3.5	6	3.5	2	1
d^2	1	2.25	4	0.25	0	0

$\Sigma d^2 = 7.5$. Spearman's rank correlation is 0.786 (3 d.p.). It would appear that there was a fair amount of agreement between the two judges.

The scatter plot of the original scores would look like this:

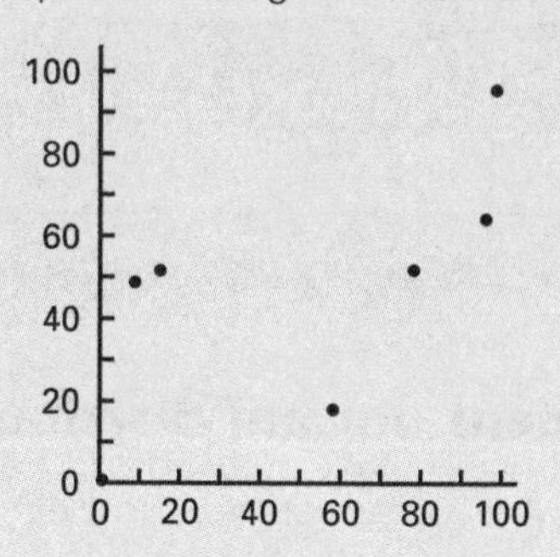

The product moment correlation coefficient, by calculator, is 0.48 (2 d.p.). The correlation coefficient, and the scatter plot, would both indicate that there is very low linear correlation between these scores. There is the possibility of the data following a quadratic trend. The correlation coefficient would therefore be unduly high, as the type of correlation it describes is not really present.

Testing correlation coefficients

Checkpoints

1 The null hypothesis here tells us what we can conclude from a test failure – in this case, that there does not exist significant correlation in the population. The alternate hypothesis tells us what type of correlation we can say exists in the population if the test is successful.

2 H_0: $\rho = 0$ (there is no correlation between age and length of trunk). H_1: $\rho > 0$ (there is positive correlation between age and length of trunk). We are given that the level of significance is 5% and the critical region starts at 0.235 3. Since $0.22 < 0.235\,3$ we state that there is not enough evidence to reject H_0 and conclude that significant correlation between age and length of trunk does not exist in this population.

Exam questions

1 H_0: $\rho = 0$ (ranks are not correlated). H_1: $\rho > 0.9$ (ranks are correlated).

1% level of significance; $n = 6$, and from tables the critical region starts at 0.942 9.

The calculated Spearman's coefficient is 0.8. Since $0.8 < 0.942\,9$, we state that there is not enough evidence to reject H_0 and conclude that significant correlation does not exist between the ranks given by each person. For 58 people $\rho = 0.934$. This is in the critical region, so correlation is significant at the 1% level; reject H_0; there is significant agreement among the 58 people.

2 H_0: $\rho = 0$ (there exists no correlation between alcohol consumed and reaction time). H_1: $\rho > 0.9$ (there exists a positive correlation between alcohol consumed and reaction time).

5% level of significance; $n = 50$, and from tables the critical region starts at 0.235 3. For the sample, $r = 0.401$ (3 d.p.). Since $0.401 > 0.235\,3$, we state that there is enough evidence to reject H_0 and conclude that significant correlation exists between the amount of alcohol consumed and reaction time. In other words, the more alcohol one consumes, the longer it takes to react.

Revision checklist

Statistics

By the end of this chapter you should be able to:

1 Understand what a mode is, and how to estimate it from tabulated and graphical data including histograms.	Confident	Not confident. **Revise** pages 112–113
2 Understand what the median represents, and how to obtain a median using interpolation.	Confident	Not confident. **Revise** pages 114–115
3 Know how to draw and interpret cumulative frequency tables/graphs, and the meaning of Inter Quartile Range (IQR).	Confident	Not confident. **Revise** pages 116–117
4 Understand and interpret the mean and standard deviation, how to get them from tables, and how to combine sample data.	Confident	Not confident. **Revise** pages 118–119
5 Know and use the probability equations, and understand when they are appropriate.	Confident	Not confident. **Revise** pages 122–123
6 Be able to construct a probability tree diagram and use it to solve a problem.	Confident	Not confident. **Revise** pages 124–125
7 Know the formulae for permutations and combinations, and when to apply them.	Confident	Not confident. **Revise** pages 126–127
8 Know the conditions for discrete random variables, the difference between probability and cumulative distribution functions, finding and manipulating E(x) and Var(x).	Confident	Not confident. **Revise** pages 128–129
9 Know the conditions for a binomial distribution, its mean and variance, how to use the formula and tables.	Confident	Not confident. **Revise** pages 130–131
10 Know the conditions for a Poisson distribution, its mean and variance, how to use the formula and tables.	Confident	Not confident. **Revise** pages 132–133
11 Know the conditions for continuous random variables, the difference between probability and cumulative distribution functions, finding and manipulating E(x) and Var(x).	Confident	Not confident. **Revise** pages 134–135
12 How to find probabilities under the normal distribution and solve problems involving simultaneous equations.	Confident	Not confident. **Revise** pages 136–137
13 List the conditions for using binomial or Poisson distributions to approximate the normal, using continuity corrections where appropriate.	Confident	Not confident. **Revise** pages 138–139
14 How to test to see if a mean is reasonable, or if a difference between two means is unusual.	Confident	Not confident. **Revise** pages 140–141
15 Know how to calculate best estimates of population statistics, and use the central limit theorem to find confidence intervals.	Confident	Not confident. **Revise** pages 142–143
16 Know the difference between repeated and multiplied samples, and using transformations of variables.	Confident	Not confident. **Revise** pages 144–145
17 Know how to calculate and interpret the product moment correlation coefficient.	Confident	Not confident. **Revise** pages 148–149
18 Be able to capture a relationship between two variables in a single equation using regression equations.	Confident	Not confident. **Revise** pages 150–151
19 Know how to use Spearman's Rank Correlation Coefficient to measure the agreement between two sets of rankings.	Confident	Not confident. **Revise** pages 152–153
20 Be able to test to see if a correlation statistic is significant for a particular sample.	Confident	Not confident. **Revise** pages 154–155

Mechanics

Most A-level students do one or two mechanics modules, and this section will take you through what you need to know. Many students find that mechanics is difficult at first, but it all comes together at the end. If you are struggling, don't despair – careful revision and practice can make all the difference! As this is a branch of applied maths, you will need to transfer skills from the pure maths modules. In particular you should be able to manipulate equations easily, understand vectors, have a knowledge of trigonometry and calculus and be able to sketch graphs. The table below shows which topic is covered by which module, but you will also need to consult a detailed specification from your awarding body, as not all aspects may be included.

Exam themes

- ➜ Kinematics – the study of acceleration, velocity and displacement
- ➜ Dynamics – the study of forces related to motion
- ➜ Statics – the study of forces in equilibrium

Topic checklist

	AQA	CCEA	EDEXCEL	OCR/A	OCR/MEI	WJEC
Modelling	M1	M1	M1	M1	M1	M1
Vectors	M1	M1	M1	M1	M1	M1
Graphs in kinematics	M1	M1	M1	M1	M1	M1
Variable acceleration	M2	M1	M2/M3	M1	M1	M2
Constant acceleration	M1	M1	M1	M1	M1	M1
Projectiles	M1	M2	M2	M2	M1	M2
Force diagrams	M1	M1	M1	M1	M1	M1
Newton's Second Law	M1	M1	M1	M1	M1	M1
Connected particles	M1	M1	M1	M1	M1	M1
Equilibrium of a particle	M1	M1	M1	M1	M1	M1
Friction	M1	M1	M1	M1	M2	M1
Using vectors	M1	M2	M1/M2		M1	M2
Moments and equilibrium	M2	M1	M1/M2	M2	M2	M1/M3
Centre of gravity 1	M2	M3	M2	M2	M2	M1
Centre of gravity 2	M2/M3	M3	M2/M3	M2	M3	
Work and power	M2	M2	M2	M2	M2	M2
Work and energy	M2	M2	M2	M2	M2	M2
Momentum and impulse	M1	M1	M1	M1	M2	M1
Momentum and restitution	M3		M2	M2	M2	M1
Circular motion 1	M2	M2	M3	M2	M3	M2
Circular motion 2	M2		M3	M3	M3	M2
Elasticity	M2	M3	M3	M3	M3	M2

Modelling

The whole point of applied maths is that you apply the skills and techniques you have learned in pure maths to 'real world' problems – in this case mechanics. On the one hand, you want the mathematical solutions to correspond to real-life experience – there's not much point in it all otherwise! On the other hand, the real world is awkward and complex, and there's no point ending up with maths that is too difficult for you to handle. 'Modelling' is all about finding the balance.

Action point

You may have had the opportunity in your maths lessons to do some practical work to give you some idea of how your calculated solutions compare to practical results. This enables you to judge whether or not the assumptions were valid.

Modelling assumptions

In order to simplify the maths to a level you can cope with, we make 'modelling assumptions'. As the range of mathematical techniques available to you increase, you can make fewer assumptions, so that your solutions correspond more closely to real experience. You need to be aware of the mathematical significance of each assumption (i.e. what difference it makes to the maths). You also need to know the likely effect on your answers.

Assumptions relating to size and mass

Checkpoint 1

Suppose you are doing a question which involves one vehicle overtaking another. What would be the significance of modelling the vehicles as particles?

Particle

A particle has mass, but no size (or, at any rate, its size is insignificant compared to other lengths in the question). If you are given length dimensions for an object, that is an indication that you are not modelling it as a particle. When you model an object as a particle, then all the forces acting on the particle act through the same point, so you don't consider that the forces might make the object rotate.

Rod

A rod has mass and length, but no thickness.

Lamina

A lamina has mass and area, but no thickness.

Checkpoint 2

Imagine a vertical string which is not light. Why would the tension in the string be greater in a section near the top of the string than in one near the bottom?

Light

If an object is modelled as 'light', it means it has no mass (or its mass is negligible compared to other masses). If a rod is light, then you can't apply Newton's Second Law to it because the mass is zero. If a light object is moving, it has no kinetic energy. Strings are often modelled as light. This means that a vertical string has a uniform tension along its length. It's usually a fair assumption.

Assumptions relating to the nature of materials

Checkpoint 3

If a man is using a rope to drag a box across a floor, does it matter whether or not the rope is inextensible if you are calculating the acceleration of the box?

Inextensible or inelastic

This relates to strings and ropes which do not extend when under tension. It is important when you are dealing with connected particles as they will move with the same velocity and acceleration if the connecting string is inextensible.

Smooth

A 'smooth' surface has no friction. In practice this is rarely (ever?) the case, so this assumption simplifies the maths, but does introduce an error. Pulleys, and pegs, are often modelled as smooth. When this is the case, you can take the tensions in the string on either side as equal.

Air resistance

In many problems you will be told to neglect air resistance, and if you are working with small speeds and objects this does not introduce much of an error. You should be aware that as speed increases so does air resistance.

Checkpoint 4

You may know that a person in free fall from a plane eventually reaches a terminal velocity. How does this fact contradict an assumption of 'no air resistance'?

Exam questions answers: page 212

Exam questions on modelling are always contained within other questions. The questions below assume that you are near the end of your revision – come back to them and use them as practice for the synoptic section.

Use $g = 9.8$ ms^{-2}.

1 Two objects of masses 1 kg and 1.5 kg are on a rough horizontal ground, 2 m apart. The 1 kg object is projected towards the other with speed 5 ms^{-1}. Modelling the objects as particles, find

(a) the speed of the 1 kg object immediately before impact, if $\mu = 0.3$;

(b) the speed of both objects after impact if the coefficient of restitution = 0.5.

What problems might you encounter if the objects were not modelled as particles? (12 min)

2 A penny is dropped from a window 12 m above ground level. Stating any necessary modelling assumptions, find the time taken for the penny to reach the ground.

Would the same assumptions be valid if a packing case measuring 2 m by 1.5 m by 1 m and mass 200 kg fell from the cargo hold of a plane 12 000 m above the ground? (10 min)

3 Two particles of masses 4 kg and 5 kg are connected by a string. The 4 kg particle is held on a smooth plane inclined at 20° to the horizontal, and the string passes over a pulley so that the 5 kg particle hangs freely. Stating any assumptions you make, find the acceleration of the particles when they are released. (12 min)

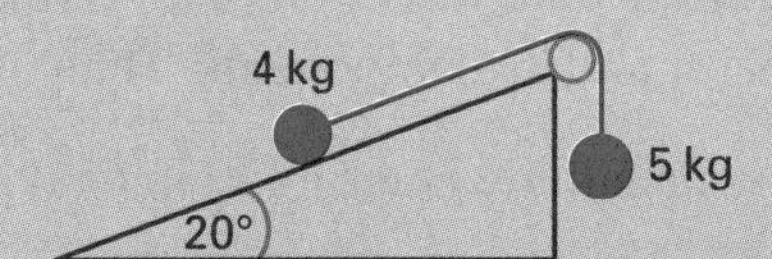

4 A man is carrying a ladder of mass 15 kg and length 3 m. At one end of the ladder there is a fixing for a tool which has mass 3 kg. In order to calculate where the man should hold the ladder for it to balance.

(a) Is it appropriate to model the fixing as a particle?

(b) Is it appropriate to model the ladder as a particle?

(c) Are any other modelling assumptions necessary?

Do the calculation. (8 min)

Examiner's secrets

Just like in a real exam, you can't tell by the title at the top of the page what topic each question is testing. *Read each question carefully*, alert for clues. Draw a diagram and decide which principles you are going to apply.

Examiner's secrets

You need to know both the common modelling assumptions, and the situations in which they are valid. You also need to know *why* you are making them!

Vectors

Many quantities you deal with in mechanics are vectors – they have both magnitude and direction. This includes force, acceleration, velocity and displacement. You need to be able to combine vector quantities to find resultants, and also use vectors in component form.

The resultant of two vectors

The sum of two (or more) vectors is called the **resultant** of the vectors. You can find the resultant by means of a vector triangle: put the two vectors 'nose-to-tail' then the resultant will form the third side of the triangle. You can calculate the sides and angles of the triangle using the sine and cosine rules.

Example: Find the resultant of two forces of magnitudes 6 N and 8 N if the angle between them is 50°.

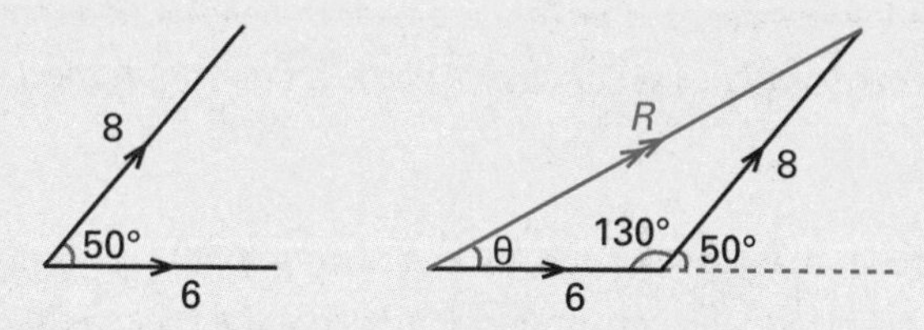

$R^2 = 6^2 + 8^2 - 2 \times 6 \times 8 \times \cos 130°$, giving $R = 12.7$ (3 s.f.)

$$\frac{\sin \theta}{8} = \frac{\sin 130}{R} \text{ giving } \theta = 28.8°$$

So the resultant has magnitude 12.7 N, and makes an angle 28.8° with the 6 N force.

Vectors in component form

You will be familiar with column vectors. In this form two components are given, in the x and y directions. Alternatively we can write the vector using **i** and **j**, where **i** and **j** are unit vectors in two perpendicular directions:

$$\begin{pmatrix} -3 \\ 2 \end{pmatrix} = -3\mathbf{i} + 2\mathbf{j}$$

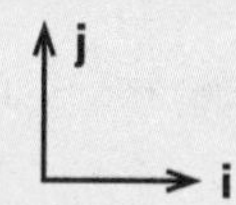

The advantage of vectors in this form is that resultants are very easy to find, as all you have to do is to add the components separately.

To find the magnitude and direction of a vector given in component form, sketch the vector and its components. Then use Pythagoras and trig.

Example: Find the magnitude and direction of $12\mathbf{i} - 5\mathbf{j}$.

$$12\mathbf{i} - 5\mathbf{j} = \begin{pmatrix} 12 \\ -5 \end{pmatrix}$$

12

5

The magnitude is $\sqrt{5^2 + 12^2} = 13$, and the direction is given by $\tan^{-1}(\frac{5}{12})$, i.e. 22.6° below the x-axis.

The jargon

When typing, you use **bold** to indicate vectors. In hand-writing, you use a squiggle underneath, or an arrow overhead:

a or $\underset{\sim}{a}$ or $\overrightarrow{AB}$

Links

For sine and cosine rules, see page 45.

Watch out!

The angle between two vectors is the angle formed when both vectors are pointing away. When you put them nose-to-tail in the vector diagram, the angle in the triangle will be the supplementary angle.

Test yourself

If the two forces have magnitudes 4 N and 5 N and the angle between them is 100°, use the cosine rule to show that the resultant has a magnitude of 5.84 N, and makes an angle of 42° to the 5 N force.

Checkpoint 1

Find the resultant of $3\mathbf{i} + 3\mathbf{j}$, $-2\mathbf{i} + \mathbf{j}$ and $-6\mathbf{i} - 5\mathbf{j}$.

Checkpoint 2

Find the magnitude and direction of the resultant you found in checkpoint 1.

Resolving vectors

The process of finding the components of a vector is called resolving. You often have to do this in mechanics problems, so it's worth while spending time practising until you can do it easily. The vector's magnitude forms the hypotenuse of a right-angled triangle, and the two components are the other two sides.

Example: Find the components in the East and North directions of a velocity 50 ms^{-1} on a bearing 072°.

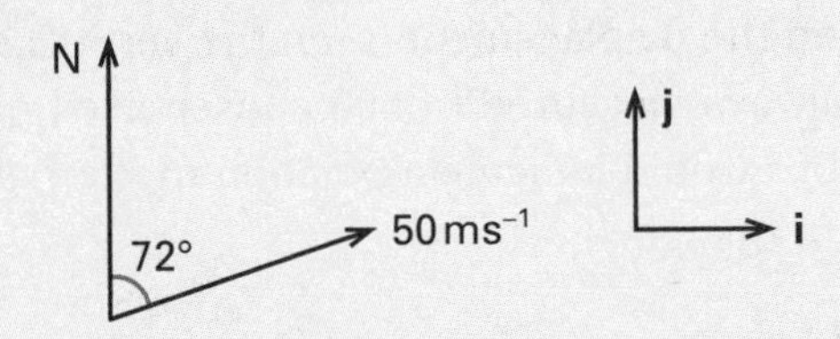

Answer: The component is 50 sin 72° = 47.6 in the East direction and 50 cos 72° = 15.5 in the North direction. So we could write the velocity as $47.6\mathbf{i} + 15.5\mathbf{j}$ ms^{-1} with **i**, **j** in East/North directions.

If you need to find the resultant of more than two vectors, it is often best to find the sum of the components in two perpendicular directions, and then convert back to magnitude-direction form if necessary.

Example: Find the resultant of the forces shown in the diagram below:

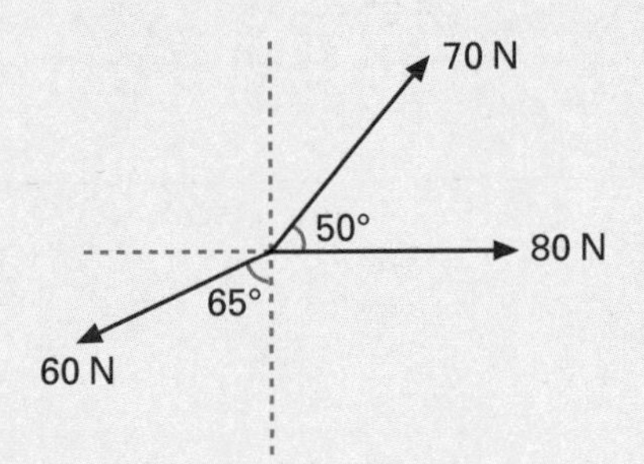

Answer: In component form, the resultant is

$$\begin{pmatrix} 80\text{N} \\ 0\text{N} \end{pmatrix} + \begin{pmatrix} 70\cos 50°\ \text{N} \\ 70\sin 50°\ \text{N} \end{pmatrix} + \begin{pmatrix} -60\sin 65°\ \text{N} \\ -60\cos 65°\ \text{N} \end{pmatrix} = \begin{pmatrix} 70.6\text{N} \\ 28.3\text{N} \end{pmatrix}$$

Exam questions answers: page 212

1 Two forces have magnitudes 5 N and P N. The resultant force has magnitude 6 N, and acts at an angle 40° to the force of magnitude 5 N. Find the value of P and the direction in which this force acts. (5 min)

2 Three dogs pull a sledge using three horizontal ropes. The tensions in the ropes are as shown in the diagram. Find the components of the resultant in the direction of **i** and **j** as indicated, and also the magnitude and direction of the resultant. (6 min)

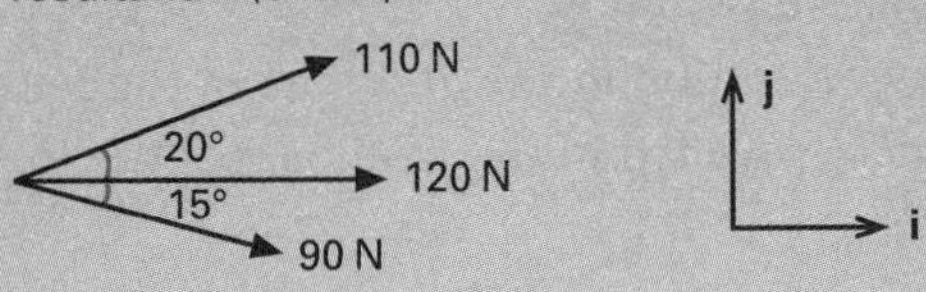

Examiner's secrets

Students often get the sin and cos the wrong way round. Remember it is *cos* if you are *turning through* the angle to get the direction you want, and *sin* if you are *turning away* from the angle. The component in the x-direction of the vector below is 8 cos 40°.

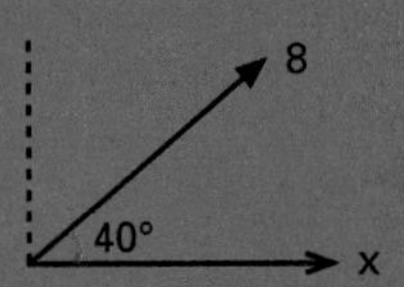

The component in the x-direction in this case is 8 sin 40°.

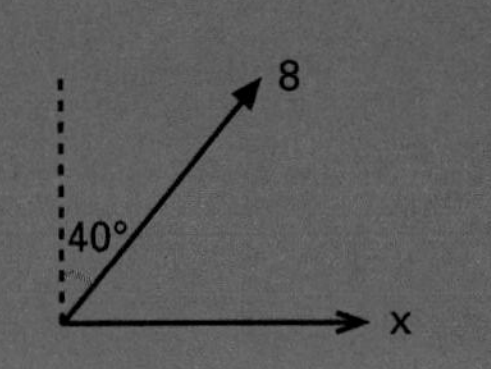

Checkpoint 3

Find the horizontal and vertical components of the velocity of a ball moving at 3 ms^{-1} at an angle 15° above horizontal.

Test yourself

Cover this working up, and on a piece of rough paper check that you can find the resultant yourself.

Checkpoint 4

Find also the magnitude and direction of the resultant in this worked example.

Graphs in kinematics

Lots of students shy away from drawing graphs, but actually they are not that difficult and sketch graphs give a helpful picture of the motion you are analysing. In this section we will assume that the motion is taking place along a straight line. We use + and – to indicate the two possible directions.

The jargon

Kinematics is the study of motion.

Displacement–time graphs 

These are also called (t,s) graphs to show that the time, t, is on the horizontal axis and the displacement, s, on the vertical axis. The displacement is always measured from a chosen fixed point, O. A positive value of s would indicate a position to the right of O and a negative value to the left.

The jargon

s is commonly used for displacement, though x is also used sometimes. We use v for velocity.

Checkpoint 1

What is the difference between displacement and distance travelled? What is the difference between velocity and speed?

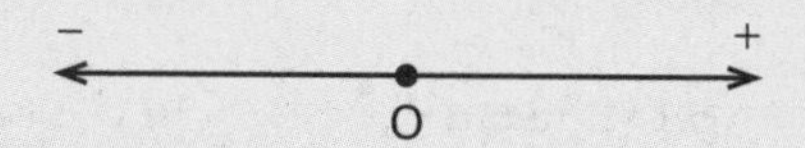

The gradient of the graph measures how fast the object is moving, its velocity.

Links

For details about calculating gradients of straight lines, see page 38.

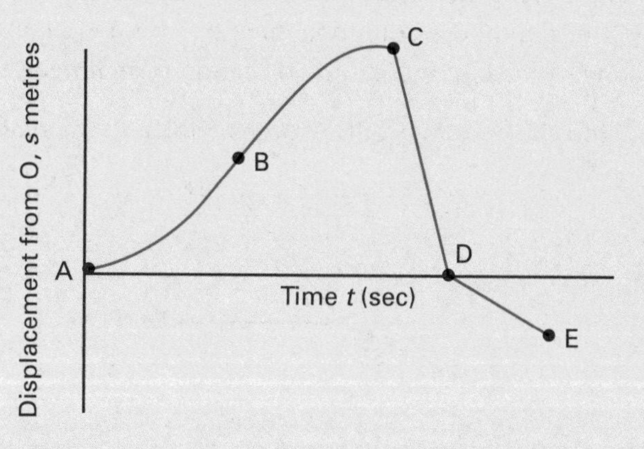

In the graph above, the object is moving away from O in the section A to C. It is moving fastest at B (why?). In the section CD it is moving back towards O, and in section DE it is moving away from O in the opposite direction.

Checkpoint 2

How can you tell that it is moving with constant velocity in section CD, and in section DE, but not in section AC? Is it moving faster in CD or DE?

Velocity–time graphs 

These are also called (t,v) – again to indicate which variable is on the horizontal axis and which on the vertical (check you know which!). This time the gradient represents the acceleration. The displacement is represented by the area between the graph and the horizontal axis. These graphs can be especially helpful if you have a multi-stage journey.

Links

We often use trapeziums to estimate the area under a curved graph. See page 74.

Example: A train is travelling between two stations A and B 1 500 m apart. It leaves A and accelerates for 15 s, reaching a top speed of 12 ms^{-1}. It travels at this speed until the final stretch when it decelerates at 0.5 ms^{-2} to come to rest at B.

Find (a) the acceleration in the first part of the journey, (b) the time taken and distance travelled in the final part of the journey and (c) the total journey time.

Checkpoint 3

Sketch a velocity–time graph for the journey and mark on it the 15 s and 12 ms^{-1}.

Answer:

(a) The acceleration in the first section is the gradient, which is $12 \div 15 = 0.8$ ms^{-2}.

(b) The final section has gradient -0.5 ms^{-2}. The time taken is 24 s (since $12 \div 24 = 0.5$). The distance covered is the area of the triangle.

(c) To find the total time, you will need to work out how far the middle section is. You know the total area is 1 500, so subtract the distances for the first and third sections. It's travelling at constant speed so the time taken should be easily found.

Common misunderstanding

Many students believe that a negative acceleration means that the object is slowing down. This is not always true! Consider the situation when a stone is thrown vertically upwards with speed 15 ms^{-1} from the top of a tower 30 m high. The acceleration is negative throughout. While the stone is on its way up, it is slowing down, but once on the way down, it is speeding up! The velocity–time graph is drawn for you below:

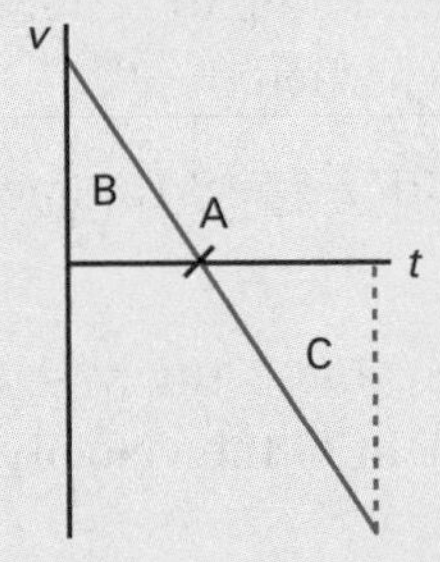

At the point marked A, the stone is at its greatest height, and is at instantaneous rest. Consider the two areas marked B and C. The difference between them is 30 m. Why is this? What does the area B represent?

Exam questions answers: page 212

1 A train travels between two stations A and B, 11 km apart. It starts from rest at A, accelerates uniformly to a speed 25 ms^{-1}, and maintains this speed until it decelerates uniformly to rest at B. The journey takes in total 10 minutes, and it takes three times as long to accelerate as it does to decelerate. Sketch the velocity–time graph for the journey, and hence, or otherwise, calculate the time taken to decelerate. (12 min)

2 The graph illustrates the displacement of a particle from a fixed point O as it moves in a straight line.

(a) Calculate its velocity during the first five seconds.

(b) Calculate its velocity during the 10th second, and write down the time at which it passes through O.

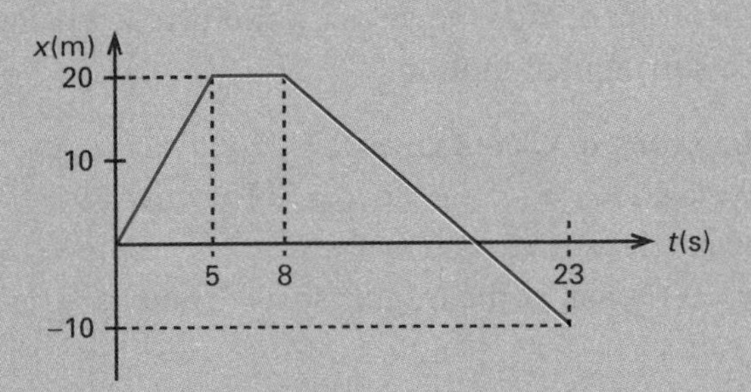

Checkpoint 4

Find the distance for the third section.

Checkpoint 5

Do the working here.

Checkpoint 6

What force(s) are acting on the particle when it is in flight?

Watch out!

The graph below is wrong!!

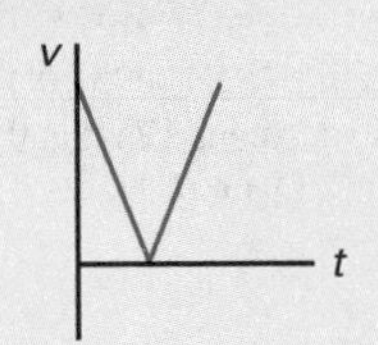

The jargon

When you say that an object is at 'instantaneous rest', you mean that just before this moment its velocity was positive, and just after it was negative (or the other way round!). So there must have been an instant when the velocity was zero. It does not mean that it stopped moving and was 'at rest'.

Variable acceleration

Links

For a reminder about calculus see pages 52–53 and 62.

The key words here are 'rate of change'. Acceleration measures the rate of change of velocity, and velocity measures the rate of change of displacement. You'll need to be comfortable with basic calculus, but in this section we'll stick to the most straightforward scenario: everything is given in terms of t – what a relief!

Using differentiation

The jargon

Some books use

$\dot{x}$ for $\frac{dx}{dt}$

$\ddot{x}$ for $\frac{d^2x}{dt^2}$

displacement → velocity → acceleration

In the previous spread we saw that the gradient of a displacement–time graph represented the velocity. From your work on calculus, you know that differentiation is used to find a gradient, so using x for the displacement and v for velocity we have

$$v = \frac{dx}{dt}$$

Checkpoint 1

If $x = 30 + 20t - 5t^2$, find the maximum positive displacement. What is the velocity at this instant?

Example: The displacement of a particle at time t is given by $x = 5t + 6t^2$. Find the velocity when $t = 10$.

Answer: Differentiating with respect to time gives $v = 5 + 12t$, so when $t = 10$, $v = 125$.

In the same way, acceleration is the rate of change of velocity and so is represented by the gradient of the velocity–time graph. This means that

$$v = \frac{dv}{dt} = \frac{d^2x}{dt^2}$$

Checkpoint 2

Find the acceleration at $t = 10$ for the example in checkpoint 1.

Example: The velocity of a particle at time t is given by $v = \frac{1}{2}t^2 + \frac{1}{4}t^3$. Find the acceleration at time $t = 10$.

Answer: Differentiating gives $a = t + \frac{3}{4}t^2$. Substituting $t = 10$ gives $a = 85$.

Integration

Watch out!

Do not forget the constant of integration. This is a common, and costly, mistake.

acceleration → velocity → displacement

If we are working back from the acceleration to find the velocity and then the displacement, we simply use the reverse of differentiation, namely integration. The important thing to remember here is the **constant of integration** at each stage. You can find the value of the constant of integration by substituting in the initial conditions.

The jargon

'Initial conditions' is just the name given to what was happening at the start, i.e. the values of x and/or v when $t = 0$.

Example: A particle is initially at the origin and is moving with speed 5 ms^{-1}. Its acceleration, a ms^{-2}, at time t is given by $a = 3t - 2$. Find expressions in terms of t for the velocity and displacement of the particle in the subsequent motion.

Test yourself

Cover up the answer here and check that you can reproduce it.

Answer: Integrating $a = 3t - 2$ gives $v = \frac{3}{2}t^2 - 2t + c$

When $t = 0$, $v = 5$ so $5 = 0 - 0 + c$, i.e. $c = 5$. This gives $v = \frac{3}{2}t^2 - 2t + 5$.

Integrating again gives $x = \frac{1}{2}t^3 - t^2 + 5t + c$.

When $t = 0$, $x = 0$ (it was at the origin) so $c = 0$ and $x = \frac{1}{2}t^3 - t^2 + 5t$.

Motion in two or three dimensions

There's not really a problem here. You still differentiate to find the velocity given the displacement, and differentiate again to find the acceleration from the velocity. The only difference is that everything is in vector form – you have got **i**, **j** and maybe **k** to contend with. Don't let that put you off!

Example: The displacement of a particle at time t is given by $\mathbf{r} = (3t - 2)\mathbf{i} + (6t - t^2)\mathbf{j}$, where **i** and **j** are unit vectors parallel to the x- and y-axes respectively. Find (a) its initial position, (b) its velocity at time t, (c) the time at which it is travelling parallel to the x-axis, and its position at this instant.

Answer:

(a) $r = \begin{pmatrix} 3t-2 \\ 6t-t^2 \end{pmatrix}$ so when $t = 0$, $r = \begin{pmatrix} -2 \\ 0 \end{pmatrix}$ giving the initial position as (–2,0).

(b) Differentiating gives $\mathbf{v} = \begin{pmatrix} 3 \\ 6-2t \end{pmatrix} = 3\mathbf{i} + (6 - 2t)\mathbf{j}$

(c) The particle is moving parallel to the x-axis when the vertical component of the velocity is zero, i.e. when $t = 3$. At this time $\mathbf{r} = 7\mathbf{i} + 9\mathbf{j}$, so the position is (7,9).

Example: The acceleration of a particle is $-10\mathbf{j}$. (a) The particle starts at (0,80) with velocity $15\mathbf{i}$. Find its velocity at time $t = 5$. (b) At time T the particle is at $(x,0)$. Find the values of x and T.

Answer:

(a) $\mathbf{a} = -10\mathbf{j}$ integrated gives $\mathbf{v} = -10t\mathbf{j} + \mathbf{u}$, where **u** is the constant of integration. When $t = 0$, the velocity is $15\mathbf{i}$ so $\mathbf{u} = 15\mathbf{i}$. This gives $\mathbf{v} = 15\mathbf{i} - 10t\mathbf{j}$, and when $t = 5$, $\mathbf{v} = 15\mathbf{i} - 50\mathbf{j}$.

(b) Integrating $\mathbf{v} = 15\mathbf{i} - 10t\mathbf{j}$ gives $\mathbf{r} = 15t\mathbf{i} - 5t^2\mathbf{j} + \mathbf{c}$.

The initial position is (0,80) so when $t = 0$, $\mathbf{r} = 80\mathbf{j}$ so $\mathbf{c} = 80\mathbf{j}$. Therefore $\mathbf{r} = 15t\mathbf{i} + (-5t^2 + 80)\mathbf{j}$. The particle is at $(x,0)$ when the **j** component of **r** is zero, i.e. $-5t^2 + 80 = 0$. This gives $T = 4$ and $x = 60$.

Watch out!

r is the displacement vector from the origin to the position at time t. It does not directly give you the distance from the origin (though you could find it – how?)

Checkpoint 3

You should be able to find the acceleration at time t too.

The jargon

Some students find it easier to work with column vectors, rather than **i** and **j**, so that is what I have done here.

Checkpoint 4

What physical situation could this example be modelling?

Checkpoint 5

Note that the constant of integration is a vector. Why? Why do you think I chose the letter **u** for the constant of integration?

Test yourself

Cover up the answers to the examples on this page and check that you can reproduce them without looking.

Exam questions

answers: pages 212–13

1 The velocity, v ms^{-1}, of a particle moving in a straight line t seconds after passing through the origin O is given by $v = 3t^2 - 8t + 4$. Find

(a) The acceleration of the particle when $t = 2$.

(b) The values of t for which the particle is at instantaneous rest.

(c) The distance travelled by the particle between these two times.

(12 min)

2 A particle moves so that its position at time t is given by $\mathbf{s} = (t^2 - 4)\mathbf{i} + 3t\mathbf{j}$. Find:

(a) The times at which the particle crosses the y-axis.

(b) An expression for the velocity vector v, and hence the particle's speed at time $t = 2$.

(c) The acceleration of the particle. (10 min)

Constant acceleration

You need to be confident in handling the constant acceleration equations as they are often tacked on to the end of other problems. Luckily they're not hard!

The jargon

The standard notation is:
a = acceleration
t = time
u = initial velocity
v = velocity at time t
s = displacement at time t

Deriving the constant acceleration equations

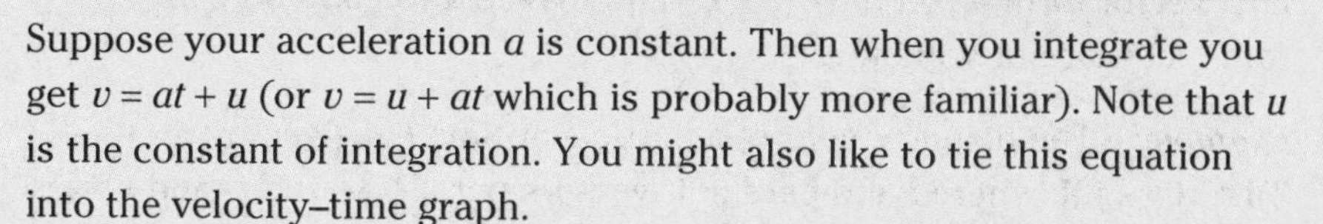

Suppose your acceleration a is constant. Then when you integrate you get $v = at + u$ (or $v = u + at$ which is probably more familiar). Note that u is the constant of integration. You might also like to tie this equation into the velocity–time graph.

Checkpoint 1

What is the gradient of the velocity–time graph? How does $v = u + at$ relate to the familiar line equation $y = mx + c$?

We can go on to integrate $v = u + at$. This gives $s = ut + \frac{1}{2}at^2 + c$. If we measure the displacement from the origin, $c = 0$, but that need not be the case. However, usually we have $s = ut + \frac{1}{2}at^2$.

The third equation can be found by considering the velocity–time graph. The displacement is equal to the area under the graph so $s = \frac{1}{2}(u + v)t$ (or average velocity × time).

Checkpoint 2

What shape is the area under the graph? What's the formula for its area?

Once we've got these three equations, we can find the other two by substituting:

$u = v - at$	into $s = ut + \frac{1}{2}at^2$	gives $s = vt - \frac{1}{2}at^2$
and $t = (v - u)/a$	into $s = \frac{1}{2}(u + v)t$	gives $v^2 = u^2 + 2as$

Test yourself

Actually do the substitution. Go on, try!

Which formula shall I use?

Look at the formulae. You'll notice that each one uses four of the five quantities and has a different quantity missing. In most questions you will find that you know three of the quantities, are asked to find one quantity and neither know nor need to know the fifth. So it's easy! Simply use the equation that doesn't contain the quantity you're not interested in.

Don't forget

$v = u + at$
$s = ut + \frac{1}{2}at^2$
$s = vt - \frac{1}{2}at^2$
$v^2 = u^2 + 2as$
$s = \frac{1}{2}(u + v)t$

Example: A car accelerates from 12 ms^{-1} to 30 ms^{-1} over a distance of 120 m. Find the acceleration of the car.

Answer: You are given u, v and s. You need to find a, but are not interested in t. So use $v^2 = u^2 + 2as$.

$30^2 = 12^2 + 2a\ 120$, giving $756 = 240a$ and therefore $a = 3.15$

Checkpoint 3

Find the time taken by the car as well.

Example: A particle leaves point A with speed 5 ms^{-1} and acceleration -0.5 ms^{-2}. Find:

(a) how far the particle is from A at the instant in which it comes to instantaneous rest;
(b) the total distance travelled in the first 12 seconds.

Answer:

(a) $u = 5$, $v = 0$, $a = -0.5$ and you need s, so use $v^2 = u^2 + 2as$.
$0 = 25 + 2 \times (-0.5) \times s$ giving $s = 25$
(b) Find the displacement after 12 seconds. Note that when $t = 12$, $v = u + at = 5 + (-0.5 \times 12) = -1$, so the particle is heading back towards A.
$u = 5$, $t = 12$, $a = -0.5$ and we want s, so use $s = ut + \frac{1}{2}at^2$
$s = 5 \times 12 + \frac{1}{2} \times (-0.5) \times 12^2 = 60 - 36 = 24$ (displacement from A)

Checkpoint 4

What clue are you being given by being asked for the *total* distance travelled?

Test yourself

Cover up this answer and check you can work it through for yourself.

Since the particle has gone to its furthest distance from A (25 m) and is now on its way back, the total distance travelled is 25 + 1 = 26 m.

Vertical motion under gravity

This is a special case of constant acceleration. The acceleration always acts downwards and has magnitude g. You will need to read your exam board's instructions about the value of g: either 10 ms^{-2}, 9.8 ms^{-2} or 9.81 ms^{-2}. In this book, we will use $g = 9.8$ ms^{-2}.

Example: A man is in a hot-air balloon which is rising vertically. At an instant when the balloon is 200 m above the ground, and rising at 4 ms^{-1}, the man knocks his binoculars off the rim of the basket. Find the time taken for the binoculars to reach the ground, and the speed with which they hit the ground.

Answer: Take the upward direction as positive.
The initial speed, u, of the binoculars is 4 ms^{-1}.
The acceleration $a = -9.8$ and $s = -200$.
Using $v^2 = u^2 + 2as$ we get $v^2 = 16 + 2 \times (-9.8) \times (-200) = 3\,936$.
So $v = -62.7$ ms^{-1} (note the negative sign as it's travelling downwards).
Then $v = u + at$ gives $-62.74 = 4 - 9.8t$, so $t = 6.81$.

Using vectors

All the examples considered so far assume that motion is taking place in a straight line. If your exam board requires you to use vectors to deal with motion in two or three dimensions, see 'Using vectors' pages 190 and 191.

Examiner's secrets

In these questions you will need to be clear about which direction (up or down) you are taking as positive. It doesn't matter which you choose, but say it clearly and stick to it!

Checkpoint 5

What modelling assumption is being made in the solution to this example?

Test yourself

You could find t first using $s = ut + \frac{1}{2}at^2$. Check that it gives the same result. Note that you get a quadratic. How might you interpret the other root of the equation?

Exam questions answers: page 213

1 A boy throws a stone vertically upwards with speed 2.8 ms^{-1} from the rim of a well. Find: (a) the greatest height reached by the stone above the well rim; (b) the depth of the well, given that the splash occurs 5 seconds after the boy throws the stone. (6 min)

2 A particle travelling in a straight line has constant acceleration 0.8 ms^{-2}. It passes two points X and Y which are 40 m apart. Given that the particle's speed increases by 4 ms^{-1} in moving from X to Y, find the speed of the particle at X. (6 min)

3 A policeman is running towards a stationary suspect at a steady speed of 2.5 ms^{-1}. When the policeman is 30 m away, the suspect sees him and starts to run away accelerating at 0.1 ms^{-2}. Show that the policeman catches the suspect, and find how far the suspect has run in this time. (8 min)

Examiner's secrets

Always quote the formula you are using to make your method clear.

Projectiles

This is a special case of motion in two dimensions under constant acceleration. In this topic you study the motion of a particle in flight, acted on by gravity only – we assume no air resistance.

Checkpoint 1

What will the constant acceleration be?

Checkpoint 2

What will be the components of the initial velocity in the horizontal and vertical directions?

The basic equations 

Suppose the particle has initial speed U, and is projected from the origin at angle θ to the horizontal. At time t after projection the particle is at (x,y), and the components of its velocity are $\dot{x}$ and $\dot{y}$.

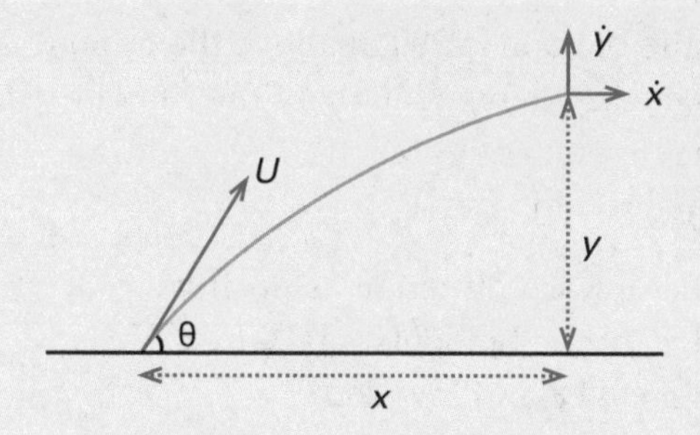

Examiner's secrets

You need to make yourself really familiar with these equations.

We consider the horizontal and vertical motion separately for the most part.

Horizontal motion: $a = 0$

Initial speed = $U \cos \theta$

Velocity = $\dot{x} = U \cos \theta$ (using $v = u + at$)

$x = Ut \cos \theta$

Vertical motion: $a = -g$

Initial speed = $U \sin \theta$

Velocity = $\dot{y} = U \sin \theta - gt$ (using $v = u + at$)

$y = Ut \sin \theta - \frac{1}{2}gt^2$ (using $s = ut + \frac{1}{2}at^2$)

Checkpoint 3

Sketch a diagram showing the information in this example.

Example: A particle is projected with initial speed 12 ms^{-1} at an upward angle of 30° to the horizontal from the top of a tower 25 m high. Find how long the particle is in flight before it hits the ground, and the horizontal distance from the foot of the tower to the landing site. (Take $g = 9.8$ ms^{-2})

Checkpoint 4

Work out the speed of the particle when it hits the ground.

Answer: It lands when $y = -25$ (note the use of negative values). So $-25 = 12 \sin 30°t - \frac{1}{2} \times 9.8t^2$. This is a quadratic in t which simplifies to $4.9t^2 - 6t - 25 = 0$. Use the quadratic formula to get $t = 2.95$ or -1.73. The positive value is the one we want. Now put this value in $x = U \cos \theta t$ to get $x = 30.7$.

Time of flight, range and greatest height 

Action point

First check that you can follow through the rearrangement to get T. Then see if you can get the same result for T by finding the time when the vertical velocity is zero. T will be double this time (why?).

Time of flight, T

The particle is at zero height at the start and end of its flight. Putting $y = 0$ and $t = T$ gives $0 = U \sin \theta T - \frac{1}{2}gT^2$. This rearranges to $T = 2U \sin \theta/g$.

Range, R

R is the value of x when $t = T$.
$R = U \cos \theta T = 2U^2 \sin \theta \cos \theta/g = U^2 \sin 2\theta/g$.
The greatest range possible occurs when $\sin 2\theta = 1$, i.e. $\theta = 45°$.

Greatest height, H

The greatest height occurs when $t = \frac{1}{2}T = U \sin \theta/g$. Putting $y = H$ and $t = U \sin \theta/g$ in the equation for y gives $H = U \sin \theta U \sin \theta/g - \frac{1}{2}g(U \sin \theta/g)^2 = (U \sin \theta)^2/2g$.

Cartesian equation for the trajectory

This is the posh name for the equation directly relating x and y. It is found by eliminating t from $x = U \cos \theta t$ and $y = U \sin \theta t - \frac{1}{2}gt^2$. You get

$$y = x \tan \theta - \frac{gx^2}{2U^2 \cos^2\theta}$$

Note that if you are given any three of x, y, U, θ you can find the fourth. The version below replaces the $1/\cos^2\theta$ by $\sec^2\theta = 1 + \tan^2\theta$. This enables you to have a quadratic in $\tan \theta$ if you need to find θ.

$$y = x \tan \theta - \frac{gx^2}{2U^2}(1 + \tan^2\theta)$$

Example: A gunner at ground level is aiming at a window 20 m above the ground in a building 200 m away. Find the possible angles of projection if the speed of projection is 70 ms^{-1}.

Answer:

$$20 = 200 \tan \theta - \frac{9.8 \times 200^2}{2 \times 70^2}(1 + \tan^2\theta)$$

$20 = 200 \tan \theta - 40(1 + \tan^2\theta)$
$40 \tan^2\theta - 200 \tan \theta + 60 = 0$
$\tan \theta = 0.320\,6$ or $4.679\,4$ (using the quadratic formula)
$\theta = 17.8°$ or $77.9°$

Links

If you have forgotten the sin 2θ formula see page 50.

Action point

Check again that you can follow this through – it is excellent practice in manipulation.

Checkpoint 5

An archer standing on level ground fires an arrow into the air with initial speed 56 ms^{-1} at 60° to the horizontal. Find the time of flight and the range of the arrow.

Test yourself

You've guessed it! I'm going to ask you to try to do this for yourself!

Checkpoint 6

Cover up the work and check you can answer the question yourself. What modelling assumption is being made here?

Exam questions answers: page 213

1 A golfer is using an iron which gives the ball an initial velocity at an angle $\sin^{-1}(\frac{3}{5})$ to the horizontal. To reach the green, he needs the range of his shot to be between 132 m and 146 m. Find the possible values for the speed of projection. (8 min)

2 A particle is projected from a point on level ground with speed 28 ms^{-1} at an angle 30° to the horizontal. Find the times between which the particle is more than 8 m above the ground. (8 min)

Examiner's secrets

A question may well tell you that you are not allowed to quote the formulae for the range, time of flight, etc. without proof. So practise this question without quoting the formula above.

Force diagrams

Drawing a good clear force diagram is the first and most important step in solving many mechanics problems. It's a skill you must acquire, so practise as much as you need to make sure that you can be successful.

Notation

Force, velocity and acceleration are all vectors and so are drawn with arrowhead lines to indicate direction. It is helpful to have different styles of arrowhead to indicate different quantities. Commonly we use

forces velocities acceleration

Drawing diagrams

Remember that this is not art – you might like to make your car or whatever look realistic, but don't waste time doing this, and don't let the details obscure the important features of the mechanics!

Make sure you are clear which object you are considering and only draw in the forces acting on that object. If there are two objects, and you need to consider the forces on both, use two colours.

Generally forces are labelled with capital letters (the only exception being weight; see below).

Watch out!

A common mistake is to draw both a force and one of its components. Either draw the force *or* both the components.

The most common forces

The jargon

If you have two or more objects to consider with different weights use mg and Mg for the weights, or m_1g and m_2g.

Weight

Newton's Law of Gravitation states that there is an attractive force between any two bodies. (If the bodies are small, e.g. two books, we don't notice this attraction. At least one of the bodies needs to be massive.) The attraction between an object and the Earth is commonly known as the weight of the object and is equal to mg, where m is the mass of the object and g is the acceleration due to gravity (approximately 9.8 ms^{-2}). So when you draw a force diagram, the first to put in is the weight, acting vertically downwards.

Tension and thrust

If two bodies are connected by means of a string, rope, chain, rod, etc. the string may be in tension (the two ends being pulled apart).

T T

A rigid connector like a rod can also be in compression (the two ends pushed together) and the force in the connector is a thrust.

T T

If the string is light, and passes over smooth light pulleys, then the tension will be the same all along its length (in particular, it will be the same at both ends).

If the object is attached to two strings you can't assume that the tensions in the string will be equal (unless the diagram is symmetrical). Label the tensions T_1 and T_2. However if you have a smooth bead or ring threaded on a string, then the tension in each part is equal.

Checkpoint 1

Why can't there be a thrust in a string?

Links

For the definition of 'light' and 'smooth' in this context see pages 168 and 169.

Checkpoint 2

A particle of mass 3 kg is attached to two strings. The free ends of the strings are attached to two fixed points A and B, on the same horizontal level. Draw a diagram showing the forces acting on the particle.

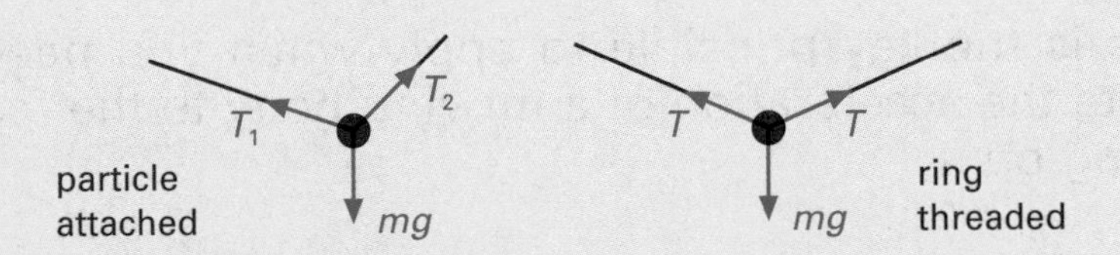

Contact forces

Where there are two surfaces in contact, there will be contact forces. These always consist of a normal reaction which is perpendicular to the contact surface. This is usually labelled N or R (for obvious reasons!) In addition, if the surface is 'rough' as opposed to 'smooth' there will be a friction force, parallel to the surface.

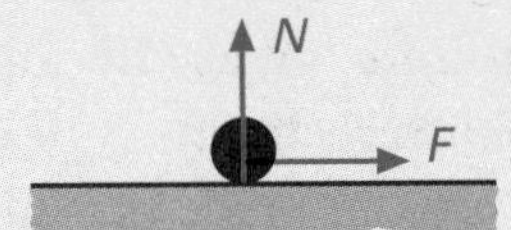

Other forces

In the question you may be told about other forces acting on the object. There may be air resistance (but this is often ignored). If there is a hinge, draw in two perpendicular components of the force acting there (unless told different).

Newton's Third Law

Essentially this says that forces come in equal and opposite pairs. This is especially relevant for contact forces. Each of the two objects in contact experiences the same pair of normal reaction and friction, but in opposite directions. Label them with the same letters to indicate this. The diagrams below show a sack of potatoes on the back of a truck. The first shows the forces on the sack due to contact with the truck, and the second shows the forces on the truck due to the sack.

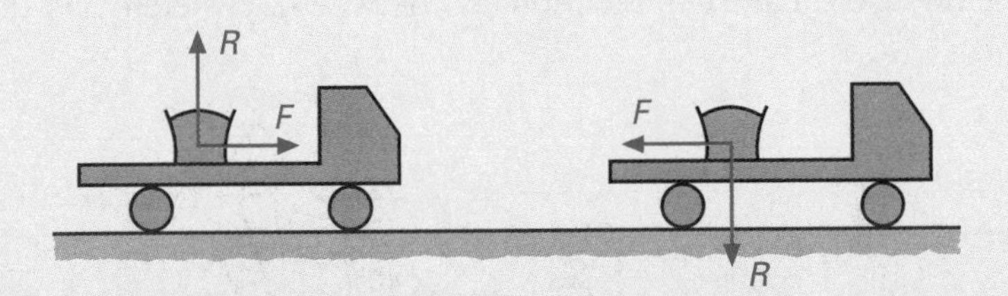

Exam questions answers: page 213

1 A man is pushing a supermarket trolley. The trolley experiences resistances to motion totalling 50 N. Draw diagrams to show the forces acting on the man and the trolley. (5 min)

2 A ladder is leaning against a smooth wall with its foot on rough ground. Two-thirds of the way up the ladder a paint tin is suspended from a rung by means of a piece of string. Draw diagrams to show the forces on the paint tin and the ladder. (4 min)

Links

For more details about friction see page 188.

Checkpoint 3

A brick of mass m kg is on a rough slope inclined at 20° to the horizontal. Draw a diagram showing the forces acting on the brick.

Checkpoint 4

A rod of mass 5 kg is hinged at one end. A rope is attached to the other end, and the free end of the rope is tied to a point directly above the hinge, so that the rod is horizontal. Draw a diagram showing the forces on the rod.

Checkpoint 5

A boy is standing in a lift which is moving upwards by means of a vertical cable. Draw diagrams to show the forces on the boy, and the forces acting on the lift.

Newton's Second Law

This is the key principle to apply when you need to relate the acceleration of a moving body to the forces acting on it.

Don't forget

$\boldsymbol{F} = m\boldsymbol{a}$

Checkpoint 1

Which of the three quantities force, mass and acceleration are vectors?

Checkpoint 2

Find the mass of a particle which is accelerating at 0.3 ms^{-2} under the action of a force of magnitude 18 N.

Examiner's secrets

Without a good clear force diagram you will be in trouble. It will be worth your while to take care over it.

Resultant force = mass × acceleration

This is a vector equation, so we use components and apply it in two **perpendicular** directions.

First consider the direction of the acceleration. Find the sum of the components of the forces in this direction and set it equal to mass × acceleration.

Second, if necessary, consider the direction perpendicular to the acceleration. Find the sum of the components of the forces in this direction and set it equal to zero. (Why? Because the acceleration has a zero component in this direction!)

Example: A particle of mass 5 g is falling vertically through oil. It experiences resistance of magnitude 0.03 N. Find its acceleration.

Answer:

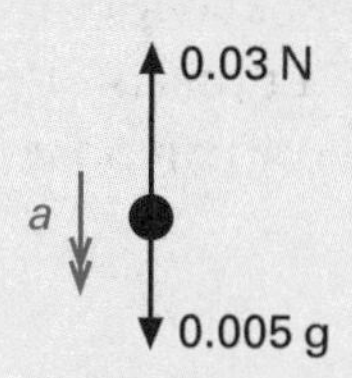

Apply Newton's Law downwards: $0.005g - 0.03 = 0.005a$

$a = 0.019 \div 0.005 = 3.8\ ms^{-2}$

Example: A man is pulling his son on a sledge over horizontal ground by means of a rope inclined at 40° to the horizontal. The total mass of the boy plus sledge is 50 kg. If the sledge has an acceleration 0.2 ms^{-2} and the resistances to motion total 60 N, find the tension (T) in the rope. Find also the normal reaction (N) between the sledge and the ground.

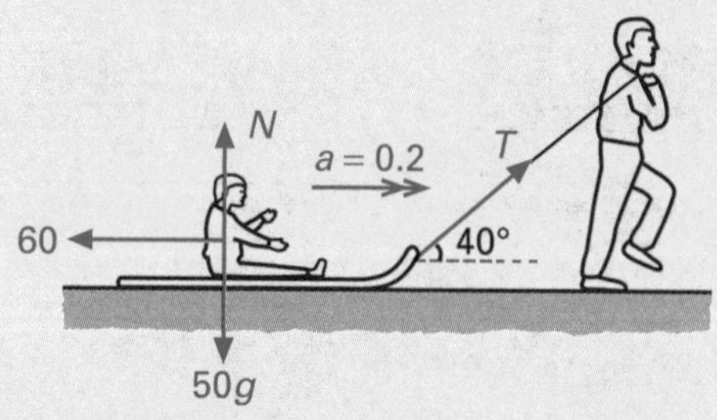

Answer:

Apply Newton's Law → (in the direction of the acceleration):

$T\cos 40° - 60 = 50 \times 0.2$

$T\cos 40° = 10 + 60 = 70$

$T = 70 \div \cos 40° = 91.4$ (3 s.f.)

Links

If you have forgotten how to resolve forces see page 171.

Apply Newton's Law ↑ (perpendicular to the direction of the acceleration):

$N + T \sin 40° = 50g$
$N = 50g - T \sin 40° = 431$ (3 s.f.)

So the tension in the string is 91.4 N and the magnitude of the normal reaction is 431 N.

Inclined planes

Lots of Newton's Law questions use inclined planes. Remember that the direction of the acceleration will be parallel to the plane. You may be asked to apply the constant acceleration equations as well.

Example: The boy now slides on his sledge down a hill inclined at 25° to the horizontal. Resistances to motion total 40 N, and the total mass of boy plus sledge is still 50 kg. Calculate:

(a) the acceleration of the boy;
(b) the speed of the boy after 5 seconds, given that he started from rest.

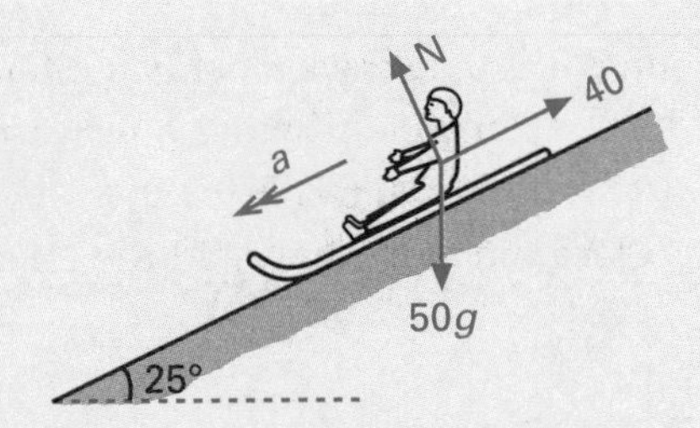

Answer:

(a) apply Newton's Law ↙ (parallel to the plane)
$50g \sin 25° - 40 = 50a$
$a = 167.18 \div 50 = 3.34$ (3 s.f.)
acceleration = 3.34 ms^{-2} (3 s.f.)

(b) Apply $v = u + at$
$v = 0 + 5a = 16.7$ (3 s.f.)
speed = 16.7 ms^{-1} (3 s.f.)

Exam questions answers: pages 213–14

1 A block of mass 20 kg is being pulled along a horizontal floor by means of a rope inclined at an angle 30° to the horizontal. Resistance to motion totals 60 N, and the tension in the rope is 100 N. Find the acceleration of the block, and the magnitude of the normal force exerted by the floor on the block. (6 min)

2 A block of mass 10 kg is projected up a smooth slope with speed 5 ms^{-1}. The angle the slope makes with the horizontal is θ where $\sin \theta = 0.2$. Find the distance travelled by the block before it comes to instantaneous rest. (8 min)

Examiner's secrets

It helps the examiner if you let her know what you are doing. Say clearly what direction you are considering. It helps you too!

Don't forget

If the angle of the plane is θ, the component of the weight parallel to the plane is *mg* sin θ, and perpendicular to the plane it is *mg* cos θ.

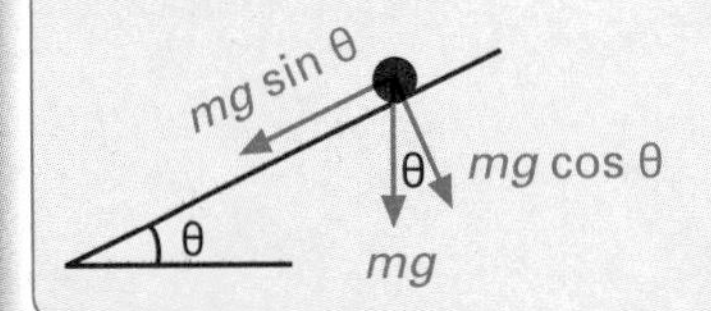

Checkpoint 3

What kinds of resistance might the sledge and rider experience? What factors might affect the magnitude of the resistance?

Test yourself

Cover the diagram, and check you can draw in all the forces yourself.

Examiner's secrets

If you are given a trig value for the angle, instead of the angle directly, it is impressive if you can use it without finding the value for the angle.

The jargon

'Is projected' just means that the question is telling you what the initial speed is. You don't need to worry about *how* it was projected.

Connected particles

This is a common application of Newton's Second Law. Two bodies are connected (by a string or coupling or some other means) and so have the same speed, and same magnitude of acceleration.

Car and caravan 

Checkpoint 1

Why would this not be true if the connecting link is elastic? If the connecting link is a rope, why will there be a problem when the car brakes?

There are lots of variations on this theme: lorry and trailer, engine and carriages, tugboat and tanker, etc. The main point here is that the car and caravan are moving in the same **direction** with the same **speed** and **acceleration**.

You have three options:

- Apply Newton's Law to the car and caravan as a whole. This does not involve the tension in the connector.
- Apply Newton's Law to the car only.
- Apply Newton's Law to the caravan only.

You will usually choose the first option to find the acceleration, then the second or third to find the tension.

Example: A car of mass 1 000 kg tows a caravan of mass 800 kg along a horizontal road. The car's engine provides a forward force of 2.3 kN, and the resistances to motion total 600 N for the car and 800 N for the caravan. Find the acceleration of the car, and the tension in the tow-bar.

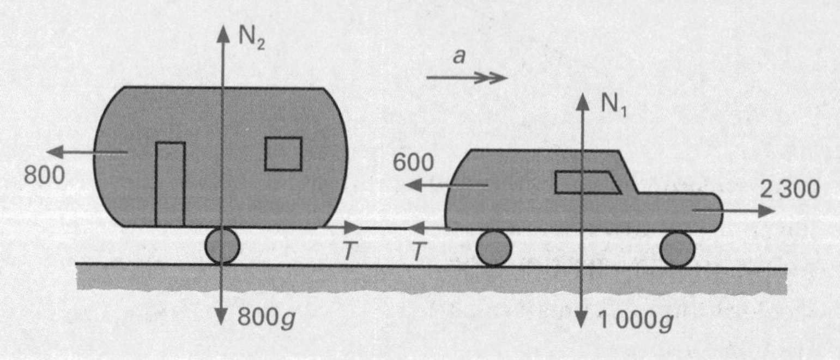

Test yourself

Cover up the diagram, and check that you can draw in the forces for yourself. Try to answer the question yourself as well.

Checkpoint 2

If the driver in the example puts the car into neutral (so that there is no forward force) calculate the deceleration of the car and the magnitude and nature of the force in the tow-bar.

Answer: Apply Newton's Law to the car and caravan as a single unit:

$$2\,300 - 1\,400 = 1\,800a$$
$$a = 900 \div 1\,800 = 0.5$$

Note: We use the total resistance 1 400 N, and the total mass 1 800 kg in this equation.

Now apply Newton's Law to the caravan only:

$$T - 800 = 800a$$
$$a = 0.5, \text{ so } T = 400 + 800 = 1\,200$$

Note: Here we use the resistance and mass of the caravan only.

The acceleration is 0.5 ms^{-2}, and the tension in the tow-bar is 1 200 N.

Pulley systems 

The main difference between this type and the car and caravan problems, is that although the magnitude of the acceleration for each particle is the same, the direction is different. So you must apply

Newton's Law to each particle separately. You cannot consider it as a single unit. You may be asked about what happens after one particle reaches the ground. Note that the string will go slack, and the other particle will continue to move but with a new acceleration.

Example: Two particles of masses 3 kg and 5 kg are joined by a light inelastic string which passes over a smooth pulley as shown in the diagram. The system is released from rest with both particles 50 cm above the ground. Find:

(a) the acceleration of the particles;
(b) the speed with which the 5 kg particle hits the ground;
(c) the greatest height attained by the 3 kg particle, assuming that it does not reach the pulley.

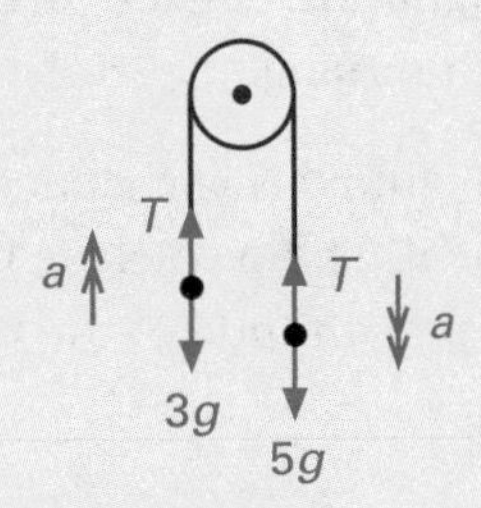

Answer:

(a) Apply Newton's Law to each particle:

3 kg ↑: $T - 3g = 3a$

5 kg ↓: $5g - T = 5a$

Adding gives: $T - 3g + 5g - T = 3a + 5a$

So: $2g = 8a$ and $a = 0.25g = 2.45 \text{ ms}^{-2}$

(b) Apply $v^2 = u^2 + 2as$ with $u = 0$ and $s = 0.5$

$v^2 = 0 + 2 \times 0.5 \times 2.45 = 2.45$, so $v = 1.57 \text{ ms}^{-1}$ (3 s.f.)

(c) The 3 kg particle is now 1 m above the ground. How much further does it go? Since the string has gone slack, its acceleration is now g downwards. Apply $v^2 = u^2 + 2as$ with $v = 0$ and $u = \sqrt{2.45}$: $0 = 2.45 + 2 \times (-g) \times s$ which gives $s = 0.125$.

So the greatest height is 1.13 m (3 s.f.).

Examiner's secrets

If the examiner asks you about your modelling assumptions, remember that the pulley must be smooth, so that the tension in the string on either side is the same. We will also assume that the string is light.

Checkpoint 3

What modelling assumptions do you make when you say that the tension in the string is the same at each end?

Watch out!

In the final part, the acceleration is different. Examiners like this twist in the tail of a question.

Exam questions answers: page 214

1 A lorry of mass 800 kg tows a trailer of mass 600 kg up a hill which is inclined at an angle 15° to the horizontal. The forward force provided by the lorry's engine is 5 kN and the resistance to motion is 900 N for the lorry and 300 N for the trailer. Calculate the acceleration up the hill, and the tension in the tow-bar. (8 min)

2 A particle of mass 6 kg rests on a rough horizontal table. It is attached to another particle of mass 4 kg by means of a light inextensible string, which passes over a smooth pulley. The 4 kg particle hangs freely, 1 m above the ground.

Given that the frictional force acting on the 6 kg particle is 9.2 N, calculate (a) the acceleration of the particles, (b) the speed with which the 4 kg particle hits the ground and (c) the total distance travelled by the 6 kg particle before coming to rest. (15 min)

Equilibrium of a particle

In this section we will look at the equilibrium of a particle. This means that all the forces acting on the object act through one point, so you don't have to worry about turning and moments. The basic principle is that the forces must have a zero resultant, because in equilibrium there is zero acceleration.

Links

If your problem is about the equilibrium of an object that is clearly *not* modelled as a particle (perhaps because its dimensions are given) refer to page 192.

Triangle of forces

If you have three forces acting on an object in equilibrium, they must form a complete triangle. So one approach is to draw the forces 'nose-to-tail' making a triangle. Then you can use trigonometry to solve the problem. Remember that the length of the appropriate side of the triangle represents the magnitude of each force. You may have to think about the angles of the triangle.

Checkpoint 1

Why do the forces form a triangle?

Example: A particle of weight 60 N is suspended from a light string which is at 30° to the vertical. The particle is held in equilibrium by a horizontal force. Find the magnitude of the tension in the string.

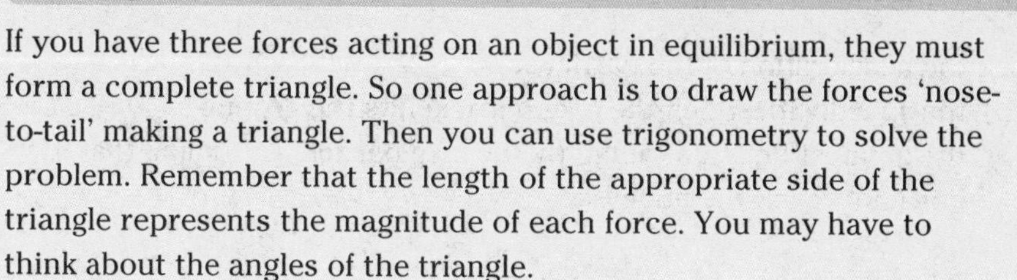

Test yourself

Cover up the diagram and working, and check that you can reproduce them.

The two diagrams show the forces on the particle, and the corresponding triangle of forces. Using the triangle you can see that $60 = T \cos 30°$ giving $T = 60 \div \cos 30° = 69.3$ N (3 s.f.).

Lami's Theorem

This is an adaptation of the sine rule for three-force problems. Each force divided by the sine of the 'opposite' angle gives an equal ratio.

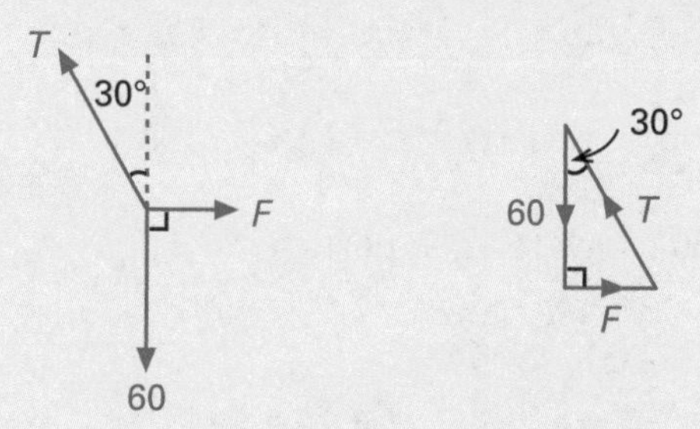

$$\frac{P}{\sin \alpha} = \frac{Q}{\sin \beta} = \frac{R}{\sin \gamma}$$

Examiner's secrets

Lami's Theorem is especially good if you know all the angles between the forces. *It only works for three-force problems* – so don't use it if there are four forces!

Checkpoint 2

A force of magnitude 400 N acts due North, and forces of magnitudes P N and Q N acts on bearings 120° and 220° respectively. Use Lami to find P.

Least force problems

These are best solved using a triangle (or polygon) of forces. First draw the forces you are given. Then you'll find that you are given the direction of a force, but not its magnitude. Sketch in this direction, and consider how you can complete the triangle (or polygon) with the shortest possible line (the *least* force). You'll find it is always perpendicular to the direction of the force with unknown magnitude.

Checkpoint 3

When would you have a *polygon*, rather than a *triangle*?

Example: A particle of weight 60 N is suspended from a light string which is at 30° to the vertical. The particle is held in equilibrium by a third force F. Find the least possible magnitude of F.

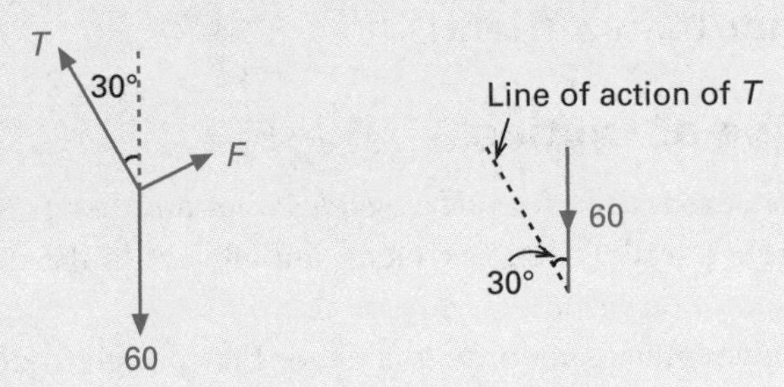

The diagram on the right above shows the weight drawn in, and the direction of the tension. Think how you can complete the triangle with the shortest possible line. Then use trig to solve the problem.

Checkpoint 4

Complete the solution of this example.

Resolving 

This often provides a quick and easy method, especially if you have more than three forces. Chose two perpendicular directions. The sum of the components in each direction must be zero.

Examiner's secrets

If there is a force you don't know, and you don't need to find it, try resolving in the perpendicular direction.

Example: Given that the forces in the diagram are in equilibrium, find the values of X and θ.

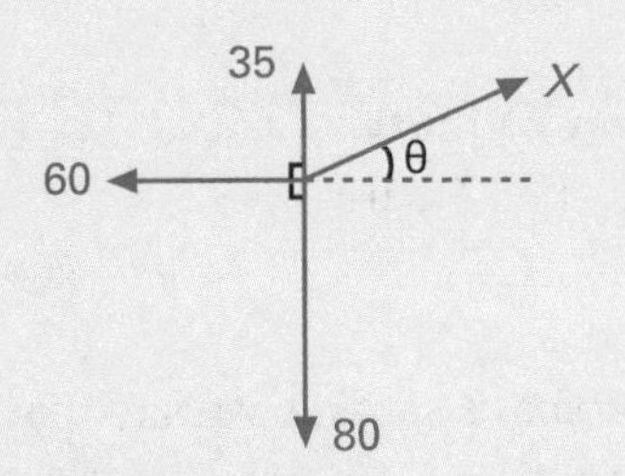

Resolving $\rightarrow$: $X\cos\theta = 60$

Resolving $\uparrow$: $X\sin\theta + 35 = 80$, so $X\sin\theta = 45$

Dividing $\dfrac{X\sin\theta}{X\cos\theta} = \dfrac{45}{60}$ gives $\tan\theta = 0.75$, so $\theta = 36.9°$

So $X = 60 \div \cos 36.9° = 75$ N.

Examiner's secrets

You often find you get one equation for $X\sin\theta$ and another for $X\cos\theta$. Divide the two equations: the X goes out, and you are left with $\sin\theta \div \cos\theta$ which gives $\tan\theta$ (see page 48).

Exam questions answers: page 214

1 (a) The forces shown in the diagram are in equilibrium. Find the values of P and Q.

(b) If the force of magnitude 4 N is removed, state the magnitude and direction of the resultant of the three remaining forces. (10 min)

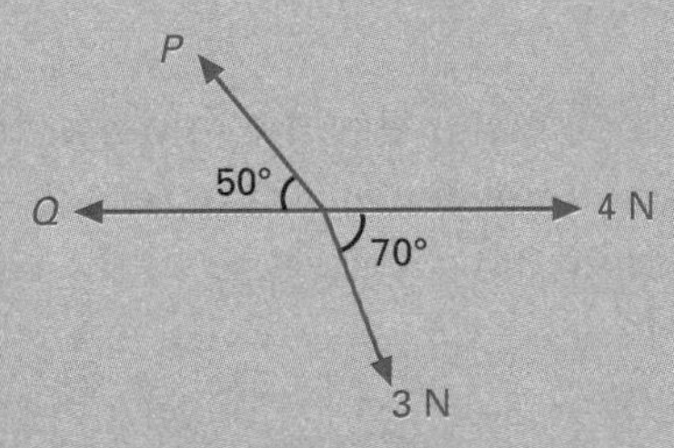

2 A particle of mass 5 kg is suspended from two strings of lengths 50 cm and 120 cm. The other ends of the strings are attached to two points A and B which are 130 cm apart on the same horizontal level. Find the tensions in the two strings. (10 min)

Examiner's secrets

'State' in question 1(b) is a clue that you can answer the question with a little thought, without any working.

Friction

Life would be impossible without friction – we couldn't even walk if friction didn't stop our feet from slipping. However in some contexts friction is a foe rather than a friend.

Basic laws of friction

- Friction occurs when one surface *slides* over another.
- Friction acts parallel to the surfaces, and always in the direction as to oppose motion or potential motion.
- Friction never causes motion. It is never big enough to do so. Up to a certain point, the magnitude of the frictional force is exactly enough as to prevent motion.
- Friction has a maximum magnitude which depends on two factors:
  - (a) the nature of the two surfaces (how rough they are);
  - (b) the magnitude of the normal reaction force.

Checkpoint 1

Imagine that you are trying to push a packing case across the floor. It is obviously easier if the case is light – but is that the only factor? Does it make it easier or harder if someone else is pushing down on the case at the same time?

Friction formula

The friction formula is unusual in that it is an inequality, rather than an equation: $F \leq \mu N$.

μ is called the coefficient of friction, and is the part which is dependent on the nature of the surfaces; N is the normal reaction force.

Checkpoint 2

Does μ have units? If so, what are they?

Friction with moving objects

If the object is moving, then $F = \mu N$.

Limiting equilibrium

If the object is *on the point of moving*, we say that it is in limiting equilibrium. Again, in this case, $F = \mu N$.

Examiner's secrets

A key step in answering a question is to read it carefully: is the object moving, or not? If not, is it in limiting equilibrium, or not?

Problems with motion

The first step is (you've guessed it!) to draw a good force diagram. Remember to include the normal reaction perpendicular to the surface, and the friction parallel to the surface in the opposite direction to motion. Then

- (a) apply Newton's Second Law in the direction of motion;
- (b) use the fact that there is no resultant force in the perpendicular direction (useful for finding N);
- (c) use $F = \mu N$.

Test yourself

Before looking at the top of the next page, have a go at drawing the force diagram and writing a few equations.

Example: A packing case of mass 50 kg is being dragged across the floor by means of a rope inclined at 40° to the horizontal. If the tension in the rope is 400 N, and the coefficient of friction between the floor and case is 0.3, calculate the acceleration of the case.

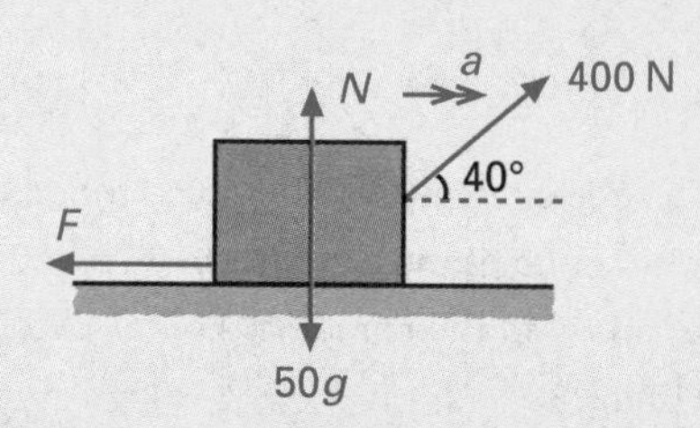

Answer:

Parallel to the floor: $400 \cos 40° - F = 50a$ ①
Perpendicular to the floor: $N + 400 \sin 40° = 50g$ ②
Friction: $F = \mu N = 0.3\ N$ ③

Equation ② gives us $N = 232.9$.
Equation ③ now gives us $F = 69.9$.
Equation ① gives $50a = 236.5$, so $a = 4.73\ ms^{-2}$.

Equilibrium problems

The key point here is to read the question to see if the equilibrium is limiting or not. As with all equilibrium questions, you will need to resolve the forces in two directions.

Example: A block of mass 12 kg is placed on a rough plane, $\mu = 0.5$, inclined at 30° to the horizontal. The block is kept in equilibrium by a horizontal force of magnitude P newtons. Find the range of possible values for P.

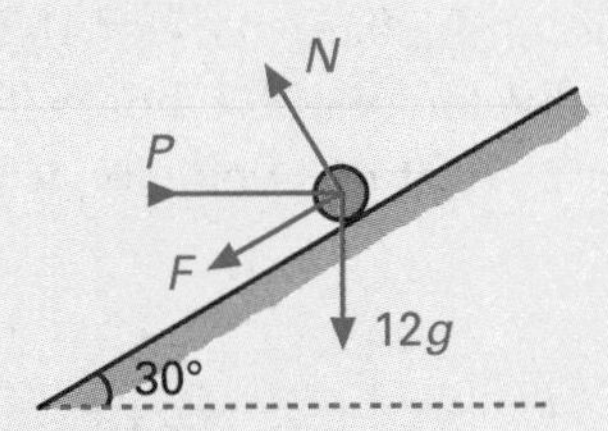

Answer: You can't be sure which way the block is tending the slip – you need to draw two diagrams, one for when it is about to slip up (drawn above), and the other when it is about to slip down.

Parallel to the plane: $P \cos 30° = F + 12g \sin 30°$ ①
Perpendicular to the plane: $N = 12g \cos 30° + P \sin 30°$ ②
Friction: $F \leq 0.5\ N$ ③

Equation ① gives $F = 0.866P - 58.8$.
Equation ② gives $N = 101.84 + 0.5P$.
Putting these results in Equation ③ gives:

$$0.866P - 58.8 \leq 0.5\ (101.84 + 0.5P).$$

And $0.616P \leq 109.72$, so $P \leq 178$.

Checkpoint 3

Suppose instead of pulling with a rope, you were pushing the case with a force of 400 N at 40° *downwards*. Which of the equations ①, ② and ③ would be different?

Checkpoint 4

What is the significance of 'limiting equilibrium'?

Examiner's secrets

Spot the clue in the question. It says 'range of possible values' which indicates that you need to work with the inequality.

Checkpoint 5

Draw the diagram and write the equations for the case when it is tending to slip down (so the friction acts up the plane). Solve them to find an inequality for P.

Examiner's secrets

Resolving parallel and perpendicular to the plane gives two separate equations for F and N which are easy to deal with. (If you feel brave, try resolving horizontally and vertically, and compare!)

Exam questions

answers: page 214

1 A parcel is on the floor of a van which is accelerating at $0.3\ ms^{-2}$. Given that $\mu = 0.25$, determine whether or not the parcel slips. (6 min)

2 A particle is projected with speed $4\ ms^{-1}$ from a point O up a rough plane inclined at 30° to the horizontal. The coefficient of friction is 0.4. Find (a) the deceleration of the particle, (b) how far up the plane from O the particle comes to instantaneous rest and (c) the speed with which the particle passes back through O. (15 min)

The jargon

'is projected' means that the particle is given an initial velocity. You do not need to consider the force that does the projecting.

Examiner's secrets

Don't be put off because you are not given the mass. Use m, and you will find that it cancels out.

Using vectors

Questions with vectors can seem quite complicated, but actually they are no more difficult than any other topic. As you read through, write down the information you are given so that you can sort it out as you go along, and see which equation you need to use. A diagram helps as well.

Using vectors with constant acceleration

a, **s**, **u** and **v** are all vector quantities so you can use the constant acceleration equation with vectors. Be wary about $v^2 = u^2 + 2as$: to make this a vector equation you need the scalar product – if this isn't in your syllabus, don't try to use it! The others become:

$$\mathbf{v} = \mathbf{u} + \mathbf{a}t \quad \mathbf{s} = \mathbf{u}t + \tfrac{1}{2}\mathbf{a}t^2 \quad \mathbf{s} = \mathbf{v}t - \tfrac{1}{2}\mathbf{a}t^2 \quad \mathbf{s} = \tfrac{1}{2}(\mathbf{u} + \mathbf{v})t$$

Example: A model assumes that a boat has initial velocity $20\mathbf{i}$ ms^{-1}, and is at the point with position vector $(-100\mathbf{i} + 60\mathbf{j})$ m. The boat is to come alongside a harbour at the point with position vector $100\mathbf{i}$, and so is given an acceleration $(-\mathbf{i} - 0.6\mathbf{j})$ ms^{-2} for 10 seconds.

(a) Find the position and velocity of the boat after 10 seconds.

(b) If the boat now continues at this velocity, will it reach the harbour?

Answer:

(a) $\boldsymbol{u} = \begin{pmatrix} 20 \\ 0 \end{pmatrix}$, $\boldsymbol{a} = \begin{pmatrix} -1 \\ -0.6 \end{pmatrix}$, $t = 10$

Using $\mathbf{v} = \mathbf{u} + \mathbf{a}t$ we get $\mathbf{v} = \begin{pmatrix} 20 \\ 0 \end{pmatrix} + 10\begin{pmatrix} -1 \\ -0.6 \end{pmatrix} = \begin{pmatrix} 10 \\ -6 \end{pmatrix}$

Using $\mathbf{s} = \mathbf{u}t + \frac{1}{2}\mathbf{a}t^2$ we get $\mathbf{s} = 10\begin{pmatrix} 20 \\ 0 \end{pmatrix} + \frac{1}{2}\begin{pmatrix} -1 \\ -0.6 \end{pmatrix} \times 10^2 = \begin{pmatrix} 150 \\ -30 \end{pmatrix}$

This is the displacement from the initial position (–100,60), so its position vector now is (50,30).

(b) The boat needs to travel a further $\begin{pmatrix} 50 \\ -30 \end{pmatrix}$ to reach the harbour. Since this vector is parallel to the velocity vector $10\mathbf{i} - 6\mathbf{j}$, the boat will reach the harbour 5 seconds later if it continues at this velocity.

Using vectors with Newton's Second Law

Here we are using $\mathbf{F} = m\mathbf{a}$ as a vector equation. There's no need to resolve, as it's all done for you – so actually it couldn't be easier!

Example: A glider of mass 400 kg is pulled by two cables. The tensions in the cables are $110\mathbf{i} + 30\mathbf{j} + 2\,100\mathbf{k}$ and $130\mathbf{i} - 20\mathbf{j} + 2\,100\mathbf{k}$, where **i** and **j** are perpendicular horizontal unit vectors, and **k** is a vertical unit vector. Find the resultant force acting on the glider, and its acceleration.

Answer: Don't forget the weight, $400g$, of the glider, acting in the negative **k** direction. Adding the two tensions to the weight we get $\mathbf{F} = (110 + 130)\mathbf{i} + (30 - 20)\mathbf{j} + (2\,100 + 2\,100 - 400g)\mathbf{k}$, then using $\mathbf{F} = 400\mathbf{a}$ we get $\mathbf{a} = 0.6\mathbf{i} + 0.025\mathbf{j} + 0.7\mathbf{k}$.

Test yourself

Sketch the information given in the question on a diagram. Cover over the work and check you can answer the question by yourself.

The jargon

Don't like **i** and **j**? Just convert to column vector form!

Checkpoint 1

How can you tell whether or not two vectors are parallel?

Watch out!

The only point you need to watch is to make sure you have noticed whether or not you have got constant acceleration.

Checkpoint 2

Have your got constant acceleration in this question, or not? Given that the glider starts from rest at the origin, find its velocity and position 20 seconds later.

Example: A force $3t^2\mathbf{i} + 2\mathbf{j}$ acts on a particle of mass 0.5 kg. Find the acceleration of the particle. Given that it starts at the origin with velocity $3\mathbf{j}$, find its position and velocity 2 seconds later.

Answer: $\mathbf{F} = 0.5\mathbf{a}$, so $\mathbf{a} = 2\mathbf{F} = 6t^2\mathbf{i} + 4\mathbf{j}$.

Note that this is *not* constant acceleration – integrate to get the velocity: $\mathbf{v} = 2t^3\mathbf{i} + 4t\mathbf{j} + 3\mathbf{j}$ and when $t = 2$, $\mathbf{v} = 16\mathbf{i} + 11\mathbf{j}$. Integrate again to get the displacement:

$\mathbf{s} = 0.5t^4\mathbf{i} + 2t^2\mathbf{j} + 3t\mathbf{j}$ and when $t = 2$, $\mathbf{s} = 4\mathbf{i} + 14\mathbf{j}$.

Equilibrium and vectors

You know that the resultant of all the forces will be zero for equilibrium. If the forces are given in component form, it's easy just to add up the components and set the result equal to zero.

Example: Three forces $(a\mathbf{i} + 3\mathbf{j} - a\mathbf{k})$ N, $(-4\mathbf{i} + b\mathbf{j} + 2b\mathbf{k})$ N and $(6\mathbf{i} + 2a\mathbf{j} + c\mathbf{k})$ N act on a particle. Given that the particle is travelling at constant speed $(6\mathbf{i} - \mathbf{j} - \mathbf{k})$ ms^{-1}, find the values of a, b and c.

Answer: The resultant force is $(a - 4 + 6)\mathbf{i} + (3 + b + 2a)\mathbf{j} + (-a + 2b + c)\mathbf{k}$.

The $\mathbf{i}$ component gives $a = -2$.
The $\mathbf{j}$ component gives $3 + b - 4 = 0$, so $b = 1$.
The $\mathbf{k}$ component gives $2 + 2 + c = 0$, so $c = -4$.

Links

If you have forgotten about variable acceleration see page 174.

Checkpoint 3

Why has the expression for $\mathbf{v}$ got $3\mathbf{j}$ added on?

Checkpoint 4

How do you know that this is an equilibrium question?

Examiner's secrets

It may be tedious, but *you do need to use proper vector notation.* The examiner is not impressed if you fail to write the twiddle underneath the vector quantities.

Exam questions answers: page 214

1 Three forces $\mathbf{F}_1 = (4\mathbf{i} - 5\mathbf{j})$ N, $\mathbf{F}_2 = (-3\mathbf{i} + 8\mathbf{j})$ N and $\mathbf{F}_3 = (a\mathbf{i} + b\mathbf{j})$ N.
(a) Given that the forces are in equilibrium, find the values of a and b.
(b) The resultant of $\mathbf{F}_1$ and $\mathbf{F}_2$ is $\mathbf{R}$. Find the magnitude of $\mathbf{R}$ and the angle that $\mathbf{R}$ makes with the $\mathbf{i}$ direction. (8 min)

2 At a certain instant, a particle of mass 0.1 kg passes through the origin with velocity $(4\mathbf{i} - \mathbf{j})$ ms^{-1}. It is acted on by a force $(-2\mathbf{i} + 3\mathbf{j})$ N.
(a) Find the velocity and position of the particle 5 seconds later.
(b) Find the time at which the particle crosses the y-axis in the subsequent motion. (6 min)

3 According to a certain model, the position at time t of a particle is given by $\mathbf{s} = (2t^3 + 3)\mathbf{i} + (6 - 5t)\mathbf{j}$, where $\mathbf{i}$ and $\mathbf{j}$ are perpendicular unit vectors.
(a) Write down the initial position vector of the particle.
(b) Obtain expressions for the velocity and acceleration at time t.
(c) Given that the particle has mass 0.2 kg, find the resultant force acting on the particle.
(d) From time $t = 5$, an additional force $-1.2t\mathbf{i}$ N acts on the particle. Describe briefly the particle's subsequent motion. (15 min)

Moments and equilibrium

The moment of a force is its turning effect. In this section you will revise finding moments, and solve problems on the equilibrium of rigid bodies. This has nothing to do with short time periods, magic or otherwise, or momentum.

Links

For the equilibrium of a particle see page 186.

Definition of the moment of a force

The moment of a force about an axis is the product of the force and the perpendicular distance of the force from the axis.

Moment of force F about axis through $A = F \times d$

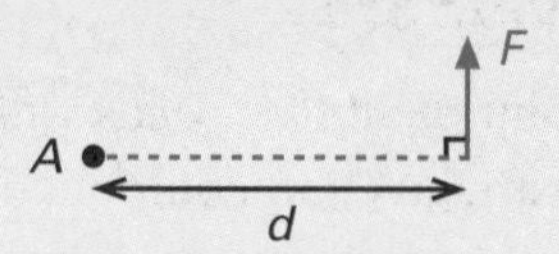

Checkpoint 1

A force 5**i** acts at the point (2,3). Find the moment of this force about the point (3,1).

Strictly speaking, we always take moments about an axis, but often we just refer to 'the moment about A' if it is clear. The units are Nm, and you need to specify the sense (clockwise or anticlockwise).

Forces at an angle

If the force is at an angle, you could find the perpendicular distance, namely $d \sin \theta$. So the moment is $Fd \sin \theta$. Alternatively, resolve the force into components as in the right-hand diagram. One component has no moment, and the other has moment $F \sin \theta d$.

Checkpoint 2

A square ABCD has sides 2 m. Forces act as follows: magnitude 4 N in direction AB, 5 N in direction CB and 2 N in direction CA. Find the resultant moment of the forces about D.

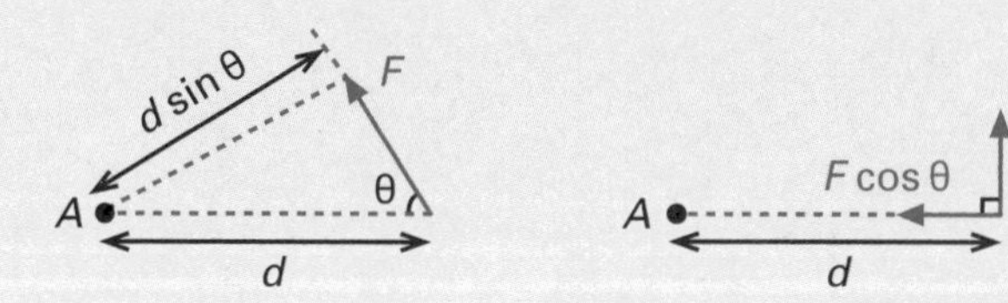

Sum of moments

If you have several forces, you can find the resultant moment by summing the moment of each force.

Checkpoint 3

A light rod of length 2 metres has particles of mass 3 kg and 5 kg attached to its two ends. Find where a support should be placed so that the rod balances horizontally. Find also the magnitude of the reaction of the support on the rod.

Equilibrium problems

If a body is in equilibrium, not only do the forces have a zero resultant, but also the resultant moment about any point is also zero. Note that you can take moments about any convenient point – there does not have to be an actual pivot there.

Your plan of campaign is as follows.

Action point

The force diagram is absolutely crucial to success. If in doubt go back to pages 180 and 181 to revise.

1. Read the question carefully, and draw a large clear diagram showing forces and dimensions. Think carefully about contact forces: the normal reaction plus friction if the surface isn't smooth. If the body is hinged, draw in two perpendicular components for the force at the hinge.
2. Choose one or two suitable points about which to take moments. In particular, if there's an unknown force you are not required to find, take moments about the point where it acts.
3. Resolve in one or two directions (think carefully about which).
4. If you have limiting equilibrium use $F = \mu N$.

Links

Forgotten about limiting equilibrium? See page 188.

5 Finally solve your equations to answer the question. The trick is to choose carefully at steps 2 and 3. Try to isolate the unknown forces, so you end up with easy equations to sort out.

Example: A rod AB of length $3a$ and mass 2 kg is hinged at the point A. It is kept in equilibrium at angle θ to the vertical by means of a string of length $4a$ attached to B. The other end of the string is attached to C, $5a$ vertically above A. Find the tension in the string.

Answer: Take moments about A:

$$T \times 3a = 2g \sin \theta \times \frac{3a}{2},$$

giving $T = 7.84$ newtons. (Make sure you know why taking moments about A was so effective.)

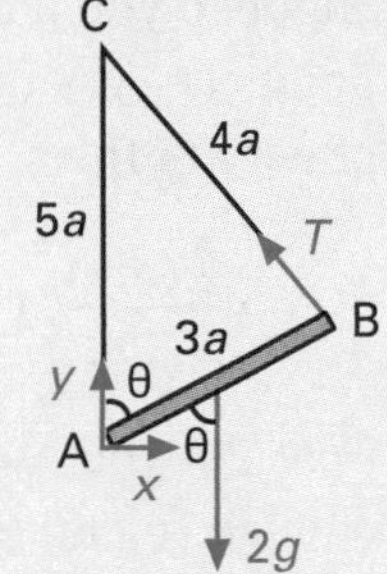

Example: A ladder of length $4d$ has mass 12 kg. One end stands on rough horizontal ground (coefficient of friction 0.25) and the other end leans against a smooth vertical wall. A boy of mass 30 kg stands on a rung one quarter of the way up the ladder. If the ladder is about to slip, find the angle the ladder makes with the vertical.

Answer:

Resolve vertically: $N = 12g + 30g = 42g$
Friction: $F = 0.25N = 0.25 \times 42g = 10.5g$
Resolve horizontally: $R = F = 10.5g$

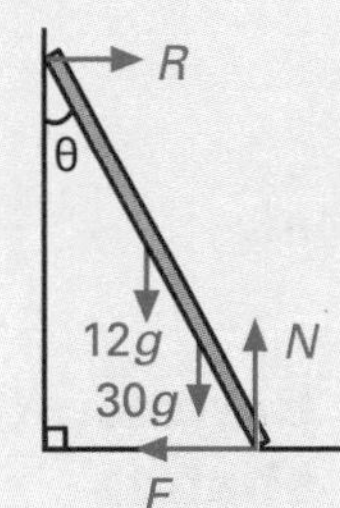

Take moments about the foot of the ladder:

$$4d \times R \cos \theta = d \times 30\, g \sin \theta + 2d \times 12g \sin \theta$$
$$= 54dg \sin \theta$$

Substitute in for R: $4d \times 10.5g \cos \theta = 54dg \sin \theta$.

This gives $\tan \theta = \frac{42}{54} = 0.777\ 78$ and $\theta = 37.9°$.

Action point

Look at the dimensions of the triangle. Deduce the size of angle B, and write down the values of $\cos \theta$ and $\sin \theta$.

Checkpoint 4

Find also the forces X and Y at the hinge by resolving.

Watch out!

A common mistake is to label both normal reactions N. They are not the same! Notice also that they are perpendicular to the wall/ground, not the ladder.

Examiner's secrets

Always say what you are doing – it helps both you and the examiner follow your method!

Test yourself

Cover up the working, and run through these examples again for yourself.

Exam questions answers: pages 214–15

1 A parcel is on the floor of a van which is accelerating at 0.3 ms^{-2}. Given that $\mu = 0.25$, determine whether or not the parcel slips. (6 min)

2 A particle is projected with speed 4 ms^{-1} from a point O up a rough plane inclined at 30° to the horizontal. The coefficient of friction is 0.4. Find
(a) the deceleration of the particle,
(b) how far up the plane from O the particle comes to instantaneous rest, and
(c) the speed with which the particle passes back through O. (15 min)

Centre of gravity 1

The centre of gravity is the point through which the weight of the body acts. It is essentially the same as the centre of mass (it would be different if the scale of the body was such that the gravitational field was not the same at every point in the body).

Checkpoint 1

Find the centre of gravity of a system of three particles of mass 2 kg at (1,3), 3 kg at (2,–3) and 1 kg at (5,0).

Centre of gravity of a system of particles

Suppose you have n particles of masses $m_1, m_2, \ldots, m_n$, situated at the points $(x_1,y_1), (x_2,y_2), \ldots, (x_n,y_n)$ respectively. Then the centre of gravity of the system is at $(\bar{x},\bar{y})$ where

$$\bar{x} = \frac{\Sigma(mx)}{\Sigma m}, \ \bar{y} = \frac{\Sigma(my)}{\Sigma m}$$

Note: If you are not given coordinates for each particle, it is easy enough to choose axes to obtain them.

One dimensional example: A collapsible rod consists of three sections, each of length 40 cm, and of masses 50 g, 30 g and 20 g. Find the distance of the centre of gravity from the heavier end.

40 cm	40 cm	40 cm
50 g	30 g	20 g

Answer: $\bar{x} = \dfrac{50 \times 20 + 30 \times 60 + 20 \times 100}{50 + 30 + 20} = \dfrac{4\,800}{100} = 48$

Centre of gravity of uniform symmetrical shapes

The jargon

A *lamina* is a shape with area but negligible thickness (like a sheet of card). *Uniform* means that it has constant density.

We can use ideas of symmetry to locate the centre of gravity of many figures. The centre of gravity of a square, rectangular or circular lamina is in the middle, exactly as you'd expect. Similarly for a cuboid, cylinder or sphere.

Centre of gravity of a triangle

This is one you have to learn. The centre of gravity lies where the medians of the triangle intersect. It is two-thirds of the way from a vertex to the midpoint of the opposite side. (It is often helpful to think of the vector from a vertex to the midpoint opposite.)

Test yourself

Check that you get the same result for the centre of gravity if you start at (4,0) and find the vector to the midpoint opposite.

Example: A uniform triangular lamina has vertices at (0,0), (4,0) and (1,3). Locate its centre of gravity.

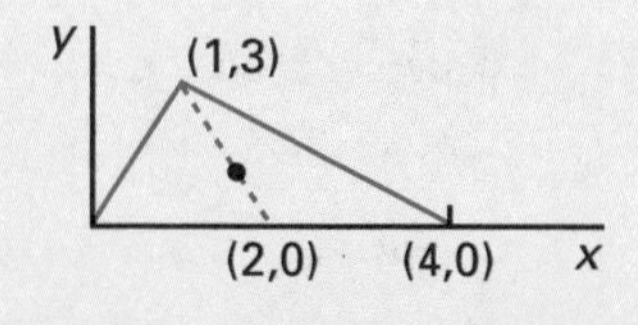

Answer: The vector from (1,3) to the midpoint opposite (2,0) is $\mathbf{i} - 3\mathbf{j}$. Two-thirds of this vector is $\frac{2}{3}(\mathbf{i} - 3\mathbf{j})$. So move $\frac{2}{3}(\mathbf{i} - 3\mathbf{j})$ from (1,3) to get $(\frac{5}{3},1)$ for the centre of gravity.

Centre of gravity of composite shapes

We are thinking here of figures made up from one or more rectangles, triangles, etc. The trick is to replace each basic shape with a particle of the same mass located at the centre of gravity of that shape. Then regard it as a system of particles, and use the formula above! The only variation is that you might be given a density and have to work out the mass from the area.

Example: The figure opposite has density m gcm^{-2}. Find its centre of gravity (dimensions in cm).

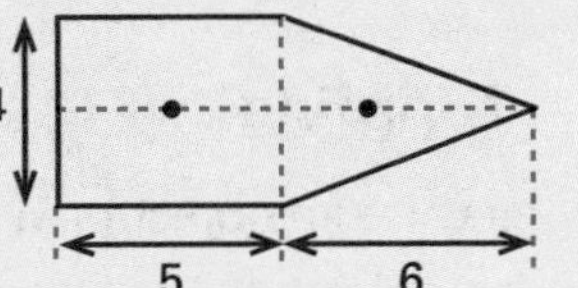

Answer: Take as origin the bottom left corner. The rectangle has mass $20m$ and its centre of gravity is at (2.5,2). The triangle has mass $12m$, and its centre of gravity is at (7,2).

So $\bar{x} = (20m \times 2.5 + 12m \times 7) \div (20m + 12m)$
$= 134m \div 32m = 4.187\,5.$

Example with 'holes': Locate the centre of mass of the figure opposite which has uniform density mg cm^{-3}.

We regard the figure as a complete circle together with a hole of negative mass. The 'complete' circle has mass $16\pi m$ and centre of gravity at (4,0). The 'hole' has mass $-4\pi m$ and centre of gravity (6,0).

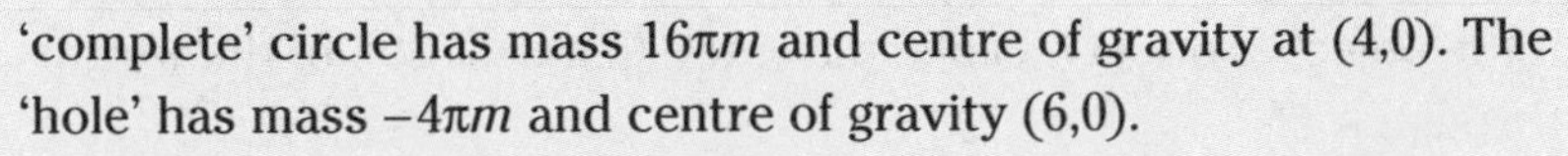

$$\bar{x} = (16\pi m \times 4 - 4\pi m \times 6) \div (16\pi m - 4\pi m) = 40\pi m \div 12\pi m = \frac{10}{3}.$$

Checkpoint 2

You can write down the y-coordinate of the centre of gravity without any working. What is it? Why?

Examiner's secrets

You can safely leave out the π in the expression for the mass. It cancels out, and leaving it in saves calculator time, and prevents rounding errors.

Using the formula sheet for other shapes

It may be that your formula sheet contains information about the location of the centre of mass of other shapes. You should check this out. Be sure you distinguish between solid figures and shells.

Checkpoint 3

A bollard is made from a solid cylinder of mass 50 kg, height 80 cm and radius 24 cm, surmounted by a solid hemisphere with mass 20 kg and radius 24 cm. Find the height of the centre of mass of the bollard from the base of the cylinder. (The centre of mass of a solid hemisphere is 3r/8 from the centre of the base circle.)

Exam questions

answers: page 215

1 A square lamina of side d has mass 4 m. Particles of masses m, $3m$ and $2m$ are attached to three of the vertices in order. Locate the centre of gravity of the lamina plus particles. (8 min)

2 The L shape shown in the diagram is made from a uniform lamina. AB = 4 cm, BC = 8 cm, AF = 10 cm and CD = x cm.

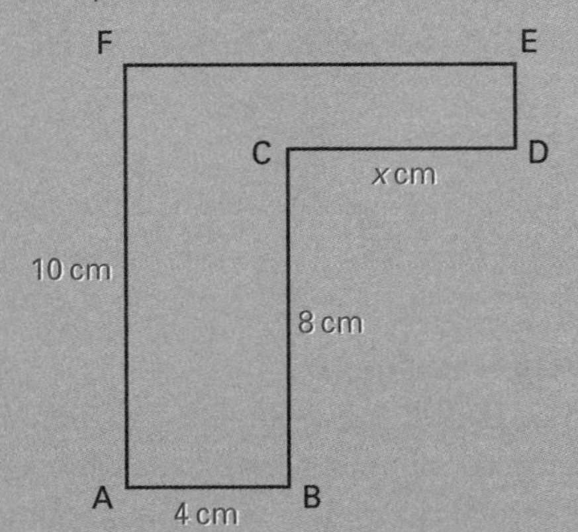

(a) Find, in terms of x, the distance of its centre of mass from AF.

(b) If the lamina is stood up on the edge AB, write down the maximum value of x if the lamina is not to topple over.

Examiner's secrets

A good diagram is essential here. Don't forget that the mass is proportional to the area of the lamina.

Centre of gravity 2

This section looks first at toppling: an object is placed on an inclined plane – will it topple over, or slide down the plane, or neither? Then we will look at the equilibrium of a suspended lamina.

Links

You will need to have revised 'Centre of gravity' first. See page 194.

Toppling

Imagine an object on a rough plane inclined at θ to the horizontal. The centre of gravity is located at C.

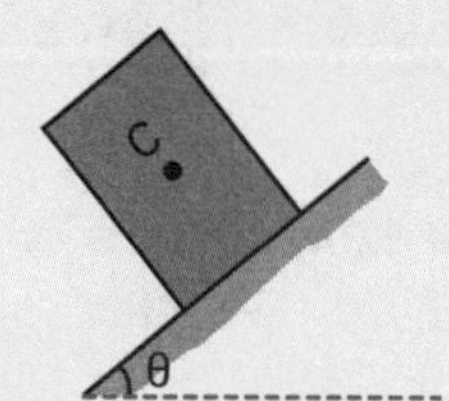

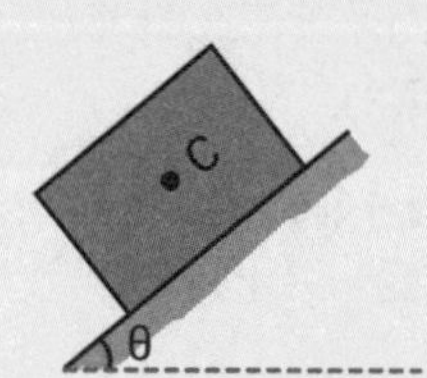

Checkpoint 1

Look at these diagrams. Will the object topple?

The critical position

In any toppling problem, there is a critical point at which it is just about to topple about a point A. When this happens the normal reaction acts at A. If the object is on an inclined plane the line of action of the weight goes through A also.

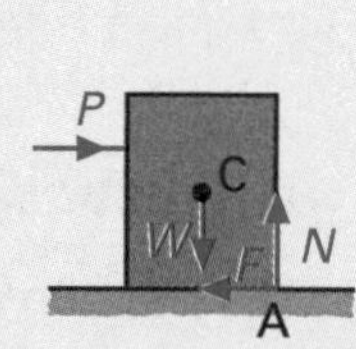

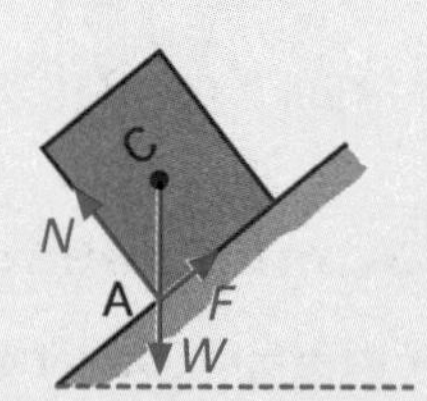

Checkpoint 2

The left-hand diagram here shows a cylinder of radius 5 cm, height 18 cm and mass 4 kg on a rough horizontal table. The force P acts 16 cm above the table, and is increased until the cylinder topples. By taking moments about A find the value of P when this happens.

Example: A cylinder of radius 4 cm and height 10 cm and mass 3 kg is placed on a rough plane inclined at angle θ to the horizontal. θ is increased until the cylinder topples. Find the value of θ for which this happens.

Answer: (Use the right-hand diagram above.) When it is about to topple, the weight and normal reaction act through A. It's really just geometry. Look at the diagram to the right, and you will see that $\tan\theta = \frac{4}{5}$ so $\theta = 38.7°$.

Checkpoint 3

For the example above, find the value of P when the cylinder slides if $\mu = 0.3$. Will it slide before it topples?

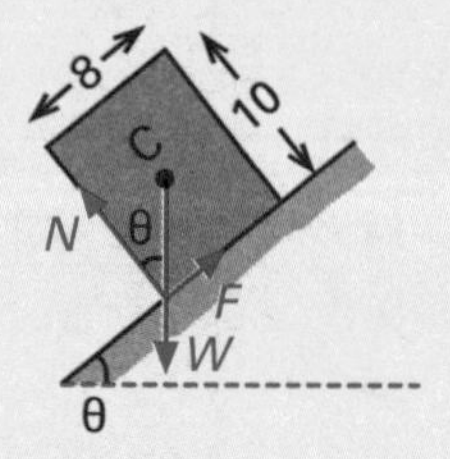

Checkpoint 4

For the cylinder on the plane, find the value of θ at which it starts to slip if $\mu = 0.85$. Will it slide before it topples?

Did it jump or was it pushed?

Actually, the question is 'will it slide or will it topple?' To answer this question, first find the condition for toppling as above using ideas of moments, and the normal reaction at the point about which toppling will take place. Then draw a force diagram, and consider the sliding possibility (using $F = \mu N$ on the point of sliding). You'll need to remember about resolving forces.

Now think which of the critical points will happen first (which is the lowest value of P or θ (or whatever).

Example: A beam AB of mass 25 kg is being dragged along rough horizontal ground by a rope attached to the end B and at 30° to the horizontal.

(a) Find the value of T if B is about to lift off the ground.
(b) Given $\mu = 0.7$, find the value of T if the beam is about to slide.
(c) If T is gradually increased from zero, how is equilibrium broken?

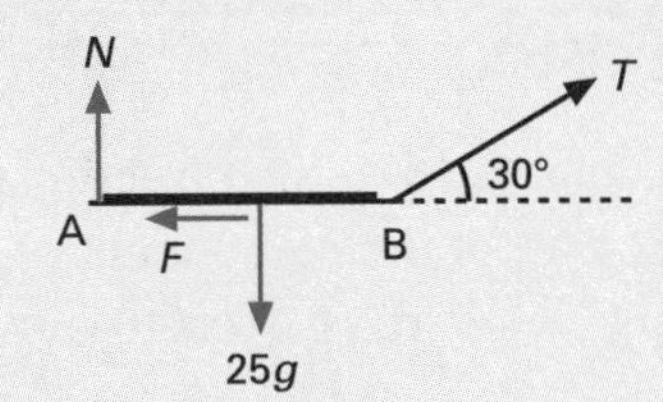

Answer: (a) If B is just about to lift off, N acts at A.
Taking moments about A: $T \sin 30° \times x = 25g \times 0.5x$
giving $T = 245$

(b) Resolving vertically: $N = 25g - T \sin 30°$
Resolving horizontally: $F = T \cos 30°$
Beam about to move so $F = 0.7N$
Giving $T \cos 30° = 0.7(25g - T \sin 30°)$
So $T = 141$

(c) It will slide because $141 < 245$.

Action point

Cover up the working, and check that you can reproduce it.

Equilibrium of a suspended lamina

A common question asks you to locate the centre of gravity of a lamina. You are then told that the lamina is suspended from a certain point, and you are asked to find the angle a certain side makes with the vertical.

The principle is that the centre of gravity must be directly below the point of suspension. The trick is to take the point of suspension as the origin, the side in question as the x-axis, and work out the centre of gravity's coordinates.

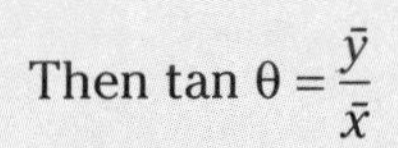

Then $\tan \theta = \dfrac{\bar{y}}{\bar{x}}$

Checkpoint 5

The figure in the example at the top of page 195 is suspended from the point A. Find the angle that AB makes with the vertical.

Exam questions answers: page 215

1 (a) Find the location of the centre of gravity of the uniform T shape shown in the diagram.
(b) If the T is placed on a rough inclined plane (with the side of length 9 units on the plane), and the angle of the plane is gradually increased, find the angle of the plane when the T starts to topple. (10 min)

2. A uniform lamina has vertices A (0,0), B (7,0), C (13,3) and D (0,3). It is suspended from the point A. Calculate the angle AD makes with the vertical. (8 min)

Examiner's secrets

A good diagram is essential for question 2. Don't forget that the mass is proportional to the area of the lamina.

Work and power

Some students do not find the concept of work very intuitive, but don't worry! It is not difficult to learn what to do.

Work done by a constant force

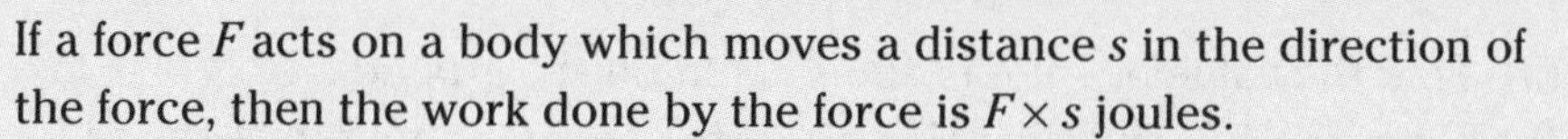

If a force F acts on a body which moves a distance s in the direction of the force, then the work done by the force is $F \times s$ joules.

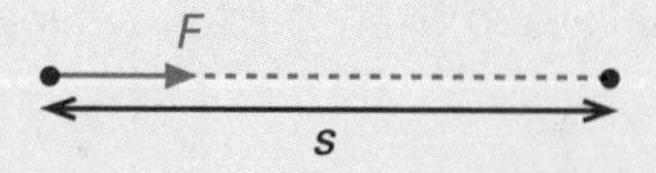

Checkpoint 1

A load is being pulled along a floor by means of a rope inclined at 30° to the horizontal. If the tension in the rope is 250 N, find the work done by the tension in dragging the load 4 m.

If the force F acts at an angle θ to the direction of motion, then the work done is the product of the component of F in the direction of motion and the distance moved, i.e. $F \cos\theta \times s = Fs \cos\theta$ joules.

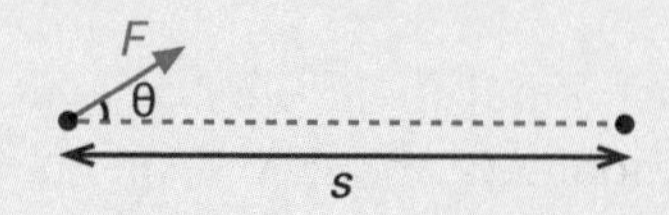

Checkpoint 2

A force $\mathbf{F} = 3\mathbf{i} + 7\mathbf{j}$ acts on a body which moves from (–1,2) to (5,3). Find the work done by the force.

The sharp among you may notice that $Fs \cos\theta$ is equivalent to the **scalar product** of the vectors **F** and **s**. Check your syllabus to see if you need to know this.

Example: A body moves from (2,6) to (5,1) under the action of a force $\mathbf{F} = 2\mathbf{i} - 3\mathbf{j}$. Find the work done by the force.

Answer: The displacement $\mathbf{s} = 3\mathbf{i} - 5\mathbf{j}$.

So the work done $= \mathbf{F} \cdot \mathbf{s} = 6 + 15 = 21$ joules.

Work done against gravity

Note that if θ is obtuse, then $\cos\theta$ is negative. In these cases we say that *work has been done against the force*. If a body of weight mg is raised a vertical distance h, then work must be done against gravity equal to mgh.

Power

Power is the rate of doing work and is measured in joules/second or watts.

Checkpoint 3

If, in the checkpoint 1 above, the rope is being pulled by a man who takes 40 s to drag the load, find the power the man is supplying.

Example: A hoist raises 1 000 bricks, each 0.9 kg, through a vertical distance 5 m every minute. Find the power of the hoist.

Answer: Work done each minute $= 1\,000 \times 0.9 \times g \times 5 = 44\,100$.

So power = work done per second $= 44\,100 \div 60 = 735$ W.

Power of a vehicle

Checkpoint 4

A car's engine is working at 8 000 watts. Find the forward force provided by the engine when the car is travelling at 20 ms^{-1}.

Suppose the engine of a vehicle develops an effective power P watts. (Note that not all the power produced by the engine is used to propel the vehicle forwards. Some is lost in heat, sound, etc.) Let F be the forward force provided by the engine (sometimes referred to as the tractive force).

Then P = work done per second

$= F \times$ distance moved per second

$= F \times v$

where v is the velocity.

Maximum velocity of a vehicle

Note that if the engine is working at a constant rate, then Fv is constant. So as the vehicle's velocity increases, F decreases and consequently the acceleration decreases in magnitude. Eventually F decreases to the value of the resistive forces acting on the vehicle. At this point, the resultant force on the vehicle is zero, the acceleration is zero, and the vehicle has reached its maximum velocity.

Example: A car of mass 750 kg is travelling along a horizontal road. Resistances to motion total 240 N, and the car's engine is working at a constant rate 12 kW. Find (a) the maximum velocity of the car, (b) the car's acceleration when its velocity is 30 ms^{-1} and (c) its maximum velocity up a hill inclined at $\sin^{-1}(\frac{1}{25})$ to the horizontal.

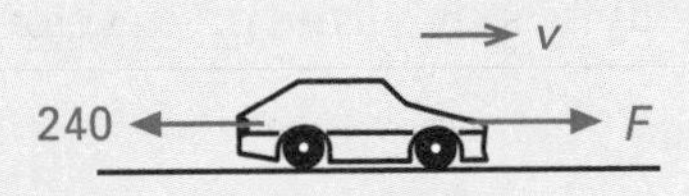

Answer:

(a) At maximum velocity, the acceleration is zero so $F = 240$. Using $P = Fv$ we have $12\,000 = 240v$ giving $v = 50\ \text{ms}^{-1}$.

(b) Using $P = Fv$ we have $12\,000 = F \times 30$, giving $F = 400$. Applying Newton's Second Law we have $400 - 240 = 750a$, so $a = 0.213\ \text{ms}^{-2}$ (3 s.f.).

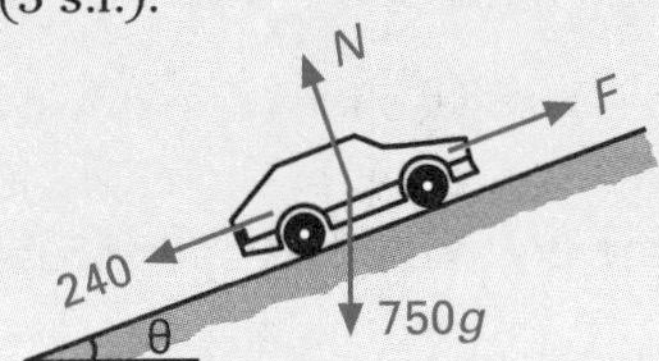

(c) At maximum velocity the acceleration is zero, so $F = 240 + 750g \sin\theta = 534$. $P = Fv$ gives $12\,000 = 534v$, so $v = 22.5\ \text{ms}^{-1}$ (3 s.f.).

Exam questions answers: page 215

1 A winch hauls a boat of mass 600 kg up a slope inclined at 15° to the horizontal against resistive forces of 300 N. The boat moves up the slope at a steady speed 0.2 ms^{-1}.

(a) Find the work done against the resistive forces per second.

(b) Find the work done against gravity per second.

(c) Hence write down the effective power output of the winch. (8 min)

2 A car's engine works at a constant rate of 12 kW. The resistive forces acting on the car are $(200 + 5v)$ N, where v is the velocity of the car. The mass of the car is 600 kg.

(a) Find the acceleration of the car at the instant when its velocity is 20 ms^{-1}.

(b) If V is the maximum velocity of the car, show that V satisfies the equation $V^2 + 40V - 2\,400 = 0$, and hence find the maximum velocity. (15 min)

Checkpoint 5

Sketch a graph of F against v for a vehicle whose engine is working at a constant rate.

Checkpoint 6

The modelling assumption here is that resistance to motion is constant. Do you think that this is a valid assumption?

Test yourself

Cover up the answer, and check you can work it through for yourself. Notice how the power equation $P = Fv$, and Newton's Law are used in tandem.

Checkpoint 7

Calculate the power developed by a cyclist riding on a horizontal road at a steady speed of 12 ms^{-1} against resistances totalling 110 N. In what way would your answer be different if he were cycling uphill?

Examiner's secrets

When you are asked to show something, don't be put off because you can't see how to begin. Start off in the way you would normally . . . 'maximum velocity' means . . . ? And see where you get to. *Whatever happens, make sure you pick up the marks for solving the quadratic.*

Work and energy

The concept of energy is very powerful, and can often be used to solve problems which are difficult to sort out using Newton's Second Law because the forces are not constant. A body which has energy can do work. We will just be considering mechanical energy here – if you study physics, you may also study heat, light, sound, electrical and chemical forms of energy.

Links

For elastic potential energy see page 210.

Gravitational potential energy

This represents the energy a body has by virtue of its position. If a body is raised, work must be done against gravity equal to mgh, where h is the vertical distance the body was raised. So the gravitational potential energy (GPE) of the body is mgh. Note that we measure h from a chosen zero level (which should be stated). This is the only form of energy which can be 'negative' if the body is below the zero level.

Checkpoint 1

A body of mass 12 kg is raised through a vertical distance h metres. Calculate the value of h if the gain in GPE by the body is equal to 58.8 joules.

Kinetic energy (KE)

This is the energy a body has by virtue of its velocity, and it is equal to $\frac{1}{2}mv^2$.

Checkpoint 2

A ball of mass 50 g is travelling at 16 ms^{-1}. Calculate its kinetic energy.

The work–energy principle

This principle states that work and energy are equivalent (you'll have noticed that they are measured in the same units, joules). If a force acting on a body does work, then the body experiences an equivalent gain in energy. If, on the other hand, work has to be done against a force acting on a body, then the body loses that amount of energy.

Example: A brick of mass 0.8 kg slides 6 m down a chute inclined at $\sin^{-1}(\frac{3}{5})$ to the horizontal. At the top of the chute it is given an initial speed 0.4 ms^{-1}, and at the bottom it has speed 5.4 ms^{-1}. Calculate

(a) the loss in GPE of the brick;
(b) the gain in KE of the brick;
(c) the work done against resistive forces;
(d) the magnitude of resistive forces (assumed constant).

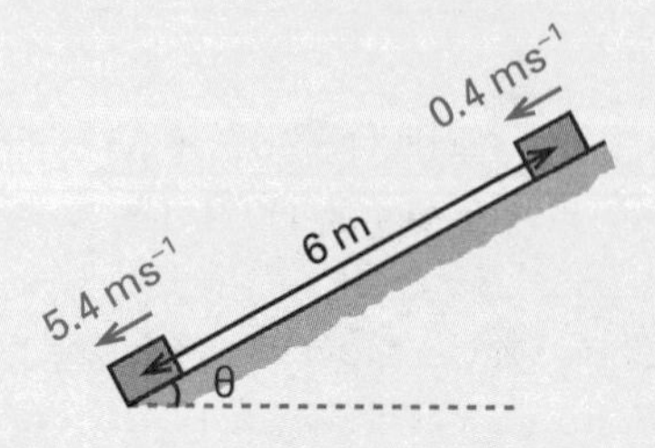

Answer:

(a) $h = 6 \sin \theta = 3.6$, so GPE lost $= 0.8 \times 9.8 \times 3.6 = 28.224$ J.
(b) KE gained $= \frac{1}{2} \times 0.8 \times 5.4^2 - \frac{1}{2} \times 0.8 \times 0.4^2 = 11.6$ J.
(c) Overall there is a net loss of energy = 16.624 J. This must be the work done against the resistive forces.
(d) Let R = resistive forces. Work done $= R \times 6 = 16.624$. So $R = 2.77$ N.

Checkpoint 3

(a) A car of mass 800 kg has an engine which is working at a constant rate of 15 kW for 30 seconds. Calculate the amount of energy given to the car by the engine.
(b) During this time the car goes 600 m up a hill inclined at $\sin^{-1}(0.05)$ to the horizontal. Calculate the gain in GPE of the car.
(c) On the way up resistive forces total 300 N. Calculate the work done against resistive forces.
(d) Deduce the change in KE experienced by the car.

Conservation of energy

This follows directly from the work–energy principle. If there are no external forces doing work on a body, then energy is conserved.

Example: A slide 130 m long in an adventure playground is completely friction-free. A child of mass 40 kg starts from rest at the top, 36 m above the ground level. Calculate the speed of the child as she reaches the bottom of the slide.

Answer: The child's loss in GPE = $40 \times 9.8 \times 36 = 14\,112$ J. As there is no work done against friction, all this energy must be converted to KE, so $\frac{1}{2} \times 40 \times v^2 = 14\,112$, giving $v = 26.6$ ms^{-1}.

Example: A bead of mass m is threaded on a smooth wire shaped as a vertical circle of radius r.

(a) If the bead is dislodged from the top of the circle, show that its speed is $\sqrt{4gr}$ when it reaches the bottom of the circle.

(b) If in fact its speed is $\sqrt{3gr}$, deduce the work done against resistive forces, and the magnitude of these forces (assumed constant).

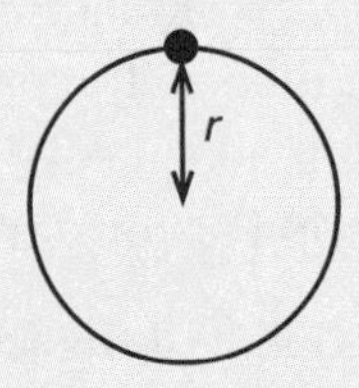

Answer:

(a) GPE lost by the bead = $mg \times 2r = 2mgr$.
This is all converted to KE if there is no work done against friction.
So $\frac{1}{2}mv^2 = 2mgr$, giving $v^2 = 4gr$ and $v = \sqrt{4gr}$.

(b) If $v = \sqrt{3gr}$ then the KE = $\frac{1}{2}m \times 3gr$ so $\frac{1}{2}mgr$ of energy is lost. This must be work done against a resistive force, R.
The distance travelled by the bead = πr (semicircle).

So $R = \dfrac{1}{2}mgr \div \pi r = \dfrac{mg}{2\pi}$.

Checkpoint 4

Does the length of the slide make any difference to your answer? If the slide were not quite friction-free, would your answer be more or less? What if the child was heavier?

The jargon

Dislodged in this context means that the bead is sent on its way with zero initial speed.

Examiner's secrets

Don't be put off by questions with algebraic answers. Remember that they are given so that you can check you are on the right lines – and they save time spent pressing calculator buttons!

Exam questions answers: page 215

1 A stone of mass 50 g is thrown vertically upwards with speed 15 ms^{-1}.

(a) Using energy considerations, find the greatest height above the point of projection reached by the particle, stating any assumptions you make.

(b) If in fact the greatest height is 10.5 m, find the work done against resistive forces, and the magnitude of the resistance (assumed constant). (7 min)

2 A sack of potatoes, mass 50 kg, is given a velocity 0.5 ms^{-1} at the top of a ramp. The ramp is inclined at 40° to the horizontal and is 9 m long. The speed of the sack at the bottom of the ramp is 1.3 ms^{-1}. Find:

(a) the loss in GPE of the sack;

(b) the work done against resistance, and the magnitude of the resistive forces. (10 min)

Examiner's secrets

Think carefully: what clues are there in the question that you need to use energy? Is energy conserved, or not?

Momentum and impulse

Checkpoint 1

What are the units of momentum?

Checkpoint 2

If in this example the bat is in contact with the ball for 0.02 seconds, find the force exerted by the bat on the ball during the impact.

Don't forget

We don't use conservation of momentum for an impact between a ball and a wall, because the wall isn't free to move.

Most students find this a relatively easy topic, so you should be able to pick up some marks here. The main focus is on impacts – what happens in that split second when bat hits ball, two balls collide, a ball bounces, etc.

Definition of momentum

The momentum of a body is the product of its mass and velocity: momentum = mass × velocity. It is a vector quantity. In everyday speech we talk about the momentum of something keeping it going. If you think about it, an object which is very heavy and/or going fast is difficult to stop – that's why both mass and velocity feature in the formula.

Definition of Impulse

There are two ways of looking at impulse.

First you can think of it as the **change in momentum**, i.e. $\mathbf{I} = m\mathbf{v} - m\mathbf{u}$, where **u** and **v** are initial and final velocities, as usual. So when a bat hits a ball, the ball receives an impulse equal to its change in momentum.

Second, impulse can be defined as the product of the force acting and the time period for which it acts, i.e. $\mathbf{I} = \mathbf{F}t$. In this section we will assume that the force in question is constant.

Note: The formulae are equivalent. If the force **F** produces constant acceleration **a**, then $\mathbf{I} = \mathbf{F}t = m\mathbf{a}t = m(\mathbf{v} - \mathbf{u})$ using $\mathbf{v} = \mathbf{u} + \mathbf{a}t$ rearranged.

Example: A ball of mass 0.3 kg is travelling with velocity $6\mathbf{i} - 5\mathbf{j}$ ms^{-1} when it is hit by a bat. Immediately after the impact it has velocity $-12\mathbf{i} + 2\mathbf{j}$ ms^{-1}. Find the impulse experienced by the ball.

Answer: $\mathbf{I} = m(\mathbf{v} - \mathbf{u})$

$$= 0.3\left\{\begin{pmatrix}-12\\2\end{pmatrix} - \begin{pmatrix}6\\-5\end{pmatrix}\right\} = \begin{pmatrix}-5.4\\2.1\end{pmatrix}$$

Example: A body of mass 10 kg has velocity $3\mathbf{i} + 2\mathbf{j}$ ms^{-1}. Find its final velocity when it has been acted on by a force $5\mathbf{i} - \mathbf{j}$ N for 3 seconds.

Answer: Impulse $= 3(5\mathbf{i} - \mathbf{j}) = 10\mathbf{v} - 10(3\mathbf{i} + 2\mathbf{j})$

$10\mathbf{v} = 3(5\mathbf{i} - \mathbf{j}) + 10(3\mathbf{i} + 2\mathbf{j}) = 45\mathbf{i} + 17\mathbf{j}$

so $\mathbf{v} = 4.5\mathbf{i} + 1.7\mathbf{j}$ ms^{-1}

Conservation of momentum

Consider an impact between two bodies which are free to move and on which either no external force is acting, or the time period of the impact is so small as to make action by a external force insignificant. By Newton's Third Law, there must be an equal and opposite contact force acting on the two bodies for the duration of the impact. Using the definition of impulse as force × time, we deduce that each body experiences an equal and opposite impulse. Now using the definition of impulse as change in momentum, we see that this change is equal and opposite for the two bodies. Therefore overall there is *no change in the total momentum of the two bodies.*

Problems on direct impacts between two particles

As always, clear diagrams help. Your diagram should show:

(a) the masses (not weights) of the particles;
(b) the velocities, both before and after impact;
(c) the direction you are taking as positive.

Example:
Two spheres of masses 0.1 kg and 0.3 kg and of equal radii are travelling directly towards each other with speeds 4 ms^{-1} and 6 ms^{-1} respectively. After the collision, the heavier sphere continues in an unchanged direction, but with speed 2 ms^{-1}. Find the velocity of the other sphere.

Answer:

before → 4 ms^{-1} ← 6 ms^{-1}

0.1 0.3 ← positive direction

after ← v ← 2 ms^{-1}

Using conservation of momentum:

$$-0.1 \times 4 + 0.3 \times 6 = 0.1 \times v + 0.3 \times 2$$

This gives $0.1v = -0.4 + 1.8 - 0.6 = 0.8$, so $v = 8$ ms^{-1}.

Loss of kinetic energy

In most collisions, some kinetic energy is lost (converted to sound and/or heat energy). The calculation of loss of energy is easy.

Example: Two trucks, of mass 500 kg and 400 kg, are travelling in the same direction at speeds 2.4 ms^{-1} and 1.05 ms^{-1} respectively. When they collide, they couple together and continue linked together. Find the loss in kinetic energy in the collision.

Answer:

before → 2.4 ms^{-1} → 1.05 ms^{-1}

500 400

after → v → v

Conservation of momentum: $500 \times 2.4 + 400 \times 1.05 = 900 \times v$.
This gives $v = 1.8$ ms^{-1}.
So the loss in KE $= (0.5 \times 500 \times 2.4^2 + 0.5 \times 400 \times 1.05^2) - 0.5 \times 900 \times 1.8^2$
$= 202.5$ J.

Exam questions answers: page 216

1 A particle of mass 0.2 kg and velocity $5\mathbf{i} + 7\mathbf{j}$ collides with a particle of mass 0.3 kg and velocity $2\mathbf{i} - 3\mathbf{j}$. If the particles coalesce, find their joint speed. (4 min)

2 Particles A and B have masses 50 g and 40 g and lie on a smooth table. A is projected towards B with speed v ms^{-1}, and after the collision A and B move in the same direction with speeds 2 ms^{-1} and 5 ms^{-1}. Find v and also the loss in kinetic energy. (6 min)

The jargon

A *direct impact* is one when the direction of motion is in the same line as the impulse. (Note that the impulse acts along the line joining the two centres.) Note that collisions between particles are always direct – a useful modelling assumption here!

Checkpoint 3

Why are you often told in questions that the spheres have equal radii? What difference does it make?

Test yourself

Cover up this answer and make sure you can do it yourself.

Links

See page 200 for work on KE.

Checkpoint 4

Find the loss in kinetic energy in the example at the top of this page.

Examiner's secrets

Don't forget that KE is *always* positive whatever the direction of motion.

The jargon

Coalesce means that the particles fuse, becoming one.

Momentum and restitution

In this section we will study the concept of restitution which measures the elasticity, or 'bounciness' of an impact. We will also tackle oblique impact, i.e. collision at an angle rather than head-on.

Newton's Law of Restitution

This is also called Newton's Experimental Law (no prizes for guessing why!). It says that:

speed of separation = $e \times$ speed of approach

where e is a constant called the coefficient of restitution and takes a value, dependent on the materials, less than or equal to 1.

The main task is to identify the speeds of separation and approach. Essentially, if the particles are going directly towards or away from each other, you add their speeds. If however they are travelling in the same direction, you subtract their speeds.

Note that the Law of Restitution can be applied in the case of a particle colliding with a surface.

The jargon

If $e = 1$, we say that the collision is *perfectly elastic*. In this case only, no kinetic energy is lost.

Checkpoint 1

Write down the speeds of separation and approach for the particles in the diagram below:

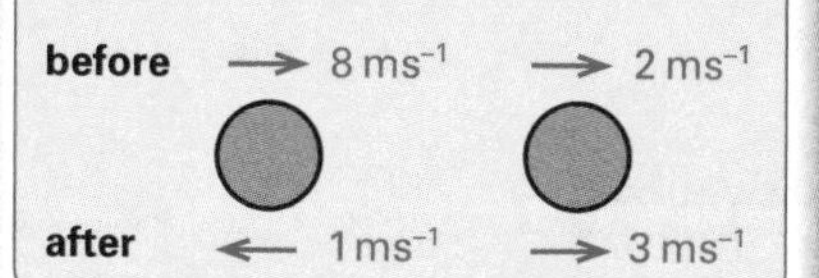

Solving problems 

Get a clear diagram first, with your positive direction indicated. Apply both conservation of momentum (if the question involves particles free to move) and the Law of Restitution. You'll end up with two equations to solve simultaneously.

Example: A particle, of mass 2 kg and travelling with speed 5 ms^{-1}, collides with a stationary particle of mass 3 kg. If the coefficient of restitution is 0.4, find their velocities after impact.

Answer:

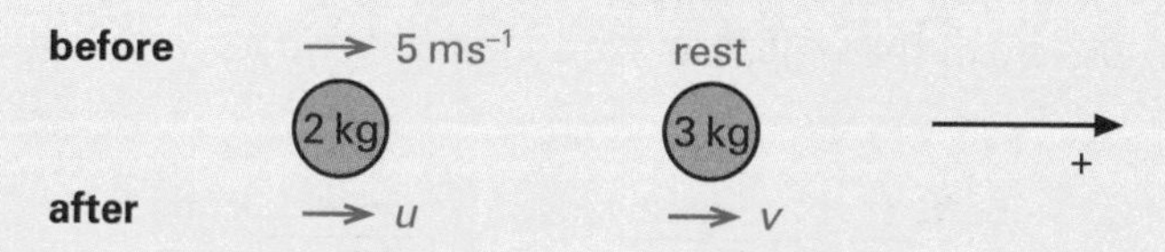

Conservation of momentum: $2u + 3v = 2 \times 5 = 10$

Law of Restitution: $v - u = 0.4 \times 5 = 2$

Solving gives $u = 0.8$ and $v = 2.8$.

Examiner's secrets

You don't know the direction of motion of the particles after impact, so make a guess for your diagram. If you are wrong, you'll just get a negative result.

Checkpoint 2

Cover up the working and do it yourself. Also find the loss in KE.

Direct impact with a fixed plane

This is easy because the plane doesn't move, so all you need to do is apply the restitution law. If the ball has speed u immediately before impact, its speed after impact is eu.

Checkpoint 3

A ball is dropped 80 cm above a surface. Given that the coefficient of restitution is $\frac{3}{4}$, find the speed of the ball immediately before impact, and the height to which it rises after impact.

Oblique impact with a surface 

We are thinking here of a ball coming in to a surface at an angle. It's not too difficult, but you must first resolve the velocity of approach of the ball into components parallel and perpendicular to the surface.

Parallel to the surface, the ball's velocity is unchanged (there are no forces acting in this direction). Use restitution in the perpendicular direction.

Recoil of a gun

When a gun is fired, a great deal of energy is released. There is an impulse on the shot giving it forward momentum, and an equal and opposite impulse on the gun giving it backward momentum. Overall, by conservation of momentum, the total momentum remains zero so the momentum of the gun is numerically equal to that of the shot.

Example: A gun of mass 1.5 kg, fires a shot of mass 0.015 kg with speed 120 ms^{-1}. Find the speed of recoil of the gun.

Answer: The momentum of the shot is $0.015 \times 120 = 1.8$ N s.

So the momentum of the gun $= 1.5v = 1.8$, giving $v = 1.2\ ms^{-1}$.

Firing at an angle

A cannon of mass 250 kg fires a shot of mass 2.5 kg at a speed 80 ms^{-1} at an angle 30° to the horizontal. Find the speed of recoil of the gun.

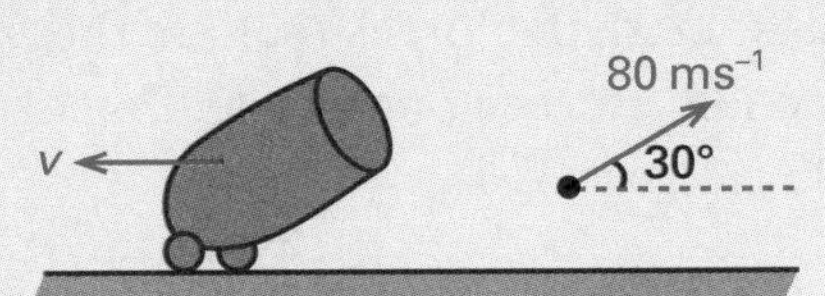

The shot experiences an impulse which has components both parallel and perpendicular to the ground. The cannon can only move parallel to the ground (if it is properly mounted – if it is on soft ground, it will dig itself in!).

Parallel to the ground we have $2.5 \times 80 \cos 30° = 250v$ so $v = 0.693$.

Checkpoint 4

A ball has speed 4 ms^{-1} immediately before impact with a fixed plane, and its direction of motion makes an angle 60° with the normal to the plane. If the coefficient of restitution is $\frac{1}{2}$, find the speed and direction of motion after impact.

Checkpoint 5

Where has the energy come from?

Checkpoint 6

Why aren't cannons fixed so that they can't recoil?

Action point

The modelling assumption here is that the length of the barrel is negligible. What is the significance of this?

Exam questions answers: page 216

1 Two particles have masses 3 kg and 5 kg. They are travelling in the same direction with speeds 0.6 ms^{-1} and 0.4 ms^{-1}, so that the lighter particle collides with the heavier. Given that the coefficient of restitution is 0.8, find the velocities of the particles after impact. (8 min)

2 A ball is projected from a point on horizontal ground with speed 14 ms^{-1} at an angle 30° to the horizontal. At the instant when the ball is travelling horizontally, the ball hits a vertical wall and rebounds. Given that the coefficient of restitution between the ball and wall is 0.7, find the distance from the wall of the point where the ball touches the ground again. (10 min)

Examiner's secrets

Examiners sometimes like to run two topics together and in Question 2 you have projectiles with an impact. You will impress the examiner if you can handle this. Just take it step by step.

Circular motion 1

In this section we will look at motion in a horizontal circle with constant speed.

Angular and linear speed

If we measure angular speed, ω, in radians per second then there is a simple relationship between angular and linear speed:

$$v = \omega r,$$

where v is the linear speed, and r is the radius of the circle. If the angular speed is measured in different unit (e.g. revolutions per minute), you need to remember than 2π rad = 360° = 1 revolution.

Checkpoint 1

Why is the velocity not constant?

Acceleration

Although the speed is constant, the velocity is not. The acceleration is directed towards the centre of the circle and has magnitude v^2/r or $\omega^2 r$. (The two forms are equivalent: check this using $v = \omega r$.)

$$a = \omega^2 r = \frac{v^2}{r}$$

Note that the direction of the acceleration is always changing, but it is always towards the centre of the circle.

Solving problems

Checkpoint 2

Why is $a = 0$ perpendicular to the circle?

You will not be surprised to learn that a good clear force diagram is the place to start! Just mark the forces that are there: weight, tension, friction, normal reaction, etc. as appropriate. Don't be tempted to add in any extra 'centrifugal' forces.

Apply Newton's Second Law ($F = ma$) towards the centre of the circle. Use $a = \omega^2 r$ or v^2/r as appropriate.
Apply Newton's Law perpendicular to the circle, with $a = 0$.

You might also need to do some trig in a triangle you are given.

Checkpoint 3

If you take a corner too fast, will you skid towards the centre, or away from the centre? Is it ever possible to take the corner too slowly?

Vehicles driving round circular bends

In these questions, it is the friction preventing side-slip that provides a force towards the centre of the circle. If the track is banked, the normal force also provides a component in this direction. Questions often focus on the maximum safe speed for cornering.

Example: A car of mass 1 000 kg is going round a horizontal corner which is part of a circle of radius 75 m. The road is banked at an angle 10° to the horizontal, and the coefficient of friction between the tyres and the road is 0.4. Find the maximum speed for which there is no tendency for the car to slip sideways.

Checkpoint 4

Which of the three forces marked have a component towards the centre of the circle? Which have a component perpendicular to the circle?

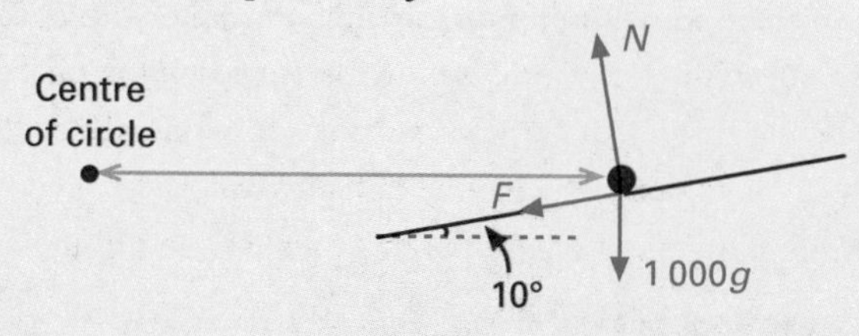

Answer:

Apply Newton's Law towards the centre of the circle:

$$F \cos 10° + N \sin 10° = \frac{mv^2}{r} = \frac{1\,000v^2}{75} \quad ①$$

Apply Newton's Law perpendicular to the circle:

$$N \cos 10° - 1\,000g - F \sin 10° = 0 \quad ②$$

Law of Friction: $F = 0.4\ N$ ③

Conical pendulum

This is the proper name given to a particle going round in horizontal circles at the end of a string. In these questions it is the tension in the string which has a component towards the centre of the circle. You will often also have to relate the radius of the circle, the length of the string and the angle of the string. More complicated questions might have two strings, or an elastic string. Don't panic – just apply the principles carefully – and start with a good force diagram!

Example: A particle of mass 0.1 kg is attached to a light inextensible string 80 cm long. It performs horizontal circles with angular speed 5 rads per second, with the string making an angle θ to the vertical. Find the tension in the string, and the value of θ.

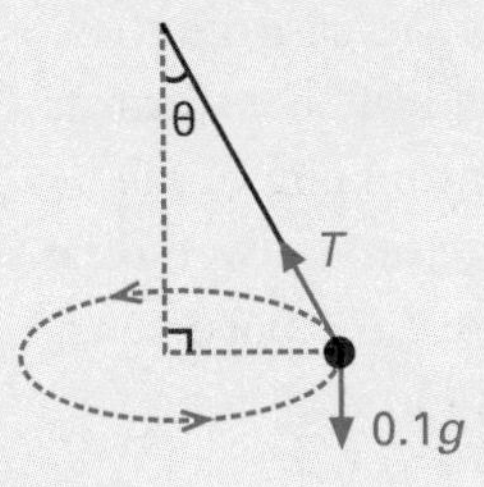

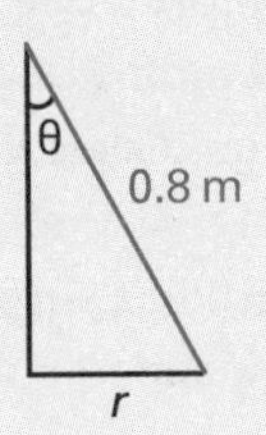

Answer:

Apply Newton's Law towards the centre of the circle:

$$T \sin \theta = m\omega^2 r = 2.5r$$

Using Checkpoint 5, we see that $r = 0.8 \sin \theta$.

So $T \sin \theta = 2.5 \times 0.8 \sin \theta$ giving $T = 2$ newtons.

To find θ, apply Newton's Law perpendicular to the circle:

$$T \cos \theta = 0.1g, \text{ so } \cos \theta = 0.1g \div 2 \text{ giving } \theta = 60.7°.$$

Extension

If the string is replaced by an elastic string with natural length 80 cm, and modulus of elasticity 3 N, what would be the tension in the string? We would have $T \sin \theta = 2.5r$ as before, but now $r = (0.8 + x)\sin \theta$. We also have Hooke's Law: $T = \lambda x/l = 3.75x$. (See page 210.)

Exam question answer: page 216

A particle of mass 2 kg is attached to two strings of lengths 30 cm and 40 cm. The free end of the longer string is attached to a fixed point A, and the free end of the shorter string is attached to B, 50 cm below A.

(a) If the particle performs horizontal circles with angular speed 8 rads^{-1} find the tension in each of the strings.

(b) If both strings remain taut, find the least possible angular speed. (15 min)

Action point

Substitute $F = 0.4\ N$ in equation ② and use your trusty calculator to find N. Then get F. Put these values into equation ① to get v. Your answer should be $v = 21.4$.

Checkpoint 5

Relate l, r and θ in the diagram below.

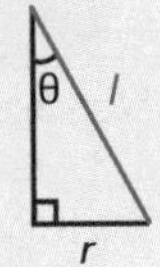

Checkpoint 6

Again, identify in the diagram the forces which have a component towards the centre, and those with a component perpendicular to the circle.

Test yourself

Cover up this working, and see if you can work it through for yourself on rough paper.

Examiner's secrets

Substitute the expressions for r and T into $T \sin \theta = 2.5r$. This gives you an equation in x only, which you can solve. You should end up with 6 N for T.

Action point

Will the tensions in the two strings be the same? Can you use the information given to work out some angles?

Circular motion 2

The focus here is on motion in a vertical circle. The main difference from the first circular motion spread is that the speed is not constant. This means that the acceleration has a component tangential to the circle as well as towards the centre. Energy considerations prove to be a helpful way towards a solution to problems.

Links

See page 206 for reminder of the acceleration towards the centre of a circle.

The jargon

The component of acceleration towards the centre is called the *radial acceleration*, and the tangential component is called the *transverse acceleration*.

Links

For reminders about work and energy see page 200.

Checkpoint 1

The weight isn't perpendicular to the motion of the particle. Does it do any work? How is it incorporated into the energy equation?

Test yourself

Before reading on, try to draw the diagram for yourself and begin to answer the question – remember: energy first, then Newton.

Checkpoint 2

In the diagram on the right, the reaction force R is drawn towards the centre. Are there any points on the circle where the reaction force is directed *away* from the centre?

Components of acceleration

Acceleration has a component towards the centre of the circle equal to v^2/r or $r\omega^2$ as before (where v is the linear speed and ω is the angular speed).

Acceleration also has a component tangential to the circle equal to

$$\frac{dv}{dt} = r\frac{d^2\theta}{dt^2}\left(\text{remember } v = r\omega = r\frac{d\theta}{dt}\right)$$

This means that we can apply Newton's Second Law ($F = ma$) both towards the centre and tangentially. Usually towards the centre proves the more fruitful, so try that unless the question gives a heavy hint otherwise.

Conservation of energy

In most situations you will encounter, apart from the weight all forces acting on the particle will act towards the centre. In most cases you will be thinking of the tension in a string, or the normal reaction between the particle and a circular surface. Because the particle is always moving perpendicular to this force, it does no work and so energy is conserved.

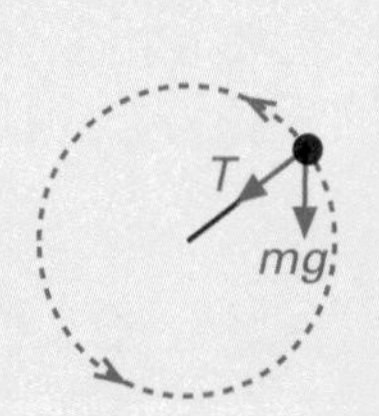

Solving problems

As always, start with a clear force diagram, and mark on it other information you are given. Generally we use energy to link the speed of the particle at two positions, and then the velocity gives the acceleration towards the centre. We then apply Newton's Second Law to find the magnitude of forces at any particular point. Note that the acceleration is not constant, so the constant acceleration equations will be a very bad idea!

Example: A smooth bead of mass m kg is threaded on a smooth circular wire of radius r metres which is held in a vertical plane. The bead is projected from the lowest point on the circle with speed $\sqrt{3rg}$. Find the speed of the bead when it has gone one sixth of the way round the circle, and the force exerted on the bead by the wire at this point.

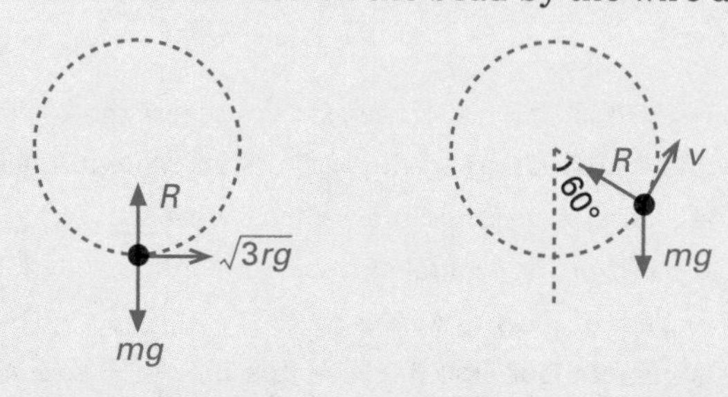

Answer:
Initially KE = $\frac{1}{2}m \times 3rg$
Gain in GPE = $mgr(1 - \cos 60°) = \frac{1}{2}mgr$
Therefore

$$\text{final KE} = \frac{1}{2}m \times 3rg - \frac{1}{2}mgr = mgr = \frac{1}{2}mv^2$$

so

$$v = \sqrt{2rg}$$

This means that the acceleration towards the centre = $v^2/r = 2g$.
Apply Newton's Second Law towards the centre:

$$R - mg \cos 60° = m \times 2g$$

which gives $R = \dfrac{5mg}{2}$

Conditions for complete circles

Is it possible for the particle to leave the circle? If your answer is no, then all that's required is that the particle has enough energy to get up to the top. If your answer is yes, as well as having sufficient energy, the tension in the string (or reaction between particle and surface) must be positive at the top.

Example: A light inelastic string of length L is fixed at one end and at the other is attached a particle of mass m. The particle is projected horizontally from its equilibrium position with speed $\sqrt{4.5Lg}$. Does the particle perform complete circles?

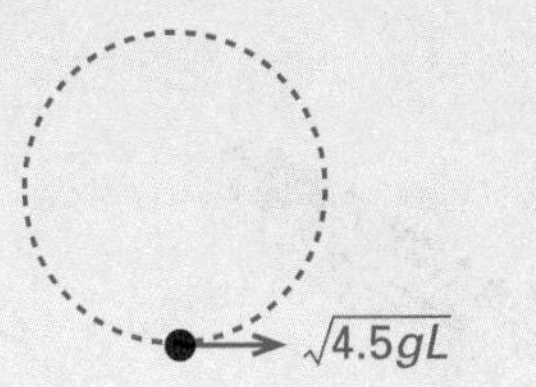

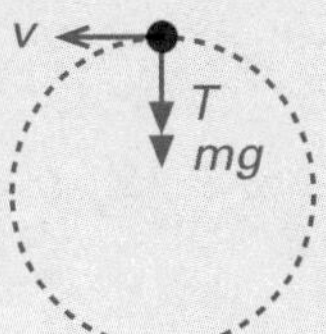

Answer:
If the particle reaches the top, it has gained GPE = $2mgL$. So its KE at the top is $\frac{1}{2}m \times 4.5gL - 2mgL = 0.25mgL = \frac{1}{2}mv^2$. This gives $v^2 = 0.5gL$. So the acceleration towards the centre at the top is $v^2/L = 0.5g$. Applying Newton's Second Law gives $mg + T = ma = 0.5mg$ which leads to $T = -0.5mg$.

But a *tension cannot be negative*, so the particle does not reach the top of the circle.

Exam question answer: page 216

A particle of mass 0.2 kg is at rest at the bottom of the inside of a smooth horizontal circular tube of radius 80 cm. The bead is projected from the lowest point with speed 7 ms^{-1} so that it starts to move in a vertical circle. Show that it performs complete circles. (15 min)

Don't forget

The first task is to read the question carefully.

Checkpoint 3

In each of the following cases, could the particle leave the circle? (In each case the circle is vertical.)
(a) A conker on the end of a string.
(b) A bead on a circular wire.
(c) A ball on the end of a light rod.
(d) A ball on the inside of a horizontal tube.

Action point

Notice that if the particle had been at the end of a rod, instead of a string, the tension could be a *thrust* and the particle would do complete circles as it had enough energy to reach the top.

Checkpoint 4

If, in the example with the string, the particle leaves the circle when the string makes an angle θ with the upward vertical, find the value of θ and describe briefly the subsequent motion of the particle. (*Hint*: draw a diagram (!) at the point when the tension is zero and apply $F = ma$ to get the acceleration.)

Elasticity

If you have a good grasp of other topics, you will not find questions on elastic strings difficult. You may get an equilibrium question for which you use standard techniques together with Hooke's Law. Alternatively, your question may have an energy focus.

Watch out!

Always read a question carefully to distinguish between elastic *strings* and *springs*. Both can be stretched, but a spring can also be compressed.

The jargon

The *natural length* of an elastic string is its unstretched length. The tension in an unstretched string is zero.

Checkpoint 1

A string is 1.2 m long unstretched, and has modulus of elasticity 3 N. Find the tension in the string when it is stretched to 1.5 m.

Hooke's Law

The main thrust here is that the tension in an elastic string is proportional to its extension. The formula is

$$T = \frac{\lambda x}{l}$$

where T = tension, x = extension, l = natural length and λ = modulus of elasticity. λ is the proportionality constant and has newtons as its unit.

Equilibrium problems

Before reading on, make sure you have revised the general equilibrium techniques on pages 186 and 192. The addition of an elastic string just means you have one more equation, Hooke's Law, to add in with the rest.

Example: A particle P of mass 400 g is attached to one end of an elastic string with natural length 80 cm, and modulus of elasticity 4 N. The other end is attached to a fixed point O on a smooth plane inclined at 30° to the horizontal. P lies in equilibrium on the plane below O, so that the string is on a line of greatest slope. Find the extension of the string.

The jargon

Line of greatest slope just means that P is directly below O on the slope, and the angle the string makes with the horizontal is 30°. *If it weren't on a line of greatest slope*, you'd have problems, but as it is don't be put off by the jargon!

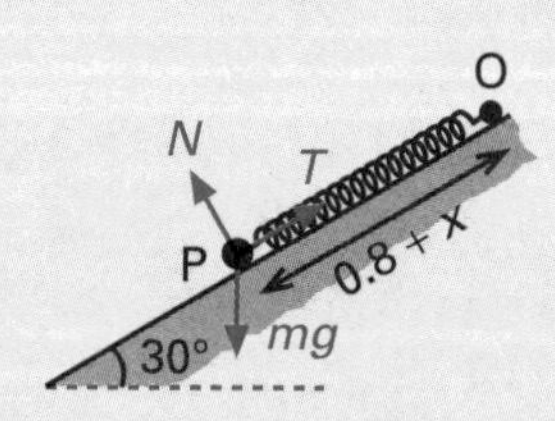

Answer:

Resolving parallel to the plane: $T = 0.4g \sin 30° = 1.96$

Hooke's Law: $T = \dfrac{4x}{0.8} = 5x$

So $5x = 1.96$ and $x = 0.392$ metres.

Elastic potential energy (EPE)

In order to stretch an elastic string, you have to do work against the tension. This means that a gain in energy is experienced, known as the elastic potential energy of the string.

$$\text{EPE} = \frac{\lambda x^2}{2l}$$

where λ is the modulus of elasticity, x is the extension and l the natural length of the string.

Don't forget

You will remember that work done = force × distance moved. The elastic potential energy is the average tension $\frac{1}{2}(\frac{\lambda x}{l})$ multiplied by the extension (which is the distance moved).

Problems involving energy

If you have a question which involves a particle moving with an elastic string, you should consider using energy. The problem with using Newton's Second Law ($F = ma$) is that the tension is changing all the time, and so is the acceleration – not so easy.

Example: A particle of mass m kg is attached to a string of natural length L and modulus of elasticity $4\,mg$. The other end is attached to a fixed point O. The particle is released from rest at O.

(a) Find the greatest distance below O reached by the particle and its acceleration at this point.

(b) Show that the speed of the particle is $1.5\sqrt{gL}$ when the acceleration is instantaneously zero.

Answer:

(a) At the lowest point, the speed of the particle is zero. All its GPE has been converted to EPE – there is no KE. Let the extension be x.

GPE lost $= mg(L + x)$

EPE gained $= \lambda x^2/2L = 2mgx^2/L$

Putting these equal (conservation of energy)

$mg(L + x) = 2mgx^2/L$,

which simplifies to $2x^2 - xL - L^2 = 0$. This factorises $(2x + L)(x - L) = 0$ giving $x = L$ as a solution.

So the greatest distance below O is $2L$.

Apply Newton's Second Law at the lowest point: $T - mg = ma$.

At this point $T = 4mgx/L = 4mg$ so $a = 3g$ (upwards).

(b) The first thing you need to work out is when the acceleration is zero. Clearly, the resultant force will be zero, so the tension is equal to the weight, i.e. $mg = T = 4mgx/L$ giving $x = \frac{1}{4}L$.

So GPE lost $= mg(L + x) = 5mgL/4$

EPE gained $= \lambda x^2/2L = mgL/8$

So KE gained $= 5mgL/4 - mgL/8 = 9mgL/8$

Therefore $\frac{1}{2}mv^2 = 9mgL/8$, so

$v^2 = 9gL/4$ and

$v = 1.5\sqrt{gL}$

Checkpoint 2

Calculate the EPE stored in an elastic string of natural length 1.5 m and modulus of elasticity 5 N, when it is stretched to 2 m.

Links

The alternative is to use simple harmonic motion equations – but only if this is included in your syllabus!

Checkpoint 3

Why is conservation of energy an appropriate principle to apply here? What modelling assumption must you make for it to be valid?

Test yourself

As previously, cover up the working, and check that you can answer the question for yourself.

Watch out!

Take care not to start with what you are supposed to show. The examiner won't be impressed if you start with the answer $v = 1.5\sqrt{gL}$, and work backwards to show that the acceleration is zero.

Exam questions

answers: pages 216–17

1 A particle of mass 3 kg is on a smooth table. It is attached to a point A on the table by means of an elastic string of natural length 75 cm, and modulus of elasticity 6 N. It is released from rest 1.25 m from A. Find the maximum speed of the particle in the subsequent motion. (8 min)

2 A smooth ring of mass 3 m is attached to one end of a light elastic string of natural length $2a$, and modulus of elasticity λ. The other end is attached to a fixed point O. The ring is threaded on a smooth vertical wire, distance $2a$ from O, and hangs in equilibrium with the string making an angle $\sin^{-1}(\frac{2}{3})$ with the vertical. Show that $\lambda = 18\sqrt{5}\,mg/5$ and find the force exerted by the wire on the ring. (12 min)

Checkpoint 4

What is the significance of the table being smooth in the first exam question? How would your answer be different if the table were rough?

Examiner's secrets

In order to get the answer required, you will need to find the value of $\cos\theta$ directly from $\sin\theta = \frac{2}{3}$. See page 48 for a hint if necessary.

Answers
Mechanics

Answers which are not exact have generally been given correct to 3 s.f. Check with your exam board the degree of accuracy you are required to give.

Modelling

Checkpoints

1 You wouldn't need to worry about the length of the vehicles and whether 'overtake' meant completely overtake or just the front of one vehicle ahead of the front of the other.
2 A section near the top has to support a greater mass of string below it.
3 Probably not.
4 If there were no air resistance, the person would continue to accelerate.

Exam questions

1 The deceleration is -2.94 ms^{-2}, and the velocity before impact is 3.64 ms^{-1}. After impact, 2.184 ms^{-1} and 0.364 ms^{-1}. If they are modelled as particles, you can treat it as a direct impact. Also you don't need to worry about whether the '2 m apart' is the distance between the centres or the edges of the objects.
2 No air resistance, particle, $t = 1.56$ seconds. Air resistance would be significant, but could still use a particle model as size negligible compared to height of aeroplane.
3 Equations of motion are $T - 4g \sin 20° = 4a$ and $5g - T = 5a$ leading to $a = 3.95$ ms^{-2}. Assumptions are: no air resistance, string light and inextensible, pulley smooth.
4 (a) yes, (b) no, (c) ladder uniform; 1.25 m from fixing.

Vectors

Checkpoints

1 The resultant is $-5\mathbf{i} - \mathbf{j}$.
2 The magnitude is $\sqrt{25 + 1} = 5.10$.
The direction is $\tan^{-1}(\frac{1}{5}) = 11.3° + 180° = 191.3°$ measured anticlockwise from the **i** direction.
3 Horizontal: $3 \cos 15° = 2.90$ ms^{-1}
Vertical: $3 \sin 15° = 0.776$ ms^{-1}
4 Magnitude is 76.1 and direction is 21.8° above horizontal.

Exam questions

1

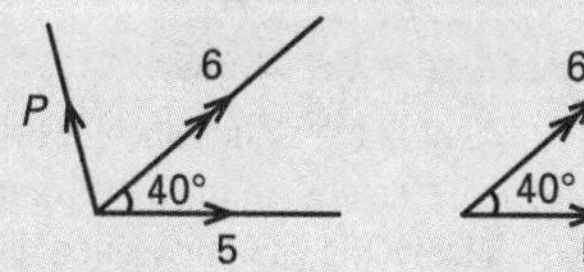

$P^2 = 25 + 36 - 2 \times 5 \times 6 \times \cos 40° = 15.0$, $P = 3.88$
$\cos \theta = (25 + P^2 - 36) \div (2 \times 5 \times P)$;
$\theta = 84.0°$
P makes angle $180° - \theta = 96.0°$ with 5 N force.

2 $(120 + 110 \cos 20° + 90 \cos 15°)\mathbf{i} + (110 \sin 20° - 90 \sin 15°)\mathbf{j} = 310.3\mathbf{i} + 14.3\mathbf{j}$. Magnitude is 310.6 N at 2.6° to **i** direction.

Examiner's secrets

If you are finding the direction of a vector, and are not sure if the angle is acute or obtuse, the cosine rule is safer than the sine rule.

Graphs in kinematics

Checkpoints

1 Displacement is a vector and is measured from a fixed point. Distance travelled is just the amount of ground you have covered in any direction. Velocity is speed in a specified direction – it is also a vector.
2 The section AC is curved, while the other sections are straight. It is travelling faster in CD.
3

4 144 m
5 Total time = 144.5 sec
6 Its weight, and possibly air resistance.

Common misunderstanding Area B represents distance travelled upwards. C is 30 m more because it starts 30 m up.

Exam questions

1

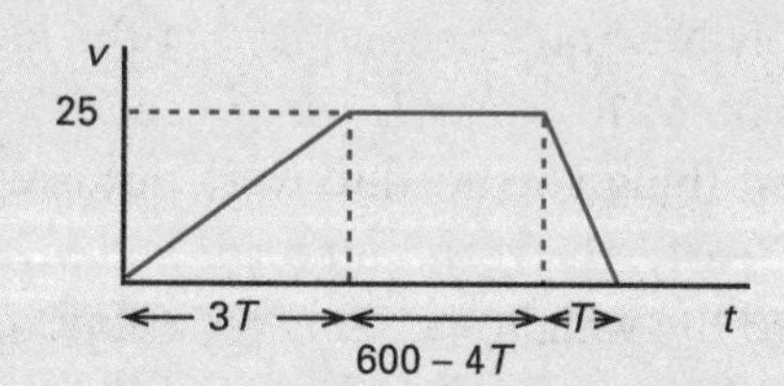

Use the fact that the area under the graph is 11 000 to show that $T = 80$.

2 (a) 4 ms^{-1}
(b) -2 ms^{-1}, 18s

Variable acceleration

Checkpoints

1 The velocity is zero at maximum displacement.
$V = 20 - 10t = 0$ when $t = 2$.
This gives maximum $x = 30 + 40 - 20 = 50$
2 $a = -10$
3 $\mathbf{a} = -2\mathbf{j}$
4 Motion under gravity
5 The constant of integration represents the initial velocity (which is a vector).

Exam questions

1 (a) $a = 6t - 8$, so when $t = 2$, $a = 4$
(b) $3t^2 - 8t + 4 = 0$ gives $t = 2$ or $\frac{2}{3}$

(c) $x = t^3 - 4t^2 + 4t + c$
$t = 2$ gives $x = c$, and $t = \frac{2}{3}$ gives $x = \frac{32}{27} + c$
The difference in the x values is $\frac{32}{27}$

2 (a) It crosses the y-axis when the **i** component is zero, i.e. $t = \pm 2$.
(b) $\mathbf{v} = 2t\mathbf{i} + 3\mathbf{j}$. When $t = 2$, $\mathbf{v} = 4\mathbf{i} + 3\mathbf{j}$, which has magnitude 5, so the speed is 5.
(c) $\mathbf{a} = 2\mathbf{i}$

Examiner's secrets

Keep a grip on whether you should differentiate or integrate, and remember not to substitute in a value for t until after you have differentiated or integrated.

Constant acceleration

Checkpoints

1. The gradient is the acceleration, a. a is like m in $y = mx + c$, while u plays the part of c
2. Trapezium; area $= \frac{1}{2}(a + b)h$
3. Time = 5.71 s.
4. It's a there-and-back distance (not just the difference in the s values).
5. No air resistance

Exam questions

1 (a) At the greatest height $v = 0$
so $0 = 2.8^2 + 2x - 9.8 \times h$, giving $h = 0.4$ metres
(b) $s = 2.8 \times 5 - \frac{1}{2} \times 9.8 \times 5^2 = -108.5$

2 $v = u + 4$ so $(u + 4)^2 = u^2 + 2 \times 0.8 \times 40$
so $u^2 + 8u + 16 = u^2 + 64$ giving $u = 6$

3 Policeman: $s = 2.5t$
Suspect: $s = \frac{1}{2} \times 0.1 \times t^2$
so $2.5t = 30 + 0.05t^2$ giving $t = 20$ or 30, giving distance for suspect = 20 m (note: at $t = 30$, the suspect would re-overtake the policeman).

Projectiles

Checkpoints

1. Magnitude g downwards
2. Horizontal: $U \cos \theta$. Vertical: $U \sin \theta$
3.

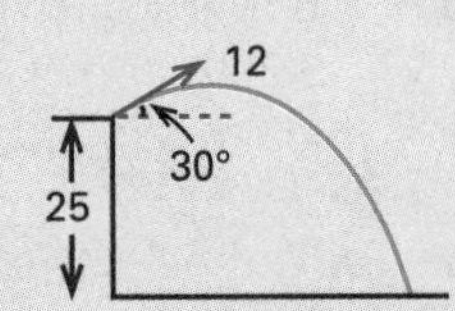

4. Horizontal velocity = 12 cos 30° = 10.4
Vertical velocity = 12 sin 30° − 9.8 × 2.95 = −22.91
Speed $= \sqrt{10.4^2 + 22.91^2} = 25.2$ ms^{-1}
5. T = 9.90s, R = 227 m
6. The modelling assumption is no air resistance.

Exam questions

1 First show that $R = u^2 \sin 2\theta / g$, then $u^2 = Rg/\sin 2\theta$ which gives u between 36.7 and 38.6 ms^{-1}

2 $8 = 28 \sin 30°\ t - \frac{1}{2} \times 9.8 \times t^2$
which gives $4.9t^2 - 14t + 8 = 0$, and $t = 0.79$ and 2.07, so t is between these times.

Force diagrams

Checkpoints

1 Because a string isn't rigid

2

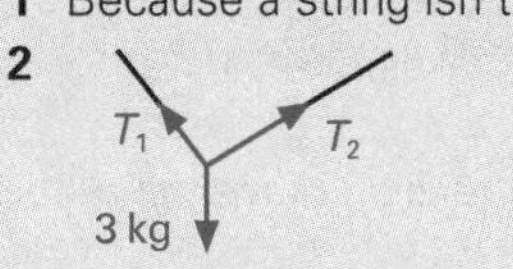

3

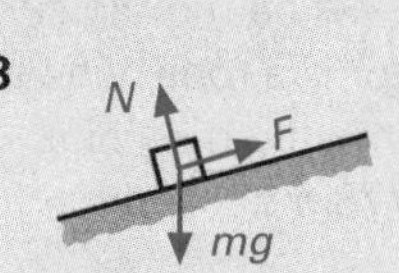

4

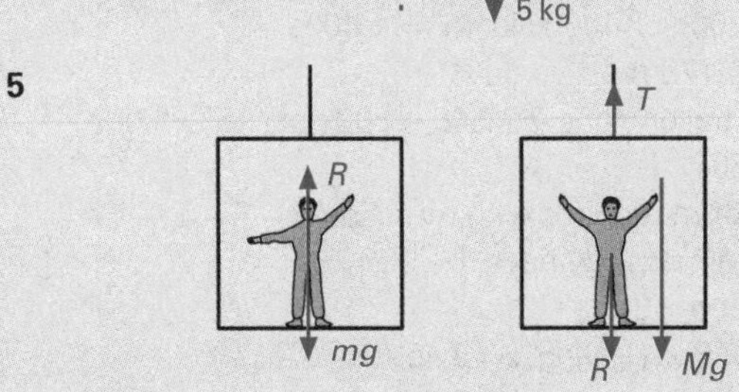

5

Exam questions

1

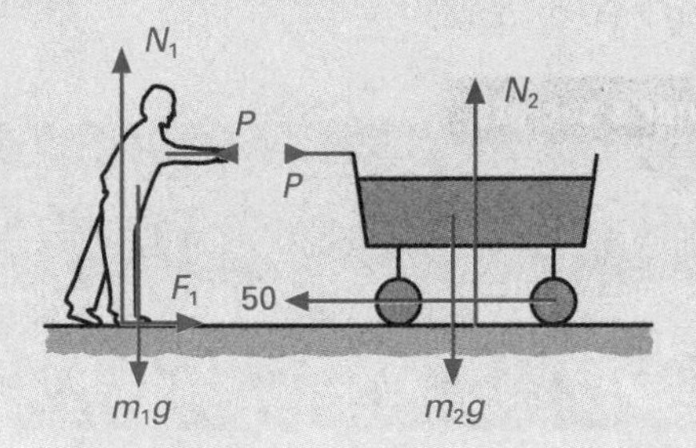

2

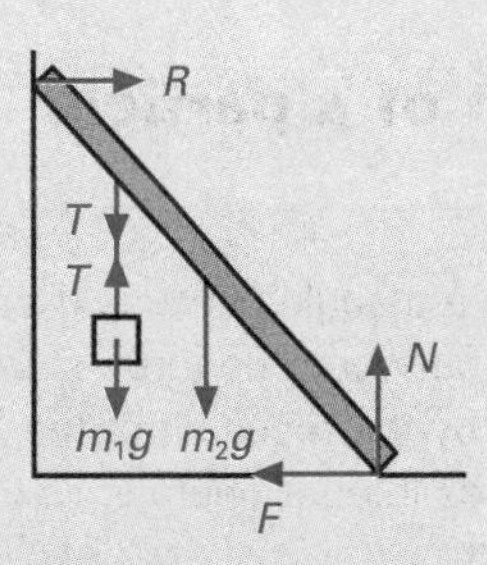

Newton's Second Law

Checkpoints

1. Force and acceleration are vectors.
Mass isn't.
2. 60 kg
3. Friction and air resistance. Friction depends on the nature of the surface, and the mass of the boy/sledge. Air resistance depends on wind speed and the shape of the boy/sledge.

Exam questions

1. $100 \cos 30° - 60 = 20a$ giving $a = 1.33$ ms^{-2}
   $N + 100 \sin 30° - 20g = 0$ giving $N = 146$
2. $-mg \sin \theta = ma$ giving $a = -1.96$
   $0 = 5^2 + 2 \times (-1.96) \times s$ giving $s = 6.38$ m

Connected particles

Checkpoints

1. If the connector is elastic, when it is extending, the caravan will be moving slower than the car. If the connector is a rope, when the car brakes, the caravan will go into the back of it.
2. $a = -0.778$ ms^{-2}, and the force is a tension of magnitude 178 N
3. The string must be light, and the pulley smooth.

Exam questions

1. Taking the lorry and trailer as a single unit
   $5\,000 - 1\,200 - 1\,400g \sin 15° = 1\,400a$
   giving $a = 0.178$ ms^{-2}
   Taking the trailer alone $T - 300 - 600g \sin 15° = 600a$, so $T = 1\,930$.
2. Equations of motion for 4 kg mass is $4g - T = 4a$.
   For the 6 kg mass we have $T - 9.2 = 6a$.
   Solve to get $a = 3$ ms^{-2}
   $v^2 = 0 + 2 \times 3 \times 1$ giving $v = 2.45$ ms^{-1}
   New equation of motion for 6 kg mass is $-9.2 = 6a$
   so $a = -1.53$, and mass travels a further 1.96 m, making 2.96 m in all.

Examiner's secrets

Make sure that you are clear to which part of the system you are applying $F = ma$. Remember that you only apply it to the system as a whole if both parts are moving in the same direction. When solving the simultaneous equations, it is usually easiest to eliminate T, and find the acceleration first.

Equilibrium of a particle

Checkpoints

1. The forces form a triangle because if they didn't, there would be a gap between the three sides you have drawn – which is filled by the resultant of the forces. But the resultant of forces in equilibrium is zero.
2. $\dfrac{P}{\sin 140°} = \dfrac{400}{\sin 100°}$ so $P = 261$
3. You'd have a polygon if you had four or more forces.
4. 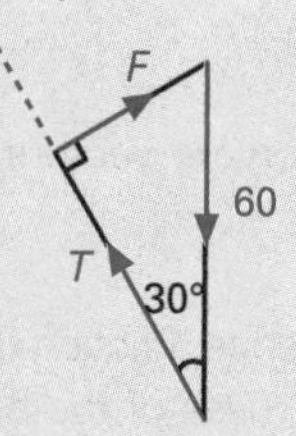

   $F = 60 \sin 30° = 30$

Exam questions

1. Resolve vertically: $P \sin 50° = 3 \sin 70°$, so $P = 3.68$.
   Horizontally: $Q + P \cos 50° = 4 + 3 \cos 70°$, so $Q = 2.66$.
   Resultant has magnitude 4 N acting to the left.
2. You need to spot that you have a Pythagorean 5, 12, 13 triangle, so the strings are perpendicular. Then you can either use Lami, or resolve first in the direction of one string, then in the direction of the other. The tensions are 18.8 N and 45.2 N.

Friction

Checkpoints

1. It is harder if someone is pushing down on the case.
2. μ has no units.
3. Equation (2) would be $N - 400 \sin 40° = 50$ g
4. Limiting equilibrium means it is on the point of moving and $F = \mu N$
5. $F = 12g \sin 30° - P \cos 30°$.
   $N = P \sin 30° + 12g \cos 30°$ giving $P > 7.06$

Exam questions

1. $N = mg$, so $\mu N = 0.25\ mg = 2.45$ m.
   $F = ma = 0.3$ m, so $F < \mu N$ and it doesn't slip.
2. (a) Perpendicular to plane; $N = mg \cos 30°$;
   $F = \mu N$, so $F = 0.4mg \cos 30°$.
   Parallel to plane: $-0.4mg \cos 30° - mg \sin 30° = ma$ so $a = -8.29$.
   (b) $v^2 = u^2 + 2as$ gives $s = 0.964$
   (c) Note there is a new acceleration – draw a new diagram. $mg \sin 30° - 0.4mg \cos 30° = ma$ giving $a = 1.505$ and $v = 1.7$.

Using vectors

Checkpoints

1. Two vectors are parallel if one is a multiple of the other.
2. Yes it is constant acceleration. $\mathbf{v} = 12\mathbf{i} + 0.5\mathbf{j} + 14\mathbf{k}$, $\mathbf{s} = 120\mathbf{i} + 5\mathbf{j} + 140\mathbf{k}$
3. $4\mathbf{j}$ is the constant of integration (the value of $\mathbf{v}$ when $t = 0$).
4. Constant speed means zero acceleration, so equilibrium.

Exam questions

1. (a) $a = -1$ and $b = -3$
   (b) magnitude is $\sqrt{10}$ and direction is 71.6°
2. (a) $\mathbf{a} = -20\mathbf{i} + 30\mathbf{j}$, $\mathbf{v} = -96\mathbf{i} + 149\mathbf{j}$, $\mathbf{s} = -230\mathbf{i} + 370\mathbf{j}$
   (b) $t = 0.4$
3. (a) $\mathbf{s} = 3\mathbf{i} + 6\mathbf{j}$
   (b) $\mathbf{v} = 3t^2\mathbf{i} - 5\mathbf{j}$, $\mathbf{a} = 6t\mathbf{i}$
   (c) $\mathbf{F} = 1.2t\mathbf{i}$
   (d) It continues with constant velocity $\mathbf{v} = 150\mathbf{i} - 5\mathbf{j}$

Moments and equilibrium

Checkpoints

1. 10 Nm clockwise
2. $5 \times 2 + 2 \sin 45 \times 2 - 4 \times 2 = 4.83$ Nm anticlockwise

3 1.25 m from 3 kg end; 78.4 N

4 $X = T\cos\theta = 4.70.$

$Y + T\sin\theta = 2g$ so $Y = 13.3$

Exam question

(a) Take moments about the hinge:
$200g \times 8 = T\sin 36.9° \times 16$, giving $T = 167g$, so $m = 167$ kg

(b) $200g \times \sin 41.4° = 12T$, giving $T = 88.2$ g, $m = 88.2$ kg, $X = 864$ N, $Y = 1\,960$ N

Examiner's secrets

Careful thought at the start about where to take moments can save a lot of time. Some candidates have correct equations, but get bogged down in sorting them out and solving them.

Centre of gravity 1

Checkpoints

1 $(\frac{13}{6}, \frac{-1}{2})$

2 It is 2 by symmetry

3 $\bar{y} = \dfrac{50 \times 40 + 20 \times (80 + 9)}{50 + 20} = 54$

Exam questions

1 From the vertex with the 3 m particle, the C of G is 0.4 d towards the 2 m particle, and 0.3 d towards the m particle.

2 Using the fact that the mass is proportional to the area, and dividing the figure into a 4 by 10 and a 2 by x rectangle:

$$\bar{x} = \frac{40 \times 2 + 2x \times (4 + 0.5x)}{40 + 2x} = \frac{80 + 8x \times x^2}{40 + 2x}$$

It will not topple provided $\bar{x} \leq 4$, which leads to $x < 8.94$ cm

Centre of gravity 2

Checkpoints

1 Left hand, yes; right hand, no

2 Critical value: $16P = 5 \times 4g$.
Topples if $P > 12.25$

3 $N = 4g$; $F = 0.3\ N = 11.76$
Slides if $P > F$, so slides before it topples as $11.76 < 12.25$

4 $N = 3g\cos\theta$; $F = 0.85\ N = 24.99\cos\theta$
Slides if $3g\sin\theta > F$, i.e. $3g\sin\theta > 24.99\cos\theta$
This gives $\tan\theta > 0.85$ (μ!) and $\theta > 40.4$, so topples.

5 $\tan^{-1}\dfrac{2}{4.1875} = 25.5°$

Exam questions

1 Taking bottom left corner as origin, centre of gravity is at (4.5,5.67). Topples if $\tan\theta > \frac{4.5}{5.67}$
$\theta > 38.5°$

2 The centre of mass is 5.15 units from AD, and 1.65 units from AD. The angle is 72.2°.

Work and power

Checkpoints

1 $250 \times \cos 30° \times 4 = 866$ J

2 $(3\mathbf{i} + 7\mathbf{j}) \cdot (6\mathbf{i} + \mathbf{j}) = 25$ J

3 21.65 W

4 400 N

5

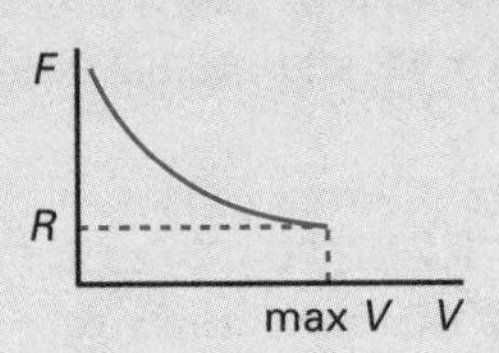

6 Not at very low speeds when you would expect the resistance to be less.

7 Constant speed so $F = 110$ and power = 1 320 W.
If cycling uphill at steady speed $F = 110 + mg\sin\theta$, so power would be greater.

Exam questions

1 (a) 60 W (b) 304 W (c) 364 W

(a) at $v = 20$, $R = 300$ and $F = 600$.
So $600 - 300 = 600a$, and $a = 0.5$ ms^{-2}

(b) At maximum velocity, $a = 0$ and $F = R$
So $12\,000/V = 200 + 5V$; $V = 32.9$ ms^{-1}

Work and energy

Checkpoints

1 $h = 0.5$ m

2 KE = 6.4 J

3 (a) 450 kJ (b) 235 200 J (c) 180 000 J (d) 34 800 J

4 The length of the slide makes no difference (it would make a difference if the slide was not friction-free). If the slide were not friction-free, the child's speed at the bottom would be less. A heavier child makes no difference on a smooth slide, but will increase the friction, and so reduce the speed, if the slide is not smooth.

Exam questions

1 (a) $\frac{1}{2} \times 0.050 \times 15^2 = 0.050 \times g \times h$, so $h = 11.5$ m
Assume no air resistance

(b) Work done $= \frac{1}{2} \times 0.050 \times 15^2 - 0.050 \times g \times 10.5 = 0.48$ J.
So $R = 0.48 \div 10.5 = 0.045\,7$ N

2 (a) $50 \times 9.8 \times 9\sin 40° = 2\,835$ J

(b) KE gained = 36 J. So work done against resistance $= 2\,835 - 36 = 2\,799$ J and the resistance is $2\,799 \div 9 = 311$ N

Examiner's secrets

It really helps if you make it clear what you are finding at each stage.
Think carefully about whether or not energy is conserved.
Think carefully about the acceleration: remember at maximum speed $a = 0$, but the acceleration may not be constant.

Momentum and impulse

Checkpoints

1 Ns
2 $\mathbf{F} = -270\mathbf{i} + 105\mathbf{j}$
3 If the spheres have unequal radii, you will have an oblique (at an angle) impact rather than a direct impact.
4 KE before = 6.2 J; KE after = 3.8 J so loss of KE = 2.4 J.

Exam questions

1 $0.5\mathbf{v} = 0.2(5\mathbf{i} + 7\mathbf{j}) + 0.3(2\mathbf{i} - 3\mathbf{j})$, so $\mathbf{v} = 3.2\mathbf{i} + \mathbf{j}$
2 $0.05v = 0.05 \times 2 + 0.04 \times 5$ so $v = 6$
Loss in KE = 0.3J

Examiner's secrets

This should be an easy question, but many candidates are not clear about which direction is positive, and so end up in a muddle.

Momentum and restitution

Checkpoints

1 Speed of approach is 6 ms^{-1}, speed of separation is 4 ms^{-1}.
2 Loss of KE = 12.6 J.
3 Speed before impact is 3.96 ms^{-1}; speed after impact is 2.97 ms^{-1} and $h = 0.45$ m.
4 Speed is 3.61 ms^{-1} and it makes angle 73.9° with the normal.
5 The energy comes from the firing device – in the old days it was chemical energy stored in the gunpowder.
6 The impulse is so large that a mounting could easily be wrecked.

Exam questions

1 Let the velocities after impact be u and v respectively (in same direction).
Conservation of momentum: $3 \times 0.6 + 5 \times 0.4 = 3u + 5v$
Restitution: $v - u = 0.8(0.6 - 0.4)$.
Solve to get $u = 0.375$, $v = 0.535$.
2 The time taken to reach the wall:
$0 = 14 \sin 30 - gt$ so $t = 0.713\,6$ seconds.
By symmetry it takes the same time to hit ground again after impact. During the flight towards the wall, the horizontal component of velocity is constant = 14 cos 30° = 12.12. After impact horizontal speed = 0.7 × 12.12 = 8.484. So horizontal distance travelled = 8.484 × 0.713 6 = 6.05 m.

Circular motion 1

Checkpoints

1 The velocity is not constant because its direction is changing. The speed is constant (magnitude).
2 $a = 0$ perpendicular to the circle because there is no change in velocity in this direction.
3 You will skid away from the centre. You can't take a corner too slowly (unless you are on a banked track, in which case you could slide down).
4 N and F towards the centre; all three have a component perpendicular to the circle.
5 $r = l \sin\theta$
6 T has a component towards the centre and perpendicular to the circle. The weight acts entirely perpendicular to the circle.

Exam question

1 (a) Did you spot the 3, 4, 5 triangle?
$T_1 \sin\theta + T_2 \cos\theta = 2 \times r \times 8^2$
$T_1 \cos\theta = 2g + T_2 \sin\theta$
$r = 0.24$, $\sin\theta = 0.6$, $\cos\theta = 0.8$
$T_1 = 34.128$ and $T_2 = 12.804$
(b) If one string goes slack, it will be T_2. With $T_2 = 0$ we have
$T_1 \sin\theta = 2 \times r \times \omega^2$ and
$T_1 \cos\theta = 2g$, giving $T_1 = 24.525$
and $\omega = 5.54$ (least value).

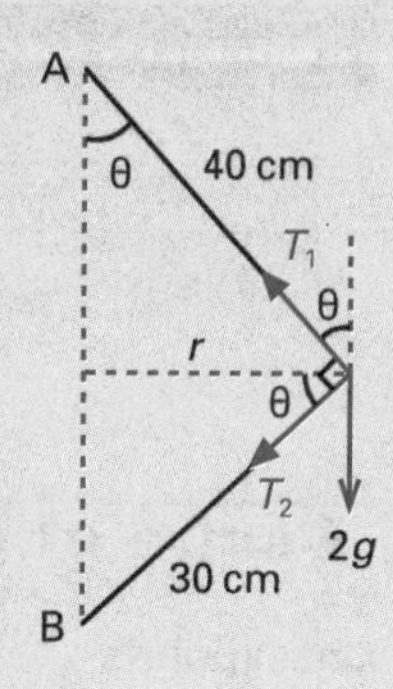

Circular motion 2

Checkpoints

1 The work done against gravity is what we call the GPE. So it's automatically included in the energy equation.
2 It depends on the speed, but there could be such points at the top of the circle.
3 (a) yes (b) no (c) no (d) yes
4 $F = ma$ gives $mg\cos\theta = mv^2/L$
So $v^2 = Lg\cos\theta$.
Initial KE = Final KE + GPE gain.
So $\frac{1}{2}m \times 4.5Lg = \frac{1}{2}mLg\cos\theta + mgL(1 + \cos\theta)$.
This gives $\cos\theta = 0.833\,3$ and $\theta = 33.6°$.

Exam question

Suppose it reaches the top with speed v.
Energy gives $\frac{1}{2} \times 0.2 \times 7^2 = \frac{1}{2} \times 0.2 \times v^2 + 0.2g \times 1.6$.
So $v^2 = 17.64$.
$F = ma$ at the top gives $0.2g + R = 0.2 \times \frac{17.64}{0.8}$.
This gives a positive value for R (2.45) so it is still in contact with the tube.

Elasticity

Checkpoints

1 0.75 J
2 0.417 J
3 No forces acting apart from the tension and weight (the work done by these forces are automatically incorporated into the energy equation). Need to assume no air resistance.
4 No work done against friction so we can use conservation of energy. Answer would be smaller if the table was rough.

Exam questions

1 EPE stored = 1 joule. At max speed this is all converted to KE so $v = 0.816\ \text{ms}^{-1}$.

2

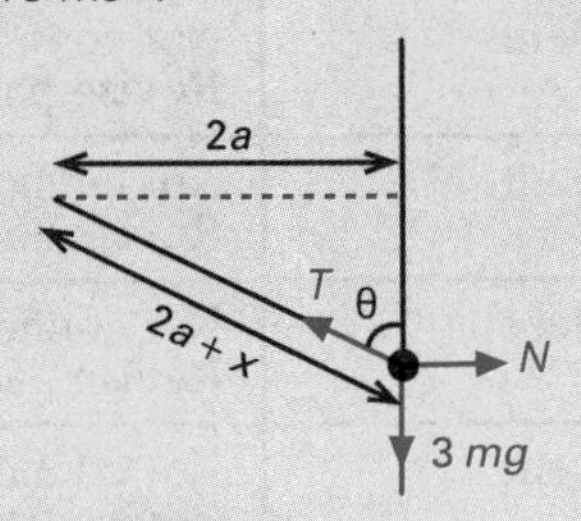

$\sin\theta = \frac{2}{3}$ leads to $x = a$; $\cos\theta = \frac{\sqrt{5}}{3}$

Vertically: $T\cos\theta = 3mg$

Hooke's Law gives $T = \frac{\lambda x}{2a} = \frac{1}{2}\lambda$

so $\frac{1}{2}\lambda\frac{\sqrt{5}}{3} = 3mg$ and $\lambda = \frac{18\sqrt{5}mg}{5}$

$N = T\sin\theta = \frac{6\sqrt{5}mg}{5}$

Examiner's secrets

Take care not to use constant acceleration equations with a string whose length is changing – if the tension is changing, so is the acceleration!

Revision checklist
Mechanics

By the end of this chapter you should be able to:

1	Understand what a modelling assumption is, and know the most common assumptions used.	Confident	Not confident. **Revise** pages 168–169
2	Know how to find the resultant of two vectors, and how to resolve a vector into components.	Confident	Not confident. **Revise** pages 170–171
3	Know how to draw and interpret displacement–time and velocity–time graphs.	Confident	Not confident. **Revise** pages 172–173
4	Know how to use calculus to solve variable acceleration problems.	Confident	Not confident. **Revise** pages 174–175
5	Know and use the constant acceleration equations, and understand when they are appropriate.	Confident	Not confident. **Revise** pages 176–177
6	Know the most common forces, and be able to draw force diagrams.	Confident	Not confident. **Revise** pages 180–181
7	Know Newton's Second Law and be able to solve problems involving forces and acceleration.	Confident	Not confident. **Revise** page 182
8	Know how to resolve forces for inclined plane problems.	Confident	Not confident. **Revise** page 183
9	Be able to solve problems involving connected particles and pulleys.	Confident	Not confident. **Revise** pages 184–185
10	Be able to solve problems about the equilibrium of a particle.	Confident	Not confident. **Revise** pages 186–187
11	Know the formula $F = \mu N$, and when it is appropriate to use it. Be able to solve problems involving friction.	Confident	Not confident. **Revise** pages 188–189
12	Be able to solve problems using vectors in column vector or **i**/**j** form.	Confident	Not confident. **Revise** pages 190–191
13	Be able to solve problems involving projectiles, finding the range and greatest height and flight time.	Confident	Not confident. **Revise** pages 178–179
14	Be able to find the moment of a force, and solve equilibrium problems using moments.	Confident	Not confident. **Revise** pages 192–193
15	Find the centre of gravity of simple shape (not using calculus).	Confident	Not confident. **Revise** pages 194–195
16	Be able to use centre of gravity to solve problems, especially toppling problems and suspended laminas.	Confident	Not confident. **Revise** pages 196–197
17	Know and use the definitions of work and power, with special reference to moving vehicles.	Confident	Not confident. **Revise** pages 198–199
18	Know the formulae for kinetic and gravitational potential energy, and use them to solve problems relating work and energy. Use conservation of energy appropriately.	Confident	Not confident. **Revise** pages 200–201
19	Know the definitions of momentum and impulse. Be able to solve impact problems using conservation of momentum.	Confident	Not confident. **Revise** pages 202–203
20	Know Newton's Law of Restitution, and be able to use it in solving impact problems.	Confident	Not confident. **Revise** pages 204–205
21	Know the formulae $v = \omega r$ and $a = r\omega^2$ for circular motion, and be able to use them to solve problems relating to circular motion with constant speed.	Confident	Not confident. **Revise** pages 206–207
22	Be able to solve problems of motion in vertical circles using energy.	Confident	Not confident. **Revise** pages 208–209
23	Know Hooke's Law, and be able to use it to solve problems involving elastic strings.	Confident	Not confident. **Revise** pages 210–211

Resources section

This section is intended to help you develop your study skills for examination success. You will benefit if you try to develop skills from the beginning of your course. Modern AS and A-level exams are not just tests of your recall of text books and your notes. Examiners who set and mark the papers are guided by assessment objectives that include skills as well as knowledge. You will be given advice on revising and answering questions. Remember to practise the skills.

Awarding body specifications

In order to organise your notes and revision you will need a copy of your awarding body's specification. You can obtain a copy by writing to the awarding body or by downloading the syllabus from its website.

- AQA (Assessment and Qualifications Alliance)
  Publications Department, Aldon House, 39 Heald Grove, Rusholme, Manchester M14 4NA – www.aqa.org.uk
- CCEA (Northern Ireland Council for Curriculum, Examinations and Assessment)
  Clarendon Dock, 29 Clarendon Road, Belfast, BT1 3BG – www.ccea.org.uk
- EDEXCEL
  190 High Holborn, London WC1V 7BH – www.edexcel.org.uk
- OCR (Oxford, Cambridge and Royal Society of Arts)
  1 Hills Road, Cambridge CB2 1GG – www.ocr.org.uk
- WJEC (Welsh Joint Education Committee)
  245 Western Avenue, Cardiff CF5 2YX – www.wjec.co.uk

Topic checklist

○ AS ● A2	AQA	CCEA	EDEXCEL	OCR/A	OCR/MEI	WJEC
Using this guide	○●	○●	○●	○●	○●	○●
Exam preparation	○●	○●	○●	○●	○●	○●
Exam technique	○●	○●	○●	○●	○●	○●
Selected formulae	○●	○●	○●	○●	○●	○●

Using this guide

Students who work hard, and carefully prepare for their exams, can improve significantly on their performance. The notes in this section may help you use revision time as productively as possible.

Maths, perhaps more than other subjects, benefits most from a 'little and often' approach, rather than the last minute all-out effort. However, if you have left it rather late in the day to start work, don't give up – you have nothing to lose by doing your best to make up lost time now!

Watch out!

Down each margin there are small snippets – they are there for a purpose!

Where shall I start?

Probably at the beginning! Start with the Pure mathematics chapter first, as you will need the techniques and skills you learn there to apply to the Discrete mathematics, Mechanics and Statistics chapters. The guide is made up of double-page sections, designed to provide enough to learn at a single sitting.

What do I need?

Pencil, paper and calculator. The most important aspect of revising maths is that you learn best by doing. It is no good just reading through your notes, or the pages in a book. It is essential that you work through each page – this means:

- **writing down important points as you read**, and saying them aloud. (If you are alone in the room!) It has been well established that you learn effectively if you see, say, write and hear the information.
- **actually do, there and then, every checkpoint, test yourself and action point as you come to them**. They are designed to help you learn, to think about important concepts, and to get a feel for how well you understand the topic. If there is a worked example on the page, you could either try to do it yourself before reading the solution, or having read it, cover it up and try for yourself. This will build up your confidence, and help you to remember standard procedures. **You can't learn maths just by reading a book or your notes – you must do it!**
- Always **check your answers**. If you were correct, fine! You know that you are getting on top of this topic. If your answer was not correct, look carefully at the correct solution and work out where you went wrong. Rewrite the solution again correctly if you made a serious mistake. This will help imprint the right way of solving the problem in your mind.
- At the end of each section there are some **exam questions** for you to try. Work through them, trying to put into practice the concepts and methods you have been learning. Again, check your answers when you have done the best you can.

Planning

You will certainly have been advised to plan your revision so that you can be sure that every topic has been covered in time. It can also give you a sense of achievement as you tick topics off your list. Your plans should be:

- realistic – don't set yourself targets you know you won't reach. It is better to set achievable goals, and stick to them, than give up because the task seems impossible.
- specific – if you have a vague plan that you will do maths on Tuesday morning, your work may end up being vague. Set yourself specific goals, e.g. differentiation and integration of trig functions from the Pure mathematics chapter, and work, energy and power from the Mechanics chapter.
- comprehensive – remember that you must include all your subjects, and every part of the syllabus. Some people are tempted to spend most of their time on the relatively easy stuff (which can make you feel good if you sail through getting everything right) at the expense of harder topics which are going to require more effort. Other students leave out topics because they are overconfident that they can do it – and then get caught out.

Grading yourself

At the end of each section, give yourself a fair assessment:

- **You were not able to get very far with the checkpoints and the exam questions**. You still have some way to go on this topic, and will need to return to it more than once. Fill in one of the three circles on the cover flaps.
- **You have made progress on this topic**, and could get quite a few of the questions correct. However, you are not yet confident that you have fully understood it all, or perhaps you are not sure that you will be able to remember it. Fill in two of the relevant circles on the cover flaps.
- **You are now confident that you have understood this topic**, and can remember the important points. Well done! Fill in all three of the circles for this topic. You should not need to go through it again, except perhaps for a brief recap.

Action point

Don't be afraid to alter your plan in the light of experience. Some topics may need longer than you expected – some less time. Use the grading scheme below.

Don't forget

Most working people get rewarded by a regular pay-packet! This isn't likely to be true in your case (yet!) but don't forget to build some treats into your planning.

Don't forget

Give yourself a break! It's better to work hard for a morning, and then have a couple of hours off, rather than too much of a stop-start approach which can leave you feeling that your whole life is dominated by work. During your work sessions, you could aim to work hard for 45–50 minutes, have a short 10–15 minute break, then get back to it.

Action point

It is important that your work is easy and clear for someone else to follow. It is a good learning experience for you to exchange work with a friend from time to time to check the answers. Your friend will enable you to see where you are setting out your work clearly, and where it was difficult to follow. When you are checking your friend's work, trying to find their mistakes is a good exercise in learning to understand what good maths is about.

Exam preparation

When you are getting closer to the exam date, you need to focus specifically on preparing for the module(s) you are taking.

The exam specification

This is a most useful resource and should be available from your teacher, from the internet or directly from the awarding body. This will give you more precise details as to what you are expected to know than this study guide is able to do. Some candidates walk into exams not knowing what to expect – don't be one of them. Look out for key words:

- 'know' – indicating something you should have learned, often a formula
- 'use' – a rule or formula you should be able to apply, e.g. the product rule for differentiation, and
- 'understand' – a concept you need to be familiar with. You should be able to demonstrate your understanding within the context of a problem.

The jargon

The 'specification' used to be called the syllabus.

Links

For awarding body addresses see page 219.

Using past papers

At the very least, each module will have a specimen exam paper, again available from your teacher or direct from the awarding body. As time passes, there will be a bank of past papers available for the new AS and A2 modules – in the meantime papers from the old syllabuses may be of some use. Some awarding bodies also sell CD-ROMs containing a range of questions.

Once you have built up some confidence in all the topics in a module, you should practise working through past papers. To begin with, don't worry about the time element, and look up anything you have forgotten. Aim first to get as much correct as you can, then focus on increasing your speed of working if necessary.

Action point

If you will need to apply to an awarding body for past papers, allow enough time for this.

Learn those formulae!

You need to know what is on your awarding body's formula sheet, and what else you need to know. Make a list for each module of the formulae you need to learn – and start learning! You are also advised to learn at least some of the formulae which are on the sheet – being really familiar with them makes it more likely that you will recognise when you need to use them. Some candidates never looked up the relevant formula on the sheet, because they didn't know of its existence.

Action point

A half-learned formula is worse than useless, so use all the necessary strategies like pinning formulae up next to the bathroom mirror, recording them on to a cassette and playing them back, writing them on small cards you can flick through – whatever works for you! Just do it!

What went wrong?

You have worked through a past paper, done a test or whatever, and got back your work marked. This is a really important and exciting time as it is the best learning opportunity if you use it.

Try to categorise your mistakes so that you know where you are weakest and can take appropriate action.

- **Inadequate revision.** You lost marks because you had forgotten a formula or forgotten a standard procedure. This can be rectified – just turn to the relevant page of the study guide and *learn*!
- **Algebraic errors.** You knew what to do, but made mistakes in the algebra. First you need to practise basic techniques until you are confident. Then get into the habit of checking your algebra every time you are not sure. You can do this by substituting numbers for your letters and checking that both expressions give the same answer.
- **Misread the question.** You gave the answer correct to 2 s.f., but the question asked for 3 d.p. You correctly found the speed, but didn't find the time taken in minutes as asked. You had not taken on board all the information you were given. All these faults can be costly – and annoying! Try using a coloured highlighter on the question paper to help focus your thinking.
- **Calculator errors.** You were doing fine until you got to the stage of wielding your trusty calculator which gave the wrong answer. It's down to you really to get familiar with the quirks of your calculator. Don't forget to check that you are in degree mode, or radians, as appropriate. Make sure that you know when you need to use brackets (especially with fractions and negative numbers).
- **Couldn't work out what to do.** If you find that you sometimes just can't work out how to start a question, especially one with an unfamiliar slant, it may be that your problem-solving skills are not good. There's no quick fix for this one – ask for help each time you get stuck, and try hard to remember what to do, and to recognise little clues that tell you where to start.
- **Other.** Everyone has their own individual weak points – learn to recognise yours, and ask advice about strategies.

Test yourself

Which of these simplifications are right, and which wrong? Put numbers in for the letters to help you decide.

(a) $\dfrac{x^4 - y^4}{x - y} = x^3 - y^3$

(b) $6^n \times (\frac{2}{3})^n = 4^n$

(c) $\dfrac{r^n - 1}{r - 1} = r^{n-1}$

Action point

Every calculator has its own little quirks, so it is no good borrowing one if yours is lost or broken or has a low battery. Make sure that yours is in good repair, and consider investing in a spare that you are also familiar with.

Making the most of your mother, your mates, etc.

You are having big trouble with an assignment your teacher has set, question 8 on the paper has you stumped, the page in the book doesn't seem to be much help . . . we've all been there!

The first step is to ask for help – burying your head in the sand and hoping the question will go away never was very successful!

The second step is harder. It is so easy, when your friend has shown you how to solve the problem, to think, 'Of course, I could have done that . . . ,' and shut your book with a sigh of relief. However, if you want to *learn*, you will wait a day or two, then try to tackle the problem again on your own. If you can now do it, you know that you have learned something, and have reinforced your learning. If you still can't do it, go back to step one!

Action point

Don't be afraid to do difficult questions several times over (with suitable time gaps between). In the end, you will find that there are not that many really different questions – they may be dressed up to look different, but once you have learned to recognise them you will realise how similar they are.

Exam technique

Many of the rules that apply to any exam also apply to maths exams. Read the question carefully, keep an eye on the time, don't get bogged down with a difficult question, be alert for key words, use checking time productively, etc. We'll go through these points below.

Examiner's secrets

Be obedient! If you are asked to give an answer to 3 decimal places, don't give it correct to 2 significant figures! Some candidates use a highlighter pen to mark these instructions on the exam paper to help them remember and check.

Read the question carefully

Sounds obvious, doesn't it, but you would be amazed how many candidates throw marks away! Some questions, especially in Pure mathematics modules, are very brief and it is immediately obvious what you are supposed to do, e.g. integrate the expression or solve the equation, etc. Others are longer, and in problem form, and here you must read carefully and check that you have got the information you are given straight.

Examiner's secrets

Be accurate! If your final answer has to be correct to 2 significant figures, remember that you will need to be *more accurate* at intermediate stages of working – as a rule of thumb keep at least *two extra figures* in at each stage until you round off your final answer. (It helps, and saves time, to use the memory functions on your calculator.)

Be alert for key words

'**State**' or '**Write down**' means that you should not need to do any working beyond simple arithmetic.

'**Giving your answer exactly**' means that you should leave π, surds like $\sqrt{2}$ or expressions like ln 5 and e^2 in your answer and not use your calculator to evaluate them. This type of question might involve an integration or trig equation and the examiner does not want you to use your calculator to find the answer (though you might well use your calculator to check!). If you find yourself writing '= 0.47 (2 s.f.)' at the end, you should know you have been disobedient! You have rounded your answer, so it is not exact!

'**Show that . . .**' Essentially the examiner is giving you the answer he wants you to reach, so don't *use* this result until you have found it yourself. However, if you *can't* show the result, then go ahead and use it to complete the rest of the question. It's better to do half a question than none.

Examiner's secrets

Remember that half a question is better than nothing! Sometimes a question gives you a result to prove, or an answer to obtain, and then continues. The purpose of this is twofold. First so you can check that you are on the right lines. Second, if you couldn't complete the first part of the question, then you have an opportunity to try the next part, and gain some marks.

'**Express . . . in the form . . .**' or '**Express . . . as a . . .**' means rewrite this in a certain format. It is essentially about rearrangement, rather than solving an equation, and almost always there is a standard procedure you should have learned.

'**Hence, or otherwise, . . .**' 'Hence' means 'using the result you have just got'. If the question permits 'otherwise', then there is an alternative method, but the 'hence' method is usually easier. The trick here is to figure out the relevance of what you have just done to the next stage of the question.

Examiner's secrets

With regard to 'hence' questions: if it is just 'hence', and not 'hence or otherwise', *you must use the result you have just got – if you don't, you won't get the marks.*

'**Sketch . . .**' A sketch diagram or graph does not need to be drawn to scale, but does need to show all the important features. A sketch graph should have labelled axes, and the coordinates of significant points given. It can be drawn freehand, but should be neat. The fact that sketches are quickly drawn does not mean that it is OK to draw carelessly!

Effective time management

Before you go into the exam you should know how long it is, and the total number of marks available. So from the outset, you should have worked out roughly how long a 5 or 10 mark question should take you. As a general rule, answer the questions in numerical order. However if you are completely baffled by a question, move on. Don't spend too long on a question that is not going well – leave it and come back to it later if you have time.

Action point

If you don't have this information, make a point of finding out!

Don't forget

The questions towards the end usually have a high mark allocation, and are not always the most difficult. Leave enough time to attempt them.

Effective checking

Towards the end of the exam you will either be battling with the questions you could not do first time, or be checking your answers.

- Read each question again – especially make sure that you have answered the question and been obedient.
- Don't just read through what you have written – it is often difficult to spot mistakes. Is there some other way of checking? If you have solved an equation, you can check by substituting the value of your answer back into the original equation. A graphical calculator can confirm some answers.
- If your algebra is weak, you can often check a possible simplification by substituting a number for your letters on a piece of rough paper. If both expressions give the same answer, you have probably simplified correctly.
- Be aware of what is a sensible answer – some candidates happily calculate a probability greater than 1 and give no indication that it can't be correct!

Action point

Some students find it helpful to mark the question paper as they go along; perhaps a tick against questions they feel confident about, a question mark if unsure or a cross if a question, or part of a question, was missed out. This helps you in the last 15 minutes or so to focus quickly on the weak areas in your answers.

Action point

Always check the first question! It's amazing how many candidates make a silly mistake right at the start when they have not yet settled.

Totally foxed?

You have done what you can, but there are still several questions you have not answered, you can't think what to do, and panic levels are rising . . . sounds familiar? There are some positive strategies that *might* work – and you've nothing to lose by trying!

- First try to work out what area of the syllabus is being tested – if it is not obvious, think through the topics you have revised, and work out what has not been tested so far.
- Then jot down the formulae which are associated with this topic – as you do this, an idea about what to do might come to mind.
- For some topics, a diagram can be really helpful in enabling you to visualise the problem.

Think back to other problems you have done which are similar in some way – there are not many completely new questions, so the chances are that you *have* done a problem like this before!

Selected formulae

Introduction

This is *not* a complete list of the formulae you need to know. It is here to provide a quick ready-reference while you are using this guide. You can also use it as a basis for your own list of formulae. You should make a list for each module – the awarding body's specification will help. List both formulae you need to learn, and those provided in the formula booklet.

Algebra

The solution of $ax^2 + bx + c = 0$ is $x = \dfrac{-b \pm \sqrt{b^2 - 4ac}}{2a}$

A.P.s: $u_n = a + (n-1)d$

$S_n = \frac{1}{2}n\{2a + (n-1)d\} = \frac{1}{2}n(a + l)$

G.P.s: $u_n = ar^{n-1}$

$S_n = \dfrac{a(r^n - 1)}{r - 1}$

$S_\infty = \dfrac{a}{1 - r}, -1 < r < 1$

Logs: $a^b = c \Leftrightarrow b = \log_a c$

$\log x + \log y = \log(xy)$

$\log x - \log y = \log\left(\dfrac{x}{y}\right)$

$k \log x = \log(x^k)$

Trigonometry

In triangle ABC: $\dfrac{a}{\sin A} = \dfrac{b}{\sin B} = \dfrac{c}{\sin C}$

$a^2 = b^2 + c^2 - 2bc \cos A$

area $= \frac{1}{2}ab \sin C$

Sectors of circles: arc length $= r\theta$, area of sector $= \frac{1}{2}r^2\theta$ } θ in radians

Identities: $\cot\theta \equiv \dfrac{1}{\tan\theta}$

$\sec\theta \equiv \dfrac{1}{\cos\theta}$

$\text{cosec}\,\theta \equiv \dfrac{1}{\sin\theta}$

$\tan\theta \equiv \dfrac{\sin\theta}{\cos\theta}$

$\sin^2\theta + \cos^2\theta \equiv 1$

$\sec^2\theta \equiv 1 + \tan^2\theta$

$\text{cosec}^2\theta \equiv 1 + \cot^2\theta$

$\sin 2A \equiv 2 \sin A \cos A$

$\cos 2A \equiv \cos^2 A - \sin^2 A \equiv 2\cos^2 A - 1 \equiv 1 - 2\sin^2 A$

$\tan 2A \equiv \dfrac{2\tan A}{1 - \tan^2 A}$

$\sin(A \pm B) \equiv \sin A \cos B \pm \cos A \sin B$

$\cos(A \pm B) \equiv \cos A \cos B \mp \sin A \sin B$

$\tan(A \pm B) \equiv \dfrac{\tan A \pm \tan B}{1 \mp \tan A \tan B}$

Small angles: $\sin\theta \approx \theta$, $\cos\theta \approx 1 - \frac{1}{2}\theta^2$, $\tan\theta \approx \theta$

Coordinate geometry

The equation of the line through (x_1, y_1) with gradient m is $(y - y_1) = m(x - x_1)$

The gradients m_1, m_2 are perpendicular if $m_1m_2 = -1$

Equation of a circle: $(x - a)^2 + (y - b)^2 = r^2$

Differentiation

y	$\dfrac{\mathbf{dy}}{\mathbf{dx}}$
x^n	$n\,x^{n-1}$
e^{kx}	$k\,e^{kx}$
$\ln x$	$\dfrac{1}{x}$
$\sin kx$	$k \cos kx$
$\cos kx$	$-k \sin kx$
$\tan kx$	$k \sec^2 kx$

Product rule:

$f(x)g(x) \qquad f'(x)g(x) + f(x)g'(x)$

Quotient rule:

$\dfrac{f(x)}{g(x)} = \dfrac{f'(x)g(x) - f(x)g'(x)}{[g(x)]^2}$

'Chain' or 'function of a function' rule:

$f\{g(x)\} \qquad f'\{g(x)\}g'(x)$

Integration

$\int x^n\,dx = \dfrac{1}{n+1}x^{n+1} + c, n \neq -1$

$\int e^{kx}\,dx = \dfrac{1}{k}e^{kx} + c$

$\int \dfrac{1}{x}\,dx = \ln|x| + c, x \neq 0$

$\int \sin kx\,dx = \dfrac{1}{k}\cos kx + c$

$\int \cos kx\,dx = +\dfrac{1}{k}\sin kx + c$

$\int \sec^2 kx\,dx = \dfrac{1}{k}\tan kx + c$

$\int \dfrac{f'(x)}{f(x)}\,dx = \ln|f(x)| + c$

$\int f(x)g'(x)\,dx = f(x)g(x) - \int f'(x)g(x)\,dx$

area under a curve $= \int_a^b y\,dx$ for $y \geq 0$

volume of revolution about x-axis $= \int_a^b \pi y^2\,dx$

volume of revolution about y-axis $= \int_c^d \pi x^2\,dy$

Vectors

$|x\mathbf{i} + y\mathbf{j} + z\mathbf{k}| = \sqrt{x^2 + y^2 + z^2}$

$(a_1\mathbf{i} + b_1\mathbf{j} + c_1\mathbf{k}) \cdot (a_2\mathbf{i} + b_2\mathbf{j} + c_2\mathbf{k}) = a_1a_2 + b_1b_2 + c_1c_2$

$\mathbf{a} \cdot \mathbf{b} = |\mathbf{a}|\,|\mathbf{b}|\cos\theta$

Equation of a line through $\mathbf{a}$ parallel to $\mathbf{b}$ is $\mathbf{r} = \mathbf{a} + t\mathbf{b}$

Mechanics

Constant acceleration:

$v = u + at$

$s = ut + \frac{1}{2}at^2$

$s = vt - \frac{1}{2}at^2$

$s = \frac{1}{2}(u + v)t$

$v^2 = u^2 + 2as$

Projectiles:

$R = \dfrac{u^2 \sin 2\theta}{g}$

$H = \dfrac{u^2 \sin^2\theta}{2g}$

Equation of trajectory: $y = x \tan\theta - \dfrac{gx^2}{2u^2 \cos\theta}$

Variable acceleration:

$v = \dfrac{dx}{dt}$

$a = \dfrac{dv}{dt}$

Newton's Second Law: $F = ma$

Law of Friction: $F \leq \mu R$

Momentum of a particle $= mv$

Impulse $= mv_2 - mv_1$

Newton's Experimental Law (Restitution):

separation speed = $e \times$ approach speed

Gravitational potential energy $= mgh$

Kinetic energy $= \frac{1}{2}mv^2$

Elastic potential energy $= \dfrac{\lambda x^2}{2l}$

Work done by a constant force $= Fd$

Circular motion: $a = r\omega^2 = \dfrac{v^2}{r}$

Hooke's Law: $T = \dfrac{\lambda x}{l}$

Probability and statistics

Mean: $\bar{x} = \dfrac{\Sigma x}{n}$ or $\dfrac{\Sigma xf}{\Sigma f}$

Standard deviation $s = \sqrt{\dfrac{\Sigma x^2}{n} - \bar{x}^2} = \sqrt{\dfrac{\Sigma(x - \bar{x})^2}{n}}$

or $s = \sqrt{\dfrac{\Sigma fx^2}{\Sigma f} - \bar{x}^2} = \sqrt{\dfrac{\Sigma f(x - \bar{x})^2}{\Sigma f}}$

Discrete random variables:

$E(X) = \mu = \Sigma p_i x_i$

$E(X^2) = \Sigma p_i x_i^2$

$\text{Var}(X) = \sigma^2 = E(X^2) - \mu^2$

$E(aX + b) = aE(X) + b$

$\text{Var}(aX + b) = a^2\text{Var}(X)$

Binominal distribution $X \sim B(n,p)$

$E(X) = np$

$\text{Var}(X) = np(1 - p)$

$P(X = r) = {}^nC_r p^r(1 - p)^{n-r}$

Geometric distribution $X \sim \text{Geo}(p)$

$E(X) = \dfrac{1}{p}$

$\text{Var}(X) = \dfrac{1 - p}{p^2}$

$P(X = r) = (1 - p)^{r-1}p$

Poisson distribution $X \sim \text{Po}(\mu)$

$E(X) = \mu$

$\text{Var}(X) = \mu$

$P(x = r) = e^{-\mu} \cdot \dfrac{\mu^r}{r!}$

Bivariate data:

$S_x^2 = \dfrac{1}{n}\Sigma x^2 - (\bar{x})^2 \quad S_y^2 = \dfrac{1}{n}\Sigma y^2 - (\bar{y})^2$

$S_{xy} = \dfrac{1}{n}\Sigma xy - \bar{x}\bar{y}$

Product moment correlation: $r = S_{xy} \div \dfrac{S_x}{S_y}$

Spearman's rank correlation: $r_s = 1 - \dfrac{6\Sigma d^2}{n(n^2 - 1)}$

Least squares regression line of y on x is: $y = a + bx$
where $b = S_{xy} \div S_x^2$ and $a = \bar{y} - b\%$

Note to students: These formulae come in various forms, so check out with the formula sheet from *your* board.

Continuous random variables

$\int_{-\infty}^{\infty} f(x)\,dx = 1$

$E(X) = \int_{-\infty}^{\infty} xf(x)\,dx$

$E(X^2) = \int_{-\infty}^{\infty} x^2f(x)\,dx$

The median m is such that $\int_{-\infty}^{m} f(x)\,dx = \dfrac{1}{2}$

Distribution of the sample mean:
If X is a random variable with mean μ and variance σ^2, then $\bar{X}$ has mean μ and variance σ^2/n, $\bar{X}$ has a normal distribution if X is normal. If n is large, $\bar{X}$ has an approximately normal distribution.

Index